Introduction to
Algebraic Geometry and
Commutative Algebra

IISc Lecture Notes Series

Introduction to Algebraic Geometry and Commutative Algebra

Dilip P Patil
Indian Institute of Science, India

Uwe Storch
Ruhr University, Germany

IISc Press

World Scientific

NEW JERSEY · LONDON · SINGAPORE · BEIJING · SHANGHAI · HONG KONG · TAIPEI · CHENNAI

Published by

World Scientific Publishing Co. Pte. Ltd.

5 Toh Tuck Link, Singapore 596224

USA office: 27 Warren Street, Suite 401-402, Hackensack, NJ 07601

UK office: 57 Shelton Street, Covent Garden, London WC2H 9HE

British Library Cataloguing-in-Publication Data
A catalogue record for this book is available from the British Library.

IISc Lecture Notes Series
INTRODUCTION TO ALGEBRAIC GEOMETRY AND COMMUTATIVE ALGEBRA
Copyright © 2010 by World Scientific Publishing Co. Pte. Ltd.

ISBN-13 978-981-4304-56-6
ISBN-10 981-4304-56-5
ISBN-13 978-981-4307-58-1 (pbk)
ISBN-10 981-4307-58-0 (pbk)

Printed in Singapore.

SERIES PREFACE
World Scientific Publishing Company - Indian Institute of Science Collaboration

IISc Press and WSPC are co-publishing books authored by world renowned scientists and engineers. This collaboration, started in 2008 during IISc's centenary year under a Memorandum of Understanding between IISc and WSPC, has resulted in the establishment of three Series: IISc Centenary Lectures Series (ICLS), IISc Research Monographs Series (IRMS), and IISc Lecture Notes Series (ILNS).

This pioneering collaboration will contribute significantly in disseminating current Indian scientific advancement worldwide.

The **"IISc Centenary Lectures Series"** will comprise lectures by designated Centenary Lecturers - eminent teachers and researchers from all over the world.

The **"IISc Research Monographs Series"** will comprise state-of-the-art monographs written by experts in specific areas. They will include, but not limited to, the authors' own research work.

The **"IISc Lecture Notes Series"** will consist of books that are reasonably self-contained and can be used either as textbooks or for self-study at the postgraduate level in science and engineering. The books will be based on material that has been class-tested for most part.

PREFACE

The present book is based on a course of lectures delivered by the second author at the Department of Mathematics, Indian Institute of Science, Bangalore during seven weeks in February/March 1998. The course met four hours weekly with tutorials of two hours in addition. The arrangement of chapters follows quite closely the sequence of these lectures and each chapter contains more or less the subject-matter of one week. In addition to the exercises covered in the tutorial sessions, further exercises are added at the appropriate places to enhance the understanding and to provide examples. We recommend to look at them while studying the text. To those exercises which are used at other places sufficient hints for straightforward solutions are given. Chapter 7 is an expanded version of the lectures given in the last week (and would at least need two weeks to deliver). The lecture notes [12] based on a series of lectures in 1971/72 and written by Dr. Michael Lippa constituted an important model.

The objective of the lectures was to introduce Algebraic Geometry and Commutative Algebra simultaneously and to show their interplay. This aspect was developed systematically and in full generality with all its consequences in the work of A. Grothendieck, cf. [4]. In Commutative Algebra we do not introduce and use the concept of completion. In geometry we start the language of sheaves and schemes from scratch, but we avoid sheaf cohomology completely. The Riemann–Roch theorem is formulated for arbitrary coherent sheaves on arbitrary projective curves over an arbitrary field. Its proof we reduce to the case of the projective line. Instead of (first) cohomology it uses the dualizing sheaf. Since the uniqueness of this sheaf is not so important for the understanding of the Riemann–Roch theorem, its proof which uses some homological algebra is postponed to the end. We have added a lot of illustrative examples and related concepts to draw many consequences, especially about the genus of a projective curve.

We start with basic Commutative Algebra and emphasize on normalization. As geometric counterpart we then introduce the K-spectrum of a finitely generated algebra over a field K. We extend these concepts to prime spectra of arbitrary commutative rings and develop the dimension theory for arbitrary commutative Noetherian rings and their spectra. After introducing the language of sheaves we develop the theory of schemes, in particular, projective schemes. The main theorem of elimination and the mapping theorem of Chevalley are proved. Regularity, normality and smoothness are discussed in detail including the theory of Kähler differentials. We give a self-contained treatment of the module of Kähler differentials and use the sheaf of Kähler differentials as a fundamental example of a coherent and quasi-coherent module on a scheme. Before we prove the Riemann–Roch theorem we describe the coherent and quasi-coherent modules on projective schemes with the help of graded modules.

With very few exceptions full proofs are given under the assumption that the reader has some experience with the basic concepts of algebra, as groups, rings, fields, vector spaces, modules etc. It should be emphasized that, for a reader who has these prerequisites at his or her fingertips, this book is largely self-contained.

This work would have been impossible without the financial support from Deutscher Akademischer Austauschdienst (DAAD). Both authors have got opportunities for visiting the Ruhr University Bochum and the Indian Institute of Science in Bangalore respectively and thank DAAD for the generous support and the encouraging cooperation. The second author was partially supported by the GARP Funds, Indian Institute of Science and Part II B-UGC-SAP grant of Department of Mathematics Phase IV-Visiting Fellows, and he would like to express his gratitude for the kind hospitality during his stays in 1998 and 2008.

A first draft of the first five chapters was written by Dr. Indranath Sengupta. Dr. Abhijit Das further pushed for the finer draft, especially for the Chapters 5 and 6, during his stay in Bochum. Both were also supported by DAAD. We express our special thanks for their interest and competent work. Dr. Hartmut Wiebe from Ruhr University Bochum has helped us in many ways. He gave us technical support and steady encouragement to come to an end. We thank him wholeheartedly.

Bangalore and Bochum, April 2008   Dilip Patil and Uwe Storch

patil@math.iisc.ernet.in
uwe.storch@ruhr-uni-bochum.de

CONTENTS

CHAPTER 6 : Regular, Normal and Smooth Points

CHAPTER 7 : Riemann–Roch Theorem

CHAPTER 1 : Finitely Generated Algebras

Throughout this book a ring will always mean a commutative ring with identity if not stated otherwise. The letter K will always denote a field and the letters A, B, C, R will be generally used for rings. As usual we use $\mathbb{Z}$, $\mathbb{Q}$, $\mathbb{R}$ and $\mathbb{C}$ to denote the ring of integers, the fields of rational, real and complex numbers respectively.

1.A. Algebras over a Ring

Let A be a ring. An A-algebra is a pair (B, φ) where B is a ring and $\varphi : A \to B$ is a ring homomorphism called the s t r u c t u r e   h o m o m o r p h i s m of the A-algebra (B, φ). We will often omit φ in the notation of (B, φ) and simply say that B is an A-algebra.

Note that an A-algebra B is also an A-module, where the scalar multiplication is defined via the structure homomorphism $\varphi : A \to B$ by $ax := \varphi(a)x$ for all $a \in A$ and $x \in B$. Conversely, if a ring B is an A-module with the property:

$$(ax)(by) = (ab)(xy) \quad \text{for all } a, b \in A \text{ and } x, y \in B,$$

then B is an A-algebra with structure homomorphism $\varphi : A \to B$ defined by $a \mapsto a1_B$.

Let (B, φ) and (C, ψ) be two A-algebras. An A - a l g e b r a   h o m o m o r p h i s m $\theta : B \to C$ is a ring homomorphism such that the diagram

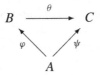

is commutative, that is, $\theta \circ \varphi = \psi$ or equivalently θ is A-linear.

1.A.1. Example (1) Let A be a subring of a ring B. Then B is an A-algebra with the natural inclusion $A \hookrightarrow B$ as the structure homomorphism.

(2) Let A be a ring and let $\mathfrak{a}$ be an ideal in A. Then the residue class ring $A/\mathfrak{a}$ is an A-algebra with the natural surjection $\pi : A \to A/\mathfrak{a}$ as the structure homomorphism.

(3) (P o l y n o m i a l   a l g e b r a) Let I be a set and let X_i, $i \in I$, be a family of indeterminates or variables over A. Then the p o l y n o m i a l   r i n g $A[X_i \mid i \in I]$ in the indeterminates X_i, $i \in I$, is an A-algebra and the natural inclusion $A \hookrightarrow A[X_i \mid i \in I]$ is the structure homomorphism.

Polynomial algebras are the free objects (in the language of categories) in the category of (commutative) A-algebras with the following universal property:

1.A.2. Universal property of polynomial algebras *Let B be an A-algebra and let x_i , $i \in I$, be a family of elements of B. Then there exists a unique A-algebra homomorphism $A[X_i \mid i \in I] \to B$ such that $X_i \mapsto x_i$ for every $i \in I$.*

In particular, we can identify $\mathrm{Hom}_{A-\mathrm{alg}}(A[X_i \mid i \in I], B)$ with B^I. Further, if $I = \{1, 2, \ldots, n\}$ then $\mathrm{Hom}_{A-\mathrm{alg}}(A[X_1, \ldots, X_n], B)$ can be identified with B^n.

Let B be an A-algebra and let $x := (x_i)_{i \in I}$ be a family of elements of B. Then the unique A-algebra homomorphism $\varepsilon : A[X_i \mid i \in I] \to B$ with $\varepsilon(X_i) = x_i$ for every $i \in I$ is called the s u b s t i t u t i o n h o m o m o r p h i s m or the e v a l u a t i o n h o m o m o r p h i s m defined by x. For $F \in A[X_i \mid i \in I]$, the image $\varepsilon(F)$ is denoted by $F(x)$ and is called the v a l u e of F at the point $x \in B^I$. Since ε is an A-algebra homomorphism, for $F, G \in A[X_i \mid i \in I]$ and $x \in B^I$, $a \in A$ we have

$$(F + G)(x) = F(x) + G(x), \quad (FG)(x) = F(x)G(x) \quad \text{and} \quad (aF)(x) = aF(x).$$

If $y \in B$ and $y = F(x)$, then x is called a y - p l a c e of F. In particular, $x \in B^I$ is called a 0 - p l a c e or z e r o of F if $F(x) = 0$.

The image of ε is the smallest A-subalgebra of B containing $\{x_i \mid i \in I\}$ and is denoted by $A[x_i \mid i \in I]$. We call it the A - s u b a l g e b r a g e n e r a t e d by the family x_i, $i \in I$. We say that B is an A - a l g e b r a g e n e r a t e d by the family x_i, $i \in I$, if $B = A[x_i \mid i \in I]$. Further, we say that B is a f i n i t e l y g e n e r a t e d A - a l g e b r a or an A - a l g e b r a o f f i n i t e t y p e if there exists a finite family $x_1, \ldots, x_n$ of elements of B such that $B = A[x_1, \ldots, x_n]$. A ring homomorphism $\varphi : A \to B$ is called a h o m o m o r p h i s m o f f i n i t e t y p e if B is an A-algebra of finite type with respect to φ.

The above discussions convey the fact that the residue class algebras $A[X_i \mid i \in I]/\mathfrak{a}$ represent all the A-algebras up to isomorphism and, therefore, a good understanding of the structure of the polynomial algebras over A is essential for the study of any A-algebra.

1.B. Factorization in Rings

We will begin by reviewing a study of division and factorization in rings. This study is modeled on properties of the ring of integers $\mathbb{Z}$.

Let R be a ring. An element $p \in R$ is called a p r i m e e l e m e n t if it is a non-zero divisor in R and if the principal ideal Rp is a prime ideal or, equivalently, if the residue class ring R/Rp is an integral domain. A non-zero divisor $a \in R$ is called i r r e d u c i b l e if a is a non-unit but not a product of two non-units. A prime element is always irreducible (but not conversely).

A ring R is called f a c t o r i a l (or a u n i q u e f a c t o r i z a t i o n d o m a i n (UFD)) if R is an integral domain and if every non-zero element $a \in R$ which is not a unit in R has a factorization $a = p_1 p_2 \cdots p_r$, where the elements $p_i \in R$ are prime elements for $i = 1, \ldots, r$. In a factorial domain every irreducible element is prime.

It is easy to show that in a factorial ring R, for every non-zero element $a \in R$ which is not a unit in R, a factorization $a = p_1 p_2 \cdots p_r$ of a into prime factors is unique up to a permutation and up to multiplication by units. Every principal ideal domain R (in particular, $\mathbb{Z}$) is factorial.

1.B.1. Proposition *Let R be a ring. If R is factorial then $R[X_i \mid i \in I]$ is also factorial.*

For $R = \mathbb{Z}$ the above proposition is a theorem due to Gauss. One important observation for proving the above proposition is the following lemma.

1.B.2. Lemma (G a u s s) *Let R be a ring and let $p \in R$ be a prime element. Then p is a prime element in $R[X_i \mid i \in I]$.*

PROOF. The rings $R[X_i \mid i \in I] / R[X_i \mid i \in I] \cdot p$ and $(R/Rp)[X_i \mid i \in I]$ are canonically isomorphic. ●

1.B.3. Corollary *Let R be a ring. If R is factorial then so is the polynomial ring $R[X_1, \ldots, X_n]$. In particular, the polynomial rings $K[X_1, \ldots, X_n]$ with K a field and $\mathbb{Z}[X_1, \ldots, X_n]$ are factorial.*

1.B.4. Example The following examples (besides (6)) are good to get a feeling about factoriality.

(1) Let $S \subseteq R$ be a multiplicatively closed set in the ring R not containing 0. Then every prime element $p \in R$ which is not a unit in $S^{-1}R$ is prime in $S^{-1}R$. Moreover, if R is factorial then so is $S^{-1}R$.

(2) Let R be a factorial domain and let $a \in R$ be either a prime element or a unit in R. Then $R[X, Y]/(XY + a)$ is a factorial domain.

(3) $\mathbb{R}[X, Y]/(X^2 + Y^2 - 1)$ is not factorial, but $\mathbb{R}[X, Y]/(X^2 + Y^2 + 1)$ is factorial.

(4) (K l e i n – N a g a t a) Let K be a field of characteristic $\neq 2$ and let $a_1, \ldots, a_n$, $n \geq 5$, be non-zero elements of K. Then $K[X_1, \ldots, X_n]/(a_1 X_1^2 + \cdots + a_n X_n^2)$ is factorial.

(5) Let K be a field of characteristic $\neq 2$ and let Q be a non-degenerate quadratic form in $K[X_1, \ldots, X_n]$. If $n \geq 5$ then $K[X_1, \ldots, X_n]/(Q)$ is factorial by (4) above.

(6) Let D be a square free integer $\neq 0, 1$ and let R_D be the ring of algebraic integers in the quadratic field $\mathbb{Q}(\sqrt{D})$. If $D < 0$ then R_D is factorial if and only if D belongs to $\{-1, -2, -3, -7, -11, -19, -43, -67, -163\}$. [1])

---

[1]) This is a very deep theorem. Gauss proved for these values of D that R_D is factorial. He also conjectured that there is no other. This much more difficult part of the theorem was finally proved in 1967, after the problem had been worked out for more than 150 years. In 1967 Stark found a proof of this theorem as did Baker soon after. The situation for positive D is not well understood. It is not known whether R_D is factorial (i.e. a principal ideal domain) for infinitely many $D > 0$.

1.B.5. Exercise (1) Let A denote a factorial domain and let K be the quotient field of A. Let $A[X_1, \ldots, X_n]$ be the polynomial ring and let $F \in A[X_1, \ldots, X_n]$ be a non-constant polynomial. Prove the following statements (use only simple arguments):

a) Let $\varphi : A[X_1, \ldots, X_n] \to A[X_1, \ldots, X_n]$ be a ring isomorphism. Then F is prime if and only if $\varphi(F)$ is prime.

b) Let B be an integral domain containing A. If the coefficients of F are relatively prime and if F is irreducible in $B[X_1, \ldots, X_n]$ then F is prime in $A[X_1, \ldots, X_n]$.

c) If the degree form F_d of F is prime then so is F.

d) F is prime if and only if its homogenization F^h is prime.

e) Let $\mathfrak{p}$ be a prime ideal in A. If the residue class $\overline{F}$ of F in $(A/\mathfrak{p})[X_1, \ldots, X_n]$ is irreducible of degree $\deg(F)$ and if the coefficients of F are relatively prime, then F is prime in $A[X_1, \ldots, X_n]$.

f) F is prime in $A[X_1, \ldots, X_n]$ if and only if F is prime in $K[X_1, \ldots, X_n]$ and the coefficients of F are relatively prime.

(2) Let A denote a factorial domain and let K be the quotient field of A. Show that the following polynomials are irreducible.

a) $XY - a \in A[X, Y]$, $\quad a \neq 0$; $\qquad\qquad (X-1)^2(X^2 + Y^2) - X^2$, $\operatorname{char} K \neq 2$;

$\quad (X^2 + Y^2)(X - 2) + X$, $\operatorname{char} K \neq 2$; $\qquad X^3 + X^2 - Y^2$.

b) $aX^m + bY^n$, $m, n \in \mathbb{N}^*$ relatively prime and $a, b \in A^* := A \setminus \{0\}$ relatively prime in A.

c) $X^{2m} + Y^{2n} \in \mathbb{R}[X, Y]$, $m, n \in \mathbb{N}^*$ relatively prime. (**Hint:** Look at the prime factorization in $\mathbb{C}[X, Y]$.)

d) $\det \left(X_{ij} \right)_{1 \leq i, j \leq n} \in A[X_{ij} \mid 1 \leq i, j \leq n]$.

e) $X^d - G(Z)/H(Z) \in K(Z)[X]$, $d \in \mathbb{N}^*$, where $G, H \in K[Z]$ are such that GH is non-constant and has no multiple factors.

f) $a_1 X_1^{\nu_1} + a_2 X_2^{\nu_2} + \cdots + a_n X_n^{\nu_n} \in A[X_1, \ldots, X_n]$, $n \geq 3$, $\nu_1, \ldots, \nu_n \in \mathbb{N}^*$ not all zero in A, $a_1, \ldots, a_n \in A^*$ relatively prime. (**Hint:** One assumes $A = K$ and using Eisenstein's criterion reduces to the case $n = 3$. Then use the fact that $a_1 X_1^{\nu_1} + a_2 X_2^{\nu_2}$ has no multiple factors if either $\nu_1 \neq 0$ or $\nu_2 \neq 0$ in K.)

g) $X_1^d + \cdots + X_n^d + G \in A[X_1, \ldots, X_n]$, $n \geq 3$, $d \neq 0$ in K and $G \in A[X_1, \ldots, X_n]$ is any polynomial of degree $< d$. What about the case $n = 2$?

h) $(X - a_1) \cdots (X - a_d) + 1 \in \mathbb{Z}[X]$, $d \geq 1$, where $a_1, \ldots, a_d \in \mathbb{Z}$ are distinct.

1.C. Noetherian Rings and Modules

Let R be a ring and let M be an R-module. We say that M is N o e t h e r i a n if it satisfies the equivalent conditions of the proposition below. A ring R is called a N o e t h e r i a n r i n g if it is Noetherian as an R-module.

1.C.1. Proposition *Let R be a ring. Then for an R-module M the following three conditions are equivalent:*

(1) Every submodule of M is finitely generated.

(2) *M satisfies the ascending chain condition for submodules, i. e. if $M_1 \subseteq M_2 \subseteq M_3 \subseteq \cdots$ is any ascending sequence of submodules of M, then there exists a positive integer n such that $M_n = M_{n+1} = M_{n+2} = \cdots$.*

(3) *Every non-empty family of submodules of M has a maximal element.*

PROOF. (1) $\Rightarrow$ (2): Let $M_1 \subseteq M_2 \subseteq M_3 \subseteq \cdots$ be an ascending sequence of submodules of M. Let $N = \bigcup_{i=1}^{\infty} M_i$. Then N is a submodule of M and is therefore generated by a finite number of elements, say $x_1, \ldots, x_r$. There exists a positive integer n such that $x_1, \ldots, x_r \in M_n$. Therefore we have $M_n = N$, so that $M_n = M_{n+1} = \cdots$.

(2) $\Rightarrow$ (3): Let $\mathcal{F}$ be a non-empty family of submodules of M. Suppose $\mathcal{F}$ does not have a maximal element. Choose any $M_1 \in \mathcal{F}$. Suppose there exist $M_2, \ldots, M_n \in \mathcal{F}$ such that $M_1 \subset M_2 \subset \cdots \subset M_n$. Then, since M_n is not maximal, there exists $M_{n+1} \in \mathcal{F}$ such that $M_n \subset M_{n+1}$. Therefore, recursively, we get an infinite sequence $M_1 \subset M_2 \subset M_3 \subset \cdots$ such that $M_n \neq M_{n+1}$ for every n. This contradicts (2).

(3) $\Rightarrow$ (1): Let N be a submodule of M. Let $\mathcal{F}$ be the family of all finitely generated submodules of N. Since $0 \in \mathcal{F}$, $\mathcal{F}$ is non-empty. Therefore $\mathcal{F}$ has a maximal element, say N'. If $N' \neq N$ then there exists $x \in N$, $x \notin N'$. The submodule $N' + Rx$ of N is finitely generated and contains N' properly. This is a contradiction. Therefore $N' = N$ and N is finitely generated. •

1.C.2. Example (1) A vector space V over a field K is Noetherian if and only if V is finite dimensional over K, that is, $\mathrm{Dim}_K V < \infty$.

(2) Every principal ideal domain is Noetherian. In particular, $\mathbb{Z}$ is Noetherian and, if K is a field, then the polynomial ring $K[X]$ and the formal power series ring $K[[X]]$ are Noetherian.

(3) If R is a Noetherian ring and $\mathfrak{a}$ is an ideal in R, then $R/\mathfrak{a}$ is a Noetherian ring.

We list some simple properties of Noetherian modules. The proofs are very easy.

1.C.3. Proposition *Let R be a ring.*

(1) *Let $0 \longrightarrow M' \longrightarrow M \longrightarrow M'' \longrightarrow 0$ be an exact sequence of R-modules. Then M is Noetherian if and only if both M' and M'' are Noetherian.*

(2) *Let N be a submodule of an R-module M. Then M is Noetherian if and only if both N and M/N are Noetherian R-modules.*

(3) *A finite direct sum of Noetherian modules is Noetherian.*

(4) *Suppose that R is a Noetherian ring. Let M be a finitely generated R-module. Then M is Noetherian.*

(5) *Let S be a multiplicatively closed subset of R and let M be a Noetherian R-module. Then $S^{-1}M$ is a Noetherian $S^{-1}R$-module.*

(6) *Let S be a multiplicatively closed subset of a Noetherian ring R. Then $S^{-1}R$ is Noetherian. In particular, the localization $R_{\mathfrak{p}}$ of a Noetherian ring R at a prime ideal $\mathfrak{p}$ is Noetherian.*

1.C.4. Hilbert's Basis Theorem *Let R be a Noetherian ring. Then the polynomial ring $R[X_1, \ldots, X_n]$ in n variables over R is also Noetherian.*

PROOF. By induction on n, it is sufficient to prove the theorem for $n = 1$, i.e. that the polynomial ring $B := R[X]$ in one variable is Noetherian. Let $\mathfrak{b}$ be any ideal of B. We will show that $\mathfrak{b}$ is finitely generated. Suppose that $\mathfrak{b}$ is not finitely generated. Then choose $f_1, f_2, f_3, \ldots$ inductively such that f_n is of smallest degree in $\mathfrak{b} \setminus \sum_{i=1}^{n-1} B f_i$. Let $d_n := \deg(f_n)$ and let a_n be the leading coefficient of f_n. Then $d_1 \le d_2 \le \cdots$. Since R is Noetherian , there exists a positive integer m such that $a_m \in \sum_{i=1}^{m-1} R a_i$. Write $a_m = \sum_{i=1}^{m-1} \alpha_i a_i$ with $\alpha_i \in R$. Let $g := f_m - \sum_{i=0}^{m-1} \alpha_i X^{d_m - d_i} f_i$. Then $g \in \mathfrak{b} \setminus \sum_{i=1}^{m-1} B f_i$ and $\deg(g) < d_m$. This contradicts the choice of f_m. Therefore $\mathfrak{b}$ is finitely generated. •

This short proof is due to H. Sarges (see: Ein Beweis des Hilbertschen Basissatzes, J. Reine und Angew. Math. **283/284** (1976), 436-437). At the end of 1.D we give a more conceptual proof of Hilbert's basis theorem.

1.C.5. Corollary *Let R be a Noetherian ring and B a finitely generated R-algebra. Then B is Noetherian.*

PROOF. Since any finitely generated R-algebra is a quotient of a polynomial algebra $R[X_1, \ldots, X_n]$, the corollary follows. •

1.C.6. Exercise Let R be a ring.

(1) Let M be an R-module . Let B be a subring of R, so that M is also a B-module. If M is Noetherian as a B-module then M is Noetherian as an R-module.

(2) Let M be a Noetherian R-module. Show that any surjective R-endomorphism of M is an isomorphism.

(3) Let M be a Noetherian R-module and let $\mathfrak{a} := \mathrm{Ann}_R M = \{a \in R \mid aM = 0\}$. Show that $R/\mathfrak{a}$ is a Noetherian ring.

(4) Let R be a non-Noetherian ring and let $\mathcal{F}$ be the set of ideals in R which are not finitely generated. Show that $\mathcal{F}$ has maximal elements and that the maximal elements of $\mathcal{F}$ are prime ideals.

(5) (I. S. C o h e n) A ring R is Noetherian if and only if every prime ideal of R is finitely generated. (**Hint:** Use (4).)

(6) Suppose that $R_\mathfrak{p}$ is Noetherian for every prime ideal $\mathfrak{p} \subseteq R$. Is R necessarily Noetherian?

(7) Let B be a faithfully flat R-algebra. If B is Noetherian, show that R is Noetherian. More generally: Let $R \subseteq B$ be a ring extension with $(\mathfrak{a}B) \cap R = \mathfrak{a}$ for all finitely generated ideals $\mathfrak{a} \subseteq R$. If B is Noetherian, then R is Noetherian. (For example, if $R \subseteq B$ and R is a direct summand of B as an R-module, then R is Noetherian if B is Noetherian.)

(8) Let $\mathfrak{P}$ be a prime ideal in the formal power series ring $R[\![X]\!]$ over R and let $\mathfrak{p} = \{f(0) \mid f \in \mathfrak{P}\}$. Show that $\mathfrak{p}$ is a prime ideal of R and if $\mathfrak{p}$ is generated by r elements then $\mathfrak{P}$ can be generated by $r + 1$ elements.

(9) If R is Noetherian then the formal power series ring $R[\![X_1, \ldots, X_n]\!]$ in n variables over R is also Noetherian. (**Hint:** Use (5) and (8).)

1.D. Graded Rings and Modules

A g r a d i n g of t y p e $\mathbb{Z}$ or $\mathbb{Z}$-g r a d i n g on a ring R is a sequence $(R_n)_{n \in \mathbb{Z}}$ of subgroups of R such that $R = \bigoplus_{n \in \mathbb{Z}} R_n$ and $R_m R_n \subseteq R_{m+n}$ for all $m, n \in \mathbb{Z}$. A ring with a $\mathbb{Z}$-grading is called a $\mathbb{Z}$-g r a d e d r i n g. A graded ring $R = \bigoplus_{n \in \mathbb{Z}} R_n$ of type $\mathbb{Z}$ is called p o s i t i v e l y g r a d e d or $\mathbb{N}$-g r a d e d if $R_n = 0$ for all $n < 0$.

1.D.1. Example Let A be any ring.

(1) The grading $A_0 := A$, $A_n := 0$ for all $n \in \mathbb{Z}$, $n \neq 0$ on A is called the t r i v i a l g r a d i n g on A.

(2) For $n \in \mathbb{Z}$, the subgroups

$$R_n := \{0\} \cup \{F \in A[X_1, \ldots, X_r] \mid F \text{ is homogeneous of degree } n\}$$

define a grading on the polynomial ring $R := A[X_1, \ldots, X_r]$. This grading is called the u s u a l or s t a n d a r d g r a d i n g on R.

(3) Let $r \in \mathbb{N}$ and $\gamma := (\gamma_1, \ldots, \gamma_r) \in \mathbb{Z}^n$. For a monomial $X^m := X_1^{m_1} \cdots X_r^{m_r} \in R := A[X_1, \ldots, X_r]$, let $\deg_\gamma X^m := m_1 \gamma_1 + \cdots + m_r \gamma_r$ be the so called γ-d e g r e e of X^m. For $n \in \mathbb{Z}$, let R_n be the A-submodule generated by all monomials of γ-degree n. Then $(R_n)_{n \in \mathbb{Z}}$ is a grading on R and is called the w e i g h t e d g r a d i n g corresponding to the w e i g h t s $\gamma_1, \ldots, \gamma_r$ on R or the γ-g r a d i n g on R. If $\gamma_i = 1$ (respectively $\gamma_i = 0$) for all $i = 1, \ldots, r$ then the corresponding weighted grading on R is the standard (respectively the trivial) grading on R.

Let $R = \bigoplus_{n \in \mathbb{Z}} R_n$ be a $\mathbb{Z}$-graded ring and let M be an R-module. A g r a d i n g of t y p e $\mathbb{Z}$ or $\mathbb{Z}$-g r a d i n g on M is a sequence $(M_n)_{n \in \mathbb{Z}}$ of subgroups of M such that $M = \bigoplus_{n \in \mathbb{Z}} M_n$ and $R_m M_n \subseteq M_{m+n}$ for all $m, n \in \mathbb{Z}$. An R-module M with a $\mathbb{Z}$-grading is called a $\mathbb{Z}$-g r a d e d R-m o d u l e.

Let $R = \bigoplus_{n \in \mathbb{Z}} R_n$ be a graded ring and let $M = \bigoplus_{n \in \mathbb{Z}} M_n$ be a graded R-module. Then R_0 is a subring of R and R_n (respectively M_n) is an R_0-submodule of R (respectively M) for every $n \in \mathbb{Z}$.

For $n \in \mathbb{Z}$, the elements of R_n (respectively M_n) are called the h o m o g e n e o u s e l e m e n t s of R (respectively M) o f d e g r e e n. The zero element of R (respectively M) is homogeneous of degree n for every $n \in \mathbb{Z}$.

Every $x \in M$ can be written uniquely in the form $x = \sum_{n \in \mathbb{Z}} x_n$ with $x_n \in M_n$ for all $n \in \mathbb{Z}$ and $x_n = 0$ for almost all n. The x_n are called the h o m o g e n e o u s c o m p o n e n t s of x of degree n.

For a non-zero element $x \in M$, the integers $\omega(x) := \inf \{n \in \mathbb{Z} \mid x_n \neq 0\}$ and $\deg(x) := \sup \{n \in \mathbb{Z} \mid x_n \neq 0\}$ are called the o r d e r and the d e g r e e of x, respectively. We put $\omega(0) := \infty$ and $\deg(0) := -\infty$. For $0 \neq x \in M$, the non-zero homogeneous elements $x_{\omega(x)}$ and $x_{\deg(x)}$ are called the i n i t i a l f o r m and the d e g r e e f o r m of x respectively.

An R-submodule N of M is called a g r a d e d or h o m o g e n e o u s submodule of M if it satisfies the following equivalent conditions: (1) $N = \sum_{n \in \mathbb{Z}} (M_n \cap N)$.

(2) $N = \bigoplus_{n \in \mathbb{Z}} (M_n \cap N)$.   (3) If $x \in N$ then every homogeneous component of x belongs to N.   (4) N is generated by a set of homogeneous elements of M.

A graded R-submodule $\mathfrak{a}$ of R is called a g r a d e d or h o m o g e n e o u s i d e a l of R.

Let $\mathfrak{a} \subseteq R$ be an ideal and let $L(\mathfrak{a})$ be the ideal generated by the degree forms $L(x) := x_{\deg(x)}$, $x \in \mathfrak{a}$, $x \neq 0$. Then $L(\mathfrak{a})$ is a homogeneous ideal in R. We say that the family f_j, $j \in J$, of elements of $\mathfrak{a}$ is a G r ö b n e r b a s i s of $\mathfrak{a}$ if $L(f_j)$, $j \in J$, generate $L(\mathfrak{a})$.

1.D.2. Lemma *Let $R = \bigoplus_{n \in \mathbb{N}} R_n$ be a positively graded ring and let $\mathfrak{a} \subseteq R$ be an ideal in R. If the family f_j, $j \in J$, of elements of $\mathfrak{a}$ is a Gröbner basis of $\mathfrak{a}$ then $\mathfrak{a}$ is generated by f_j, $j \in J$.*

PROOF. Let $x \in \mathfrak{a}$, $x \neq 0$. We shall prove by induction on $\deg x$ that $x \in \sum_{j \in J} R f_j$. Write $x = x_0 + \cdots + x_d$ with $d := \deg x$. Then $L(x) = x_d \in L(\mathfrak{a})$ and so there exist homogeneous elements $a_i \in R$, $i = 1, \ldots, n$, such that $x_d = \sum_{i=1}^{n} a_i L(f_{j_i})$ and $\deg(a_i) = d - \deg(f_{j_i})$. In particular, $\deg \left(x - \sum_{i=1}^{n} a_i f_{j_i} \right) < d$. If $d = 0$ then $x = x_0 = \sum_{i=1}^{n} a_i f_{j_i} \in \sum_{j \in J} R f_j$, since R is positively graded. Now assume that $d \geq 1$. Then $x - \sum_{i=1}^{n} a_i f_{j_i} \in \sum_{j \in J} R f_j$ by induction and therefore $x \in \sum_{j \in J} R f_j$.   •

We use the above lemma to give a conceptual p r o o f of the H i l b e r t B a s i s T h e o r e m, using the language of Gröbner bases: Let R be a Noetherian ring and let $\mathfrak{a} \subseteq R[X]$ be an ideal. Since $L(\mathfrak{a})$ is a homogeneous ideal in $R[X]$ (we take the standard grading on $R[X]$), we have $L(\mathfrak{a}) = \bigoplus_{m \in \mathbb{N}} \mathfrak{a}_m X^m$, where $\mathfrak{a}_m$, $m \in \mathbb{N}$, are ideals in R with $\mathfrak{a}_0 \subseteq \mathfrak{a}_1 \subseteq \cdots \subseteq \mathfrak{a}_n \subseteq \cdots$. Now, since R is Noetherian, there exists an $m_0 \in \mathbb{N}$ such that $\mathfrak{a}_m = \mathfrak{a}_{m_0}$ for all $m \geq m_0$. Then $L(\mathfrak{a})$ is generated by $\mathfrak{a}_0, \mathfrak{a}_1 X, \ldots, \mathfrak{a}_{m_0} X^{m_0}$ and since the ideals $\mathfrak{a}_0, \mathfrak{a}_1, \ldots, \mathfrak{a}_{m_0}$ are finitely generated in R the ideal $L(\mathfrak{a})$ in $R[X]$ is finitely generated. This proves that $\mathfrak{a}$ has a finite Gröbner basis and therefore $\mathfrak{a}$ is finitely generated by the previous lemma.   •

1.E. Integral Extensions

In this section we collect few basic facts on integral extensions which we will come across quite often in these lectures. Let R be an algebra over a ring A.

We say that R is a f i n i t e A-algebra if R is a finitely generated as an A-module, i. e. if there exist finitely many elements $x_1, \ldots, x_n \in R$ such that $R = A x_1 + \cdots + A x_n$. A ring homomorphism $\varphi : A \to R$ is called f i n i t e if R is a finite A-algebra with respect to φ.

Obviously, we have the t r a n s i t i v i t y o f f i n i t e n e s s : *If R is a finite A-algebra and S is a finite R-algebra then S is a finite A-algebra.*

It is clear that a finite algebra is of finite type, but the converse is not true. For example, the polynomial algebra $A[X_1, \ldots, X_n]$, $n \geq 1$, $A \neq 0$, is of finite type over A, but not a finite A-algebra.

An element x of an A-algebra R is said to be integral over A if it is a zero of a monic polynomial with coefficients in A, that is,

$$x^d + a_{d-1}x^{d-1} + \cdots + a_0 = 0, \quad a_0, \ldots, a_{d-1} \in A,$$

equivalently, the kernel of the substitution homomorphism $A[X] \to R$, $X \mapsto x$, contains a monic polynomial (in $A[X]$). Such a monic polynomial equation is called an integral equation of x over A. In case, A is a field, the concept of integral elements and the concept of algebraic elements are equivalent.

1.E.1. Example (1) (Theorem of Cayley–Hamilton) Let R be a finite free A-algebra of rank d with A-basis $x_1, \ldots, x_d$ and let $x \in R$. Consider the left translation $\lambda_x : R \to R$, $y \mapsto xy$. For each $j = 1, \ldots, d$, write $xx_j = \sum_{i=1}^{d} a_{ij}x_i$ with $a_{ij} \in A$. Then $\sum_{i=1}^{d} (x\delta_{ij} - a_{ij}) x_i = 0$ for all $j = 1, \ldots, d$ and therefore, by Cramer's rule, we have $\det (x\mathfrak{E}_d - \mathfrak{A}) x_i = 0$ for all $i = 1, \ldots, d$, where $\mathfrak{E}_d$ is the $d \times d$ identity matrix and $\mathfrak{A}$ is the $d \times d$ matrix (a_{ij}). *Therefore* $\chi_x(x) = \det (x\mathfrak{E}_d - \mathfrak{A}) = 0$ *is a canonical integral equation of degree d of x over A*, where $\chi_x = \det (X\mathfrak{E}_d - \mathfrak{A})$ is the characteristic polynomial of the A-linear endomorphism λ_x of R.

(2) Let b be a non-zero divisor in A which is not a unit. Then the element $1/b$ in the total quotient ring $Q = Q(A)$ of A is not integral over A. The kernel of the substitution homomorphism $A[X] \to Q$, $X \mapsto 1/b$, is generated by the linear polynomial $bX - 1$ and contains no monic polynomial.

The next proposition gives the connection between integral elements and finite algebras. First recall that an A-module M is said to be faithful if $\mathrm{Ann}_A M = 0$.

1.E.2. Proposition *Let R be an A-algebra and let $x \in R$. The following statements are equivalent*:

(1) *x is integral over A.*

(2) *$A[x]$ is a finite A-algebra.*

(3) *$A[x]$ is contained in a finite A-subalgebra S of R.*

(4) *There is a faithful $A[x]$-module M which is finite as an A-module.*

PROOF. $(1) \Rightarrow (2)$: By (1) $x^d + a_{d-1}x^{d-1} + \cdots + a_0 = 0$ for some $a_0, \ldots, a_{d-1} \in A$. Then $x^d \in A + Ax + \cdots + Ax^{d-1}$ and so by induction $x^m \in A + Ax + \cdots + Ax^{d-1}$ for all $m \geq 0$. Therefore $A[x] = A + Ax + \cdots + Ax^{d-1}$ is a finite A-algebra.

The implication $(2) \Rightarrow (3)$ is trivial.

$(3) \Rightarrow (4)$: Since $A[x]$ is a subring of S, the $A[x]$-module S is faithful and therefore $M = S$ serves the purpose.

$(4) \Rightarrow (1)$: Let $M = Ax_1 + Ax_2 + \cdots + Ax_n$ be a faithful $A[x]$-module. For each $j = 1, \ldots, n$, write $xx_j = \sum_{i=1}^{n} a_{ij}x_i$ with $a_{ij} \in A$. Then $\sum_{i=1}^{n} (x\delta_{ij} - a_{ij}) x_i = 0$

for all j and so $\det\left(x\delta_{ij} - a_{ij}\right) M = 0$ by Cramer's rule. Therefore, since M is a faithful $A[x]$-module, the equation $\det\left(x\delta_{ij} - a_{ij}\right) = 0$ is an integral equation of x over A (of degree n).   •

1.E.3. Remark The proof of the implication (4) $\Rightarrow$ (1) shows: *If* $xM \subseteq \mathfrak{a}M$ *for an ideal* $\mathfrak{a} \subseteq A$, *then there is an integral equation* $x^n + a_{n-1}x^{n-1} + \cdots + a_0 = 0$ *with* $a_j \in \mathfrak{a}^{n-j}$, $j = 0, \ldots, n-1$.

1.E.4. Corollary *Let R be an A-algebra and let $x_1, \ldots, x_r \in R$. If $x_1, \ldots, x_n$ are integral over A, then $A[x_1, \ldots, x_r]$ is a finite A-algebra.*

PROOF. We use induction on r. For $r = 1$ the assertion is a part of 1.E.2. For $r \geq 2$, $R' := A[x_1, \ldots, x_{r-1}]$ is a finite A-algebra by induction hypothesis. Now x_r being integral over A, it is also integral over R' and so $R'[x_r] = A[x_1, \ldots, x_r]$ is a finite R'-algebra by 1.E.2. Therefore $A[x_1, \ldots, x_r]$ is a finite A-algebra, too. •

The set of elements of R which are integral over A is called the **integral closure** of A in R, and is usually denoted by $\overline{A}$.

1.E.5. Corollary *For any A-algebra R, $\overline{A}$ is a subalgebra of R.*

PROOF. For any $x, y \in \overline{A}$, $A[x, y] \subseteq R$ is a finite A-algebra by 1.E.4 and therefore $A[x, y] \subseteq \overline{A}$ by Proposition 1.E.2.   •

1.E.6. Corollary *Let S be an R-algebra and R be an A-algebra. If S is integral over R and R is integral over A then S is integral over A.*

PROOF. Let $x \in S$ and let $x^d + b_{d-1}x^{d-1} + \cdots + b_0 = 0$ be an integral equation of x over R. Then x is integral over $R' := A[b_0, \ldots, b_{d-1}]$ and so $R'[x]$ is a finite R'-algebra by 1.E.2 and hence a finite A-algebra by 1.E.4. Therefore x is integral over A by 1.E.2.

1.E.7. Corollary *For an A-algebra R, the following statements are equivalent:*
(1) *R is a finite A-algebra.*   (2) *R is of finite type and integral over A.*

1.E.8. Example (1) For a monic polynomial $F = X^d + a_{d-1}X^{d-1} + \cdots + a_0 \in A[X]$ the A-algebra $R = A[X]/(F) = A[x]$ is a finite A-algebra. The residue class x of X is integral over A because $x^d + a_{d-1}x^{d-1} + \cdots + a_0 = F(x) = 0$. In fact, R is a free A-algebra of rank d with A-basis $1, x, \ldots, x^{d-1}$.

(2) Let $F \in A[X_1, \ldots, X_n]$ be a monic polynomial in X_n, i.e. $F = X_n^d + F_{d-1}X_n^{d-1} + \cdots + F_0$ with $F_i \in A[X_1, \ldots, X_{n-1}]$ and $x_1, \ldots, x_n$ be the residue classes of $X_1, \ldots, X_n$ modulo the principal ideal (F). Then $A[X_1, \ldots, X_n]/(F) = A[x_1, \ldots, x_n]$ is a finite free algebra of rank d over the polynomial algebra $A[X_1, \ldots, X_{n-1}] \cong A[x_1, \ldots, x_{n-1}]$. This is a special case of (1). In case $A = K$ is a field, *every non-zero polynomial $G \in K[X_1, \ldots, X_n]$, after a change of variables, can be expressed, up to a constant, as a monic polynomial in X_n.* We prove this in the next section, see 1.F.1.

(3) (N o r m a l i z a t i o n) Let A be a ring and let $Q := Q(A)$ be its total quotient ring. Then the integral closure $\overline{A}$ of A in Q is called the n o r m a l i z a t i o n of A. An integral domain A is called n o r m a l if $\overline{A} = A$. The normalization of an integral domain A is the smallest subring of its quotient field Q which is normal and contains A.

For example: *Every factorial domain is normal. In particular, polynomial rings over* $\mathbb{Z}$ *or a field* K *are normal.* For the *proof* note that if $x = a/b \in Q$ with $\mathrm{GCD}(a, b) = 1$ and x is a zero of a polynomial $a_n X^n + \cdots + a_0 \in A[X]$ then a divides a_0 and b divides a_n. More generally, $bX - a$ is a generator of the kernel of the subsitution homomorphism $A[X] \to Q$, $X \mapsto a/b$.

(4) (C o n d u c t o r) Let $A \subseteq R$ be a ring extension. Then the ideal $\mathfrak{C}_{R|A} := \mathrm{Ann}_A R/A = \{a \in A \mid aR \subseteq A\}$ is called the c o n d u c t o r of R o v e r A. It is the largest ideal in A which is also an ideal in R. If $\mathfrak{C}_{R|A}$ contains a non-zero divisor of A then $R \subseteq Aa^{-1}$ can be embedded in the total quotient ring $Q(A)$ of A. Moreover, R is then finite over A if A is Noetherian. The c o n d u c t o r of A is the ideal $\mathfrak{C}_A := \mathfrak{C}_{\overline{A}|A}$ where $\overline{A}$ is the normalization of A (see (3)).

(5) The following lemma gives an important property of a normal domain.

1.E.9. Lemma *Let A be a normal domain with quotient field Q. If an element x of a Q-algebra L (not necessarily a field) is integral over A then the minimal polynomial μ_x of x over Q has coefficients in A.*

PROOF. By extending L, we may assume that μ_x splits into linear factors over L (see also Exercise 1.E.10 (1) below). Let $f(x) = 0$, $f \in A[X]$, be an integral equation of x over A. Then μ_x divides f in $Q[X]$ and so, every zero y of μ_x is integral over A (with integral equation $f(y) = 0$). Therefore the coefficients of μ_x are integral over A by 1.E.5 and hence elements of A, since A is normal. •

(6) Let M be a n u m e r i c a l m o n o i d, i.e. M is a submonoid of $\mathbb{N} = (\mathbb{N}, +)$ such that $\mathbb{N} \setminus M$ is finite. Let $A = K[M] := \{\sum_{m \in M} a_m T^m \in K[T]\} \subseteq K[T]$ be the monoid algebra of M over a field K. Then the polynomial algebra $K[T]$ is finite over $K[M]$, indeed, $\mathrm{Dim}_K K[T]/K[M] = \mathrm{Card}(\mathbb{N} \setminus M)$, and so $K[T]$ is integral over $K[M]$. Since T belongs to the quotient field of $K[M]$ and $K[T]$ is normal, $K[T]$ is the normalization of $K[M]$. The K-algebra $K[M]$ is called the coordinate algebra of the m o n o m i a l c u r v e over K defined by M. (See Exercise 2.B.14 (3).)

(7) *Let K be a field and let A be a normal K-subalgebra of $K[T]$, $A \neq K$. Then A is a polynomial algebra $K[f]$ for some $f \in A$.* (f is necessarily a non-constant polynomial in A of least degree.) For the *proof*, let $\mu_T = X^n + f_{n-1}X^{n-1} + \cdots + f_0 \in Q(A)[X]$ be the minimal polynomial of T over $Q(A)$. By the lemma in (5), the coefficients $f_0, \ldots, f_{n-1} \in A$. But every non-constant coefficient f of μ_T generates the field $Q(A)$ over K (see the proof of L ü r o t h ' s t h e o r e m in van der Waerden, B. L.: Algebra, Part I, § 73, p. 222). Then $K[f] \subseteq A \subseteq Q(A) = K(f)$ and $K[f] = A$, since $K[T]$ and hence A is integral over $K[f]$ and $K[f]$ is normal. (For Lüroth's theorem see also the end of Example 7.E.18.) •

As a consequence we get: *Let A be a K-subalgebra of $K[T]$, $A \neq K$. Then the normalization $\overline{A}$ of A is a polynomial algebra $K[f]$ for some non-constant polynomial $f \in K[T]$.* (Note that every K-subalgebra of $K[T]$ is a K-algebra of finite type.)

In general a K-algebra A of finite type is called r a t i o n a l if it is an integral domain and if the quotient field $Q(A)$ of A is K-isomorphic to a rational function field $K(T_1, \ldots, T_m)$ in m variables $T_1, \ldots, T_m$. The integer m is nothing but the transcendence degree of the field extension $K \subseteq Q(A)$. By Lüroth's theorem, any K-subalgebra A of $K(T)$, $A \neq K$, of finite type is rational with $m = 1$.

As an example, consider the K-algebra homomorphism $\varphi : K[X, Y] \to K[T]$ defined by $x := \varphi(X) = T^2 - 1$ and $y := \varphi(Y) = T(T^2 - 1)$ and the K-subalgebra $A := \operatorname{im} \varphi$ of $K[T]$. Obviously, if $f \in K[X, Y]$ with $f \neq 0$ and $\deg_Y f \leq 1$ then $f \notin \operatorname{Ker} \varphi$, therefore $\operatorname{Ker} \varphi$ is the principal ideal generated by $Y^2 - X^2 - X^3$. Since $T = y/x$ belongs to the quotient field of A, *the polynomial algebra $K[T]$ is the normalization of A* $\cong K[X, Y]/\left(Y^2 - X^2(X + 1)\right)$.

1.E.10. Exercise (1) Let $A \subseteq B$ be a ring extension and let $H, G \in B[X]$ be monic polynomials such that $HG \in A[X]$. Then the coefficients of H and G are integral over A. If A is integrally closed in B then $H, G \in A[X]$. (**Hint:** There is a finite ring extension C of B (even a free one) such that H and G factor into monic linear factors in $C[X]$.)

(2) Let R be an A-algebra. Then the integral closure of $A[T_1, \ldots, T_n]$ in $R[T_1, \ldots, T_n]$ is $\overline{A}[T_1, \ldots, T_n]$, where $\overline{A}$ is the integral closure of A in R. (**Hint:** Assume $n = 1$. Let $g \in R[T]$ be integral over $A[T]$ with integral equation $0 = f(g) = g^n + f_{n-1}g^{n-1} + \cdots + f_1 g + f_0$. Let r be an integer larger than n and the degrees of $f_{n-1}, \ldots, f_0$ and let $g_1 := g - T^r$. From

$$(g_1 + T^r)^n + f_{n-1}(g_1 + T^r)^{n-1} + \cdots + f_1(g_1 + T^r) + f_0 = 0$$

one gets $(-g_1)(g_1^{n-1} + h_{n-1}g_1^{n-2} + \cdots + h_1) = (T^r)^n + (T^r)^{n-1} f_{n-1} + \cdots + (T^r) f_1 + f_0$ which is a monic polynomial in $A[T]$. Now use (1) to conclude that g_1 and hence g has coefficients in $\overline{A}$.) In particular, if A is a normal integral domain then $A[T_1, \ldots, T_n]$ too.

(3) (Graded integral extensions) a) Let $A \subseteq B$ be an extension of $\mathbb{Z}$-graded rings. Then the integral closure $\overline{A}$ of A in B is a graded A-subalgebra of B. (**Hint:** Note that, by (2), $\overline{A}[T, T^{-1}]$ is the integral closure of $A[T, T^{-1}]$ in $B[T, T^{-1}]$. Now, use the fact that, for any graded ring $C = \bigoplus_{m \in \mathbb{Z}} C_m$, the map $C \to C[T, T^{-1}]$, $\sum_m c_m \mapsto \sum_m c_m T^m$, is an injective graded ring homomorphism.)

b) Let A be a $\mathbb{Z}$-graded integral domain. Then the normalization $\overline{A}$ of A is a graded subalgebra of $S^{-1}A$ where S is the multiplicative system of non-zero homogeneous elements in A. If A is positively graded (i.e. $A_m = 0$ for $m < 0$) then $\overline{A}$ is also positively graded.

1.E.11. Exercise Let $A \subseteq B$ be an extension of integral domains such that A is a direct summand of B as an A-module. Show that:

(1) If B is normal, then A is normal too. (**Hint:** Let $f, g \in A$, $g \neq 0$, and let f/g be integral over A. Then $f \in A \cap Bg = Ag$.)

(2) $\overline{B} \cap Q(A) = \overline{A}$, where $\overline{A}$ and $\overline{B}$ are the normalizations of A and B respectively. (**Hint:** Let f, g be as in the hint of (1). If $f/g \in \overline{B}$, then $f^n \in A \cap (Bf^{n-1}g + \cdots + Bfg^{n-1} + Bg^n) = Af^{n-1}g + \cdots + Afg^{n-1} + Ag^n$ for some $n \in \mathbb{N}^*$.)

1.F. Noether's Normalization Lemma and Its Consequences

First we prove the classical version of Noether's normalization lemma.

1.F.1. Lemma *Let K be a field and $F \in K[X_1, \ldots, X_n]$ be a non-constant polynomial. Then there exists a K-automorphism $\varphi : K[X_1, \ldots, X_n] \to K[X_1, \ldots, X_n]$ such that $\varphi(X_n) = X_n$ and $F = aX_n^d + f_{d-1}X_n^{d-1} + \cdots + f_0$ where $a \in K^\times$ and $f_j \in K[Y_1, \ldots, Y_{n-1}]$, $0 \leq j \leq d - 1$, $Y_i := \varphi(X_i)$, $1 \leq i \leq n - 1$.*

PROOF. First assume that K is infinite. Then there exist $a_1, \ldots, a_{n-1} \in K$ such that $Y_i = \varphi(X_i) := X_i - a_i X_n$, $1 \le i \le n-1$, serves the purpose. To prove this let $F = F_0 + F_1 + \cdots + F_d$, where $F_m \in K[X_1, \ldots, X_n]$ is the homogeneous component of degree m of F, $0 \le m \le d := \deg F$. For any $a_1, \ldots, a_{n-1} \in K$, put $Y_i := X_i - a_i X_n$, $1 \le i \le n-1$. Then

$$F = \sum_{m=0}^{d} F_m(Y_1 + a_1 X_n, \ldots, Y_{n-1} + a_{n-1} X_n, X_n)$$

$$= \sum_{m=0}^{d} \left(F_m(a_1, \ldots, a_{m-1}, 1) X_n^m + \sum_{j=0}^{m-1} f_{mj}(Y_1, \ldots, Y_{n-1}) X_n^j \right)$$

where $f_{mj} \in K[Y_1, \ldots, Y_{n-1}]$ are homogeneous polynomials of degree $m-j$. Now, since $F_d(X_1, \ldots, X_{n-1}, 1) \ne 0$ and K is infinite, we can choose $a_1, \ldots, a_{n-1} \in K$ such that $a := F_d(a_1, \ldots, a_{n-1}, 1) \ne 0$.

In the general case, there exist positive integers $\gamma_1, \ldots, \gamma_{n-1}$ such that $Y_i = \varphi(X_i) := X_i - X_n^{\gamma_i}$, $1 \le i \le n-1$, serves the purpose. Let $F = \sum_{\alpha \in \Lambda} a_\alpha X^\alpha$ where Λ is a finite subset of $\mathbb{N}^n$ and $a_\alpha \in K^\times$ for every $\alpha = (\alpha_1, \ldots, \alpha_n) \in \Lambda$. For any positive integers $\gamma_1, \ldots, \gamma_{n-1}$, put $Y_i := X_i - X_n^{\gamma_i}$, $1 \le i \le n-1$. Then

$$F = \sum_{\alpha \in \Lambda} a_\alpha X_1^{\alpha_1} \cdots X_n^{\alpha_n} = \sum_{\alpha \in \Lambda} a_\alpha (Y_1 + X_n^{\gamma_1})^{\alpha_1} \cdots (Y_{n-1} + X_n^{\gamma_{n-1}})^{\alpha_{n-1}} X_n^{\alpha_n}.$$

For $\gamma = (r, r^2, \ldots, r^{n-1}, 1)$, where r is an integer bigger than all the components of all $\alpha = (\alpha_1, \ldots, \alpha_n) \in \Lambda$, we have $\deg_\gamma X^\alpha \ne \deg_\gamma X^\beta$ for all $\alpha, \beta \in \Lambda$, $\alpha \ne \beta$. Therefore, there exists a unique $\nu \in \Lambda$ such that $d := \deg_\gamma F = \deg_\gamma X^\nu (> 0)$ and so $F = a_\nu X_n^d + f_{d-1} X_n^{d-1} + \cdots + f_0$, $f_j \in K[Y_1, \ldots, Y_{n-1}]$. •

An affine transformation of a polynomial algebra $A[X_1, \ldots, X_n]$ is an A-algebra automorphism φ defined by $\varphi(X_j) = \sum_{i=1}^{n} a_{ij} X_i + b_j$, $1 \le j \le n$, where $(a_{ij}) \in GL_n(A)$ and $(b_j) \in A^n$. If (a_{ij}) is the identity matrix then φ is called a translation, if $(b_j) = 0$ then φ is called linear. In the proof of Lemma 1.F.1 for an infinite field K, we have used a simple linear transformation of $K[X_1, \ldots, X_n]$.

Now we prove Noether's normalization lemma with the help of the above lemma.

1.F.2. Noether's Normalization Lemma *Let K be a field and $R = K[x_1, \ldots, x_n]$ be a K-algebra of finite type. Then there exist $z_1, \ldots, z_m \in R$ which are algebraically independent over K such that R is integral (and hence finite) over the K-subalgebra $K[z_1, \ldots, z_m]$. – If $x_1, \ldots, x_n$ are algebraically dependent over K then $m < n$.*

PROOF. We prove the assertion by induction on n. In case $x_1, \ldots, x_n$ are algebraically independent over K we are through. Otherwise, let $F \in K[X_1, \ldots, X_n]$, $F \ne 0$, be such that $F(x_1, \ldots, x_n) = 0$. By the previous lemma, we can write

$$F = a X_n^d + f_{d-1} X_n^{d-1} + \cdots + f_0$$

with $f_j \in K[Y_1, \ldots, Y_{n-1}]$ and $a \in K^\times$, where $Y_1, \ldots, Y_{n-1}$ are as in 1.F.1. Therefore,

$$0 = a^{-1} F(x_1, \ldots, x_n) = x_n^d + a^{-1} \sum_{j=1}^{d} f_{d-j}(y_1, \ldots, y_{n-1}) x_n^{d-j}$$

where $y_j := Y_j(x_1, \ldots, x_n)$. This shows that x_n is integral over $K[y_1, \ldots, y_{n-1}]$ and so $K[x_1, \ldots, x_n] = K[y_1, \ldots, y_{n-1}, x_n]$ is integral over $K[y_1, \ldots, y_{n-1}]$. By induction hypothesis there exist $z_1, \ldots, z_m \in K[y_1, \ldots, y_{n-1}]$, $m \le n - 1$, which are algebraically independent over K such that $K[y_1, \ldots, y_{n-1}]$ is integral over $K[z_1, \ldots, z_m]$. Now, the assertion follows from Corollary 1.E.6 .   •

1.F.3. Remark If in Noether's normalization lemma R is an integral domain then m is the transcendence degree of the field of fractions $Q(R)$ over K and therefore uniquely determined. Even for an arbitrary K-algebra R of finite type, the non-negative integer m is uniquely determined. It is in fact the K r u l l - d i m e n s i o n of the (Noetherian) ring R. See Theorem 3.B.8 (and 3.B.14).

The normalization lemma has many consequences as we will see in these lectures. One example is the following:

1.F.4. Example A h y p e r s u r f a c e   a l g e b r a over a field K is a K-algebra of the form $K[X_1, \ldots, X_n]/(f)$ for some $n \in \mathbb{N}$ and a non-constant polynomial $f \in K[X_1, \ldots, X_n]$. Besides polynomial algebras these are the simplest K-algebras of finite type.

For any algebra R of finite type over a field K of characteristic zero which is an integral domain, there exists a hypersurface algebra $H \subseteq R$, $H \cong K[X_1, \ldots, X_n]/(f)$, with quotient field $Q(R)$ and the same normalization as R, that is, $\overline{R} = \overline{H}$.

PROOF. Let $P = K[z_1, \ldots, z_m] \subseteq R$ be a Noether's normalization of R as in 1.F.2. Since the charateristic of K is zero, the quotient field $Q(R)$ is finite separable over the function field $K(z_1, \ldots, z_m)$. Therefore by the primitive element theorem, there exists an element $\alpha \in Q(R)$ such that $Q(R) = K(z_1, \ldots, z_m)[\alpha]$. By the following Lemma 1.F.5 (2) $Q(R) = S^{-1}R$ where $S := K[z_1, \ldots, z_m] \setminus \{0\}$ and hence we may assume that $\alpha \in R$. Then $H := P[\alpha] \cong K[z_1, \ldots, z_m][X]/(\mu_\alpha)$ where μ_α is the minimal polynomial of α over $Q(P)$ (cf. Lemma 1.E.9) which is a hypersurface algebra contained in R with $Q(H) = Q(R)$. Further, $\overline{H} = \overline{R}$ is the integral closure of P in $Q(H) = Q(R)$.   •

The above statement is also true for a perfect field K of characteristic $p > 0$. (Use 6.D.12 (3) and Exercise 6.D.26.)

1.F.5. Lemma *Let $A \subseteq B$ be an algebraic extension of integral domains, i. e. the field extension $Q(A) \subseteq Q(B)$ is algebraic. Then:*

(1) *If $\mathfrak{b}$ is an ideal $\neq 0$ in B, then $\mathfrak{b} \cap A \neq 0$.*

(2) *If A is a field, then B is a field.*

(3) *If B is integral over A and if B is a field, then A is a field.*

PROOF. (1) Let $b \in \mathfrak{b}$, $b \neq 0$, and $a_n b^n + a_{n-1} b^{n-1} + \cdots + a_0 = 0$ be a non-trivial algebraic equation of b over A. If necessary, cancelling a power of b, we may assume that $a_0 \neq 0$. Then $a_0 \in (Bb) \cap A \subseteq \mathfrak{b} \cap A$.

(2) Let A be a field and let $b \in B$, $b \neq 0$. Then $(Bb) \cap A \neq 0$ by (1) and so $(Bb) \cap A = A$ and $Bb = B$.

(3) Let B be a field and let $a \in A$, $a \neq 0$. Then $a^{-1} \in B$ and therefore we have an equation $a^{-n} + a_{n-1} a^{-(n-1)} + \cdots + a_0 = 0$ for some $a_0, \ldots a_{n-1} \in A$. Multiplying by a^n, we get $1 = -\left(a_{n-1} + \cdots + a_0 a^{n-1}\right) a \in A$ and so $a^{-1} \in A$. •

Now we deduce the famous Hilbert's Nullstellensatz from Noether's Normalization Lemma 1.F.2.

1.F.6. Hilbert's Nullstellensatz (algebraic version) *Let $K \subseteq L$ be a field extension. If L is a K-algebra of finite type then L is a finite extension of K.*

PROOF. By the normalization lemma there exist $z_1, \ldots, z_m \in L$ which are algebraically independent over K such that $K[z_1, \ldots, z_m] \subseteq L$ is a finite extension. We have to show that $m = 0$, which follows from 1.F.5 (3). •

A reformulation of 1.F.6 is the following corollary.

1.F.7. Corollary *Let K be a field and let R be a K-algebra of finite type. Then, for every maximal ideal $\mathfrak{m} \subseteq R$, the field $R/\mathfrak{m}$ is a finite extension of K.*

For a ring R, the set of all maximal ideals of R is called the maximal spectrum of R and is denoted by

$$\mathrm{Spm}\, R .$$

If $R \neq 0$ then $\mathrm{Spm}\, R \neq \emptyset$. This is Krull's theorem which is an easy consequence of Zorn's lemma. Of course, in the Noetherian case, it is an immediate consequence of the Noetherian condition for the ideals of R.

For an algebraically closed field K, the maximal spectrum of a polynomial algebra over K has a simple description:

1.F.8. Corollary *Let K be an algebraically closed field. Then the map*

$$a = (a_1, \ldots, a_n) \mapsto \mathfrak{m}_a := (X_1 - a_1, \ldots, X_n - a_n)$$

from K^n to $\mathrm{Spm}\, K[X_1, \ldots, X_n]$ is bijective.

PROOF. For any $a \in K^n$, the ideal $\mathfrak{m}_a$ is the kernel of the substitution homomorphism $K[X_1, \ldots, X_n] \to K$, $X_i \mapsto a_i$, and therefore maximal.

For $\mathfrak{m} \in \mathrm{Spm}\, K[X_1, \ldots, X_n]$, $K[X_1, \ldots, X_n]/\mathfrak{m} = K$ by Corollary 1.F.7. Therefore, there exists $a = (a_1, \ldots, a_n) \in K^n$ such that $X_i \equiv a_i \bmod \mathfrak{m}$ for $i = 1, \ldots, n$ and so $\mathfrak{m}_a \subseteq \mathfrak{m}$. Therefore $\mathfrak{m} = \mathfrak{m}_a$. •

Of course, if K is not algebraically closed and $n \geq 1$, then there are maximal ideals in $K[X_1, \ldots, X_n]$ which are not point ideals $\mathfrak{m}_a$, $a \in K^n$. For example, the principal ideal $(X^2 + 1) \subseteq \mathbb{R}[X]$ is a maximal ideal in $\mathbb{R}[X]$, but not of the form $(X - a)$ for any $a \in \mathbb{R}$.

In general, the contraction $\varphi^{-1}(\mathfrak{m})$ of a maximal ideal $\mathfrak{m} \in \operatorname{Spm} S$ with respect to a ring homomorphism $\varphi : R \to S$ is not a maximal ideal in R. However, for K-algebras of finite type, we have:

1.F.9. Theorem *Let K be a field and let $\varphi : R \to S$ be a homomorphism of K-algebras of finite type. Then, for every $\mathfrak{m} \in \operatorname{Spm} S$, $\varphi^{-1}(\mathfrak{m}) \in \operatorname{Spm} R$.*

PROOF. Since $K \hookrightarrow R/\varphi^{-1}(\mathfrak{m}) \hookrightarrow S/\mathfrak{m}$ and $S/\mathfrak{m}$ is a finite field extension of K by 1.F.7, $R/\varphi^{-1}(\mathfrak{m})$ is also a field. $\quad\bullet$

1.F.10. Remark For an uncountable field, there is a very simple proof of 1.F.6, more precisely we prove:

1.F.11. Proposition *Let K be an uncountable field and let $K \subseteq L$ be a field extension. If L is countably generated as a K-algebra then L is algebraic over K.*

PROOF. If $x \in L$ is transcendental over K then $K(x) \subseteq L$ is a rational function field over K and the elements $1/(x - a)$, $a \in K$, in $K(x)$ are linearly independent over K. In particular, L is a K-vector space of uncountable dimension, but any countably generated K-algebra has countable K-vector space dimension. $\quad\bullet$

We have the following partial generalization of Noether's normalization lemma:

1.F.12. Proposition *Let $A \subseteq R$ be an extension of integral domains such that R is an A-algebra of finite type. Then there exist an element $f \in A$, $f \neq 0$, and elements $z_1, \ldots, z_m \in R$ such that $z_1, \ldots, z_m$ are algebraically independent over A and R_f is finite over $A_f[z_1, \ldots, z_m]$.*

PROOF. Let $R = A[x_1, \ldots, x_n]$ and let $K := Q(A)$ and $L := Q(R)$ be the quotient fields of A and R respectively. By 1.F.2 there are elements $z_1, \ldots, z_m \in K[x_1, \ldots, x_n] \subseteq L$ which are algebraically independent over K such that the algebra $K[x_1, \ldots, x_n]$ is finite over $K[z_1, \ldots, z_m]$. We may assume that $z_1, \ldots, z_m \in R$. If $f \in A$, $f \neq 0$, is a common denominator of the coefficients of integral equations of $x_1, \ldots, x_n$ over $K[z_1, \ldots, z_m]$ then $R_f = A[x_1, \ldots, x_n, 1/f] = A_f[x_1, \ldots, x_n]$ is integral over $A_f[z_1, \ldots, z_m]$. $\quad\bullet$

As an application, we consider the polynomial algebra $R := A[X_1, \ldots, X_n]$, $n \geq 1$, over an integral domain A. *Then there exists a maximal ideal $\mathfrak{M} \in \operatorname{Spm} R$ with $\mathfrak{M} \cap A = 0$ if and only if $Q(A) = A_f$ for some $f \neq 0$ in A. Proof.* If $Q(A) = A_f$ then $(fX_1 - 1, X_2, \ldots, X_n)$ is such a maximal ideal. Conversely, if $\mathfrak{M}$ is such a maximal ideal then by the proposition there exists an element $f \in A$, $f \neq 0$, and elements $z_1, \ldots, z_m$ in the field $L := R/\mathfrak{M}$ such that $z_1, \ldots, z_m$ are algebraically independent over A and L is finite over $A_f[z_1, \ldots, z_m]$. However, by 1.F.5 (3) $A_f[z_1, \ldots, z_m]$ is also a field which means $m = 0$ and $A_f = Q(A)$.

For example, in a principal ideal domain A, there exists an element $f \in A$, $f \neq 0$, with $Q(A) = A_f$ if and only if A has only a finite number of prime ideals. Therefore we have: *Let A be a principal ideal domain with infinitely many prime ideals. Then for every maximal ideal $\mathfrak{M} \in \mathrm{Spm}\, A[X_1, \ldots, X_n]$ the ideal $\mathfrak{m} := \mathfrak{M} \cap A$ is maximal in A and $A[X_1, \ldots, X_n]/\mathfrak{M}$ is finite over $A/\mathfrak{m}$.* In particular, for every maximal ideal $\mathfrak{M}$ in $\mathbb{Z}[X_1, \ldots, X_n]$, the residue class field $\mathbb{Z}[X_1, \ldots, X_n]/\mathfrak{M}$ is finite. In other words, a field of characteristic zero is never a $\mathbb{Z}$-algebra of finite type.

CHAPTER 2: The K-Spectrum and the Zariski Topology

2.A. The K-Spectrum of a K-Algebra

Let K be a field and let $K[X_1, \ldots, X_n]$ be the polynomial algebra in n variables over K. We have seen in Chapter 1 (see the remarks following 1.A.2) that the affine space K^n can be identified with the set of K-algebra homomorphisms $\text{Hom}_{K\text{-alg}}(K[X_1, \ldots, X_n], K)$ by identifying $a = (a_1, \ldots, a_n) \in K^n$ with the substitution homomorphism $\xi_a : K[X_1, \ldots, X_n] \to K$, $X_i \mapsto a_i$. The kernel of ξ_a is the maximal ideal $\mathfrak{m}_a = (X_1 - a_1, \ldots, X_n - a_n)$ in $K[X_1, \ldots, X_n]$. Moreover, every maximal ideal $\mathfrak{m}$ in $K[X_1, \ldots, X_n]$ with $K[X_1, \ldots, X_n]/\mathfrak{m} = K$ is of the type $\mathfrak{m}_a$ for a unique $a = (a_1, \ldots, a_n) \in K^n$; the component a_i is determined by the congruence $X_i \equiv a_i \bmod \mathfrak{m}$. The subset $\{\mathfrak{m}_a \mid a \in K^n\}$ of $\text{Spm}\, K[X_1, \ldots, X_n]$ is the K-spectrum of $K[X_1, \ldots, X_n]$ and is denoted by $K\text{-Spec}\, K[X_1, \ldots, X_n]$. We have the identifications

$$K^n \longleftrightarrow \text{Hom}_{K\text{-alg}}(K[X_1, \ldots, X_n], K) \longleftrightarrow K\text{-Spec}\, K[X_1, \ldots, X_n],$$

$$a \longleftrightarrow \xi_a \longleftrightarrow \mathfrak{m}_a = \text{Ker}\, \xi_a.$$

More generally, for any K-algebra R we use the map $\xi \mapsto \text{Ker}\, \xi$ to identify $\text{Hom}_{K\text{-alg}}(R, K)$ with the set of maximal ideals $\mathfrak{m}$ in R with $R/\mathfrak{m} = K$. Therefore we make the following:

2.A.1. Definition Let R be any K-algebra. The set of maximal ideals $\mathfrak{m}$ in R with $R/\mathfrak{m} = K$ is called the K-spectrum of R and is denoted by

$$K\text{-Spec}\, R.$$

Under the above identification we have $K\text{-Spec}\, R = \text{Hom}_{K\text{-alg}}(R, K)$.

By the algebraic version of Hilbert's Nullstellensatz, for an arbitrary maximal ideal $\mathfrak{m}$ in an algebra R of finite type over any field K, the residue field $R/\mathfrak{m}$ is a finite field extension of K (see 1.F.7). In particular:

2.A.2. Theorem *For an algebra R of finite type over an algebraically closed field K, we have $K\text{-Spec}\, R = \text{Spm}\, R$.*

For example, $\text{Spm}\, \mathbb{C}[X_1, \ldots, X_n] = \mathbb{C}\text{-Spec}\, \mathbb{C}[X_1, \ldots, X_n]$, but $\text{Spm}\, \mathbb{R}[X_1, \ldots, X_n] \supset \mathbb{R}\text{-Spec}\, \mathbb{R}[X_1, \ldots, X_n]$ for $n \geq 1$. In fact, the maximal ideal $\mathfrak{m} := (X_1^2 + 1, X_2, \ldots, X_n)$ does not belong to $\mathbb{R}\text{-Spec}\, \mathbb{R}[X_1, \ldots, X_n]$. For this $\mathfrak{m}$ the residue field is $\mathbb{C}$, therefore it is called a **c o m p l e x   p o i n t** of $\text{Spm}\, \mathbb{R}[X_1, \ldots, X_n]$. See also Example 2.C.2.

Let $F \in K[X_1, \ldots, X_n]$ be a polynomial. Then the function $\varphi_F^* : K^n \to K$, $a \mapsto F(a)$, is called the p o l y n o m i a l   f u n c t i o n defined by F. The above identifications allow us to write $F(a) = \xi_a(F) \equiv F \bmod \mathfrak{m}_a$ for any $a \in K^n$; $F(a)$ is called the v a l u e   o f F at a, or at ξ_a, or at $\mathfrak{m}_a$.

For an infinite field K, the polynomial function φ_F^* defined by F determines the polynomial F. This is the following well-known i d e n t i t y   t h e o r e m   f o r p o l y n o m i a l s:

Let K be an infinite field and let $F, G \in K[X_1, \ldots, X_n]$. If $\varphi_F^* = \varphi_G^*$ then $F = G$.

Let $\varphi : K[Y_1, \ldots, Y_m] \to K[X_1, \ldots, X_n]$ be a K-algebra homomorphism and let $F_i := \varphi(Y_i)$, $1 \le i \le m$. Then the map $\varphi^* : K^n \to K^m$ defined by $\varphi^*(a_1, \ldots, a_n) = (F_1(a), \ldots, F_m(a))$ is called the p o l y n o m i a l   m a p associated to φ. Under the above identifications the polynomial map φ^* is obviously described as follows: $\xi_a \mapsto \varphi^* \xi_a = \xi_a \circ \varphi$ or by $\mathfrak{m}_a \mapsto \varphi^* \mathfrak{m}_a = \varphi^{-1}(\mathfrak{m}_a) = \mathfrak{m}_{F(a)}$, $a \in K^n$. For every $G \in K[Y_1, \ldots, Y_m]$, we have $\varphi_G^* \circ \varphi^* = \varphi_{\varphi(G)}^*$.

More generally, for any K-algebra homomorphism $\varphi : R \to S$, we define the map $\varphi^* : K\text{-Spec } S \to K\text{-Spec } R$ by $\varphi^* \xi := \xi \circ \varphi$ or by $\varphi^* \mathfrak{m} = \varphi^{-1}(\mathfrak{m})$, $\mathfrak{m} = \operatorname{Ker} \xi \in K\text{-Spec } S = \operatorname{Hom}_{K\text{-alg}}(S, K)$. Further, if $\psi : S \to T$ is an another K-algebra homomorphism then $(\psi \circ \varphi)^* = \varphi^* \circ \psi^*$. With this *the assignments* $R \rightsquigarrow K\text{-Spec } R$, $\varphi \rightsquigarrow \varphi^*$ *is a contravariant functor from the category of K-algebras to the category of sets.*

2.A.3. Example Let $\varphi : R \to S$ be a K-algebra homomorphism. If φ is an isomorphism then by functoriality φ^* is bijective (with $(\varphi^*)^{-1} = (\varphi^{-1})^*$). However, if φ^* is bijective then φ need not be an isomorphism. We give some easy examples to illustrate this fact.

(1) Let K be a perfect field of characteristic $p > 0$. Then the Frobenius map $F : K \to K$, $x \mapsto x^p$, is bijective, but the corresponding K-algebra homomorphism $K[X] \to K[X]$, $X \mapsto X^p$, is not an automorphism.

(2) For an odd integer $n > 1$, the map $\mathbb{R} \to \mathbb{R}$, $x \mapsto x^n$, is bijective, but the corresponding $\mathbb{R}$-algebra homomorphism $\mathbb{R}[X] \to \mathbb{R}[X]$, $X \mapsto X^n$, is not bijective. If we replace $\mathbb{R}$ by $\mathbb{C}$ then the map $x \mapsto x^n$ is not bijective. This is in accordance with the following result (which can be generalised in many ways, but we do not prove it): *Let K be an algebraically closed field of characteristic zero and let φ be a K-algebra endomorphism of $K[X_1, \ldots, X_n]$. If $\varphi^* : K^n \to K^n$ is bijective then φ is an automorphism.*

The example (1) above shows that the assumption about the characteristic in this theorem is necessary. The group of K-algebra automorphisms of a polynomial algebra $K[X_1, \ldots, X_n]$, $n > 1$, is not yet well understood. In this connection, let us state a famous J a c o b i a n c o n j e c t u r e which is still open in general.

Let K be a field of characteristic zero, φ be a K-algebra endomorphism of $K[X_1, \ldots, X_n]$ and let $F_i := \varphi(X_i)$, $1 \le i \le n$. Then φ is bijective if (and only if) the Jacobian determinant

$$\frac{\partial(F_1, \ldots, F_n)}{\partial(X_1, \ldots, X_n)} := \det \left(\frac{\partial F_i}{\partial X_j} \right)_{1 \le i, j \le n}$$

is a non-zero constant.

The general case would follow from the special case $K = \mathbb{C}$.

2.A.4. Exercise Let K be a finite field with q elements.

a) Show that *every* map $K^n \to K^m$ is a polynomial map.

b) Let $F : K^n \to K^m$ and $G : K^n \to K^m$ be two maps defined by the polynomials $F_1, \ldots, F_m \in K[X_1, \ldots, X_n]$ and $G_1, \ldots, G_m \in K[X_1, \ldots, X_n]$, respectively. Give necessary and sufficient conditions on $F_1, \ldots, F_m, G_1, \ldots, G_m$ for the equality of the maps F, G. ($F = G$ if and only if $F_j - G_j \in (X_1^q - X_1, \ldots, X_n^q - X_n)$ for $j = 1, \ldots, m$.)

2.B. Affine Algebraic Sets

Let $\varphi : K[Y_1, \ldots, Y_m] \to K[X_1, \ldots, X_n]$ be a K-algebra homomorphism with $\varphi(Y_i) = F_i, i = 1, \ldots, m$, and let $\varphi^* : K^n \to K^m$ be the corresponding polynomial map defined as $a \mapsto (F_1(a), \ldots, F_m(a)), a \in K^n$. Then the fibre of φ^* over 0 is precisely the set of all common zeros (in K^n) of the polynomials $F_1, \ldots, F_m$, that is,

$$(\varphi^*)^{-1}(0) = \{ a \in K^n \mid F_1(a) = \cdots = F_m(a) = 0 \}.$$

More generally, the fibre of φ^* over $b = (b_1, \ldots, b_m)$ is the set of all common zeros of the polynomials $F_1 - b_1, \ldots, F_m - b_m$, that is,

$$(\varphi^*)^{-1}(b) = \{ a \in K^n \mid F_1(a) = b_1, \ldots, F_m(a) = b_m \}.$$

As we will see later, these fibres can be nicely described by using the so called
fibre algebra

$$K[X_1, \ldots, X_n]/\varphi(\mathfrak{m}_b)K[X_1, \ldots, X_n] = K[X_1, \ldots, X_n]/(F_1 - b_1, \ldots, F_m - b_m)$$

of the map φ at the point $b \in K^m$. The set of common zeros (in K^n) of the polynomials $F_1, \ldots, F_m \in K[X_1, \ldots, X_n]$ is denoted by

$$V_K(F_1, \ldots, F_m)$$

and is called the affine algebraic set in K^n defined by the polynomials $F_1, \ldots, F_m$. This motivates us to make the following definition:

2.B.1. Definition A subset $V \subseteq K^n$ is called an affine algebraic set in K^n if there is a family $F_j, j \in J$, of polynomials in $K[X_1, \ldots, X_n]$ such that

$$V = \{ a \in K^n \mid F_j(a) = 0 \text{ for all } j \in J \} = \bigcap_{j \in J} V_K(F_j).$$

The affine algebraic set defined by the family $\mathcal{F} = (F_j)_{j \in J}$ is denoted by

$$V_K(\mathcal{F}) = V_K(F_j, j \in J).$$

Using the identifications

$$K^n \longleftrightarrow \text{Hom}_{k-\text{alg}}(K[X_1, \ldots, X_n], K) \longleftrightarrow K\text{-Spec}(K[X_1, \ldots, X_n]),$$

$$a \longleftrightarrow \xi_a \longleftrightarrow \mathfrak{m}_a,$$

(see 2.A) we note that

$$a \in V(\mathcal{F}) \Leftrightarrow F_j(a) = 0, j \in J \Leftrightarrow \xi_a(F_j) = 0, j \in J \Leftrightarrow F_j \in \mathfrak{m}_a, j \in J.$$

This is equivalent to the condition that the ideal $\mathfrak{a} \subseteq K[X_1, \ldots, X_n]$ generated by F_j, $j \in J$, is contained in the maximal ideal $\mathfrak{m}_a$. By Hilbert's basis theorem 1.C.4, the ideal $\mathfrak{a}$ is finitely generated and so there exists a finite subset $J' \subseteq J$ such that $\mathfrak{a} = \sum_{j \in J'} K[X_1, \ldots, X_n]F_j$. This shows that $V_K(\mathcal{F}) = V_K(F_j, j \in J') = \bigcap_{j \in J'} V_K(F_j)$. In other words, *every affine algebraic set in K^n is a set of common zeros of a finite number of polynomials.*

Affine algebraic sets in K^n satisfy the properties of closed sets in a topological space. We state this in the following proposition and leave its proof to the reader.

2.B.2. Proposition *Let $\mathcal{F}_i, i \in I$, be a family of subsets of $K[X_1, \ldots, X_n]$ and let $\mathfrak{a}_j, j \in J$, be a family of ideals in $K[X_1, \ldots, X_n]$. Then:*

(1) $V_K(\bigcup_{i \in I} \mathcal{F}_i) = \bigcap_{i \in I} V_K(\mathcal{F}_i)$.

(2) $V_K(\sum_{j \in J} \mathfrak{a}_j) = \bigcap_{j \in J} V_K(\mathfrak{a}_j)$.

(3) $V_K(FG) = V_K(F) \cup V_K(G)$ *for polynomials* $F, G \in K[X_1, \ldots, X_n]$.

(4) $V_K(\mathfrak{a}\mathfrak{b}) = V_K(\mathfrak{a} \cap \mathfrak{b}) = V_K(\mathfrak{a}) \cup V_K(\mathfrak{b})$ *for ideals* $\mathfrak{a}, \mathfrak{b} \subseteq K[X_1, \ldots, X_n]$.

(5) $V_K(\mathcal{F}) \subseteq V_K(\mathcal{G})$ *for subsets* $\mathcal{F}, \mathcal{G} \subseteq K[X_1, \ldots, X_n]$ *with* $\mathcal{F} \supseteq \mathcal{G}$.

(6) $V_K(1) = \emptyset$ *and* $V_K(0) = K^n$.

Therefore the affine algebraic sets in K^n form the closed sets of a topology on K^n. This topology is called the Z a r i s k i   t o p o l o g y on K^n. The open sets are the complements

$$K^n \setminus V_K(F_j, j \in J) =: D_K(F_j, j \in J) = \bigcup_{j \in J} D_K(F_j).$$

In particular, $D_K(F) = \{a \in K^n \mid F(a) \neq 0\} = K^n \setminus V_K(F)$ for every polynomial $F \in K[X_1, \ldots, X_n]$. These open subsets are called the d i s t i n g u i s h e d   o p e n   s u b s e t s in K^n. *They form a basis for the Zariski topology on K^n.* Proposition 2.B.2 translates into:

2.B.3. Proposition *Let $\mathcal{F}_i, i \in I$, be a family of subsets of $K[X_1, \ldots, X_n]$ and let $\mathfrak{a}_j, j \in J$, be a family of ideals in $K[X_1, \ldots, X_n]$. Then:*

(1) $D_K(\bigcup_{i \in I} \mathcal{F}_i) = \bigcup_{i \in I} D_K(\mathcal{F}_i)$.

(2) $D_K(\sum_{j \in J} \mathfrak{a}_j) = \bigcup_{j \in J} D_K(\mathfrak{a}_j)$.

(3) $D_K(FG) = D_K(F) \cap D_K(G)$ *for polynomials* $F, G \in K[X_1, \ldots, X_n]$.

(4) $D_K(\mathfrak{a}\mathfrak{b}) = D_K(\mathfrak{a} \cap \mathfrak{b}) = D_K(\mathfrak{a}) \cap D_K(\mathfrak{b})$ *for ideals* $\mathfrak{a}, \mathfrak{b} \subseteq K[X_1, \ldots, X_n]$.

(5) $D_K(\mathcal{F}) \supseteq D_K(\mathcal{G})$ *for subsets* $\mathcal{F}, \mathcal{G} \subseteq K[X_1, \ldots, X_n]$ *with* $\mathcal{F} \supseteq \mathcal{G}$.

(6) $D_K(1) = K^n$ *and* $D_K(0) = \emptyset$.

We recall the definition of the radical of an ideal: For an ideal $\mathfrak{a}$ in a (commutative) ring R, the ideal $\sqrt{\mathfrak{a}} := \{ f \in R \mid f^r \in \mathfrak{a} \text{ for some integer } r \geq 1 \}$ is called the **radical** of $\mathfrak{a}$. Clearly $\mathfrak{a} \subseteq \sqrt{\mathfrak{a}}$. If $\sqrt{\mathfrak{a}} = \mathfrak{a}$, then $\mathfrak{a}$ is called a **radical ideal**. Obviously, $\sqrt{\sqrt{\mathfrak{a}}} = \sqrt{\mathfrak{a}}$. Therefore the radical of an ideal is a radical ideal. The radical $\mathfrak{n}_R := \sqrt{0}$ of the zero ideal is the ideal of nilpotent elements and is called the **nilradical** of R. *The nilradical $\mathfrak{n}_R$ is the intersection of all prime ideals in R. Proof.* If $f \in R$ is not nilpotent, then the ring of fractions R_f is non-zero and has a maximal ideal $\mathfrak{M}$. Then $\mathfrak{M} \cap R$ is a prime ideal in R not containing f.

Since for any $F \in K[X_1, \ldots, X_n]$ the polynomials F and $F^r, r \geq 1$, have the same zero set in K^n, we have the following proposition:

2.B.4. Proposition *For an ideal* $\mathfrak{a} \subseteq K[X_1, \ldots, X_n]$ *we have* $\mathrm{V}_K(\mathfrak{a}) = \mathrm{V}_K(\sqrt{\mathfrak{a}})$ *and* $\mathrm{D}_K(\mathfrak{a}) = \mathrm{D}_K(\sqrt{\mathfrak{a}})$.

2.B.5. Example (1) For the polynomial map $f : \mathbb{R}^2 \to \mathbb{R}$, $(x_1, x_2) \mapsto x_1^2 + x_2^2$, and $b \in \mathbb{R}$, the fibre $f^{-1}(b) = \mathrm{V}_{\mathbb{R}}(X_1^2 + X_2^2 - b)$ is a circle if $b > 0$, the origin $(0, 0)$ if $b = 0$ and empty if $b < 0$.

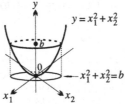

Note that the prime ideals $(X_1^2 + X_2^2)$ and (X_1, X_2) in $\mathbb{R}[X_1, X_2]$ are different, but they define the same affine algebraic set in $\mathbb{R}^2$. Further, all the prime ideals $(X_1^2 + X_2^2 - b), b < 0$, in $\mathbb{R}[X_1, X_2]$ define the same affine algebraic set, namely $\emptyset = \mathrm{V}_{\mathbb{R}}(1)$.

Note that $\mathrm{V}_{\mathbb{R}}(F_1, \ldots, F_m) = \mathrm{V}_{\mathbb{R}}(F_1^2 + \cdots + F_m^2)$ for arbitrary polynomials $F_1, \ldots, F_m \in \mathbb{R}[X_1, \ldots, X_n]$. Thus, *every affine algebraic set in $\mathbb{R}^n$ is the zero set of a single polynomial.*

(2) Let $g : \mathbb{R} \to \mathbb{R}$ be the polynomial map $x \mapsto x^2$ which is the restriction of the map f of the example (1) to the line $\mathrm{V}_{\mathbb{R}}(X_2) = \{ x_2 = 0 \}$. The fibre $g^{-1}(b) = \mathrm{V}_{\mathbb{R}}(X^2 - b)$ has exactly two points if $b > 0$; it has exactly one point if $b = 0$ and it is empty if $b < 0$.

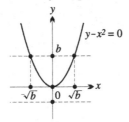

For these three cases the corresponding fibre algebras $\mathbb{R}[X]/(X^2 - b), b \in \mathbb{R}$, are isomorphic to the product algebra $\mathbb{R} \times \mathbb{R}$, to the algebra $\mathbb{R}[\varepsilon] := \mathbb{R}[X]/(X^2)$ of dual numbers and to the algebra $\mathbb{C}$, respectively. The different nature of the fibres is well represented by the different nature of these fibre algebras, cf. Example 4.D.4.

2.B.6. Example (Plane algebraic curves) Let $F \in K[X, Y]$ be a non-constant polynomial over an infinite field K. If $F_1, \ldots, F_m \in K[X, Y]$ are all distinct prime factors of F, then $V_K(F) = V_K(F_1 \cdots F_m) = V_K(F_1) \cup \cdots \cup V_K(F_m)$, that is, $V_K(F)$ depends only on the r e d u c t i o n red $F := F_1 \cdots F_m$ of F which generates the radical $\sqrt{(F)}$ of the principal ideal (F). Moreover, if each $V_K(F_i)$ is infinite, then the ideal $\mathfrak{J}_K(V_K(F))$ of all polynomials in $K[X, Y]$ which vanish on $V_K(F)$ coincides with the ideal (red F) because, by Exercise 2.B.14 (4), $\mathfrak{J}_K(V_K(F)) = \bigcap_{i=1}^{m} \mathfrak{J}_K(V_K(F_i)) = \bigcap(F_i) = (\text{red } F)$. In this case the zero set $V_K(F) = V_K(\text{red } F) \subseteq K^2$ is called a (a f f i n e) p l a n e a l g e b r a i c c u r v e o v e r K and red F is called the d e f i n i n g p o l y n o m i a l of this curve, since it is uniquely determined by the curve, up to a non-zero constant. From now on assume that F is a square-free polynomial, that is, $F = \text{red } F = F_1 \cdots F_m$. The algebraic curves $V_K(F_i)$ are called the components of $V_K(F)$. If $m = 1$ then the curve is called i r r e d u c i b l e. Sometimes we denote the curve $V_K(F)$ by $F(x, y) = 0$ or by $\{F(x, y) = 0\}$. The degree of the polynomial $F_1 \cdots F_m$ is called the d e g r e e o f t h e c u r v e $V_K(F)$. Obviously the complexity of a curve increases with its degree. It is easy to see that under an affine transformation of $K[X, Y]$ the equation of a curve is transformed into another equation of the same degree. Therefore the degree of a curve is a first quantitative measure for the geometric, qualitative complexity of a curve. Plane algebraic curves of degrees 1, 2, 3, 4, 5, 6 are called l i n e s, q u a d r i c s, c u b i c s, q u a r t i c s, q u i n t i c s, s e x t i c s, respectively. The curves of degree 1 are just straight lines: $ax + by + c = 0$, $(a, b) \neq (0, 0)$. The curves of degree 2 are also called c o n i c s e c t i o n s [1]):

$$F(x, y) = ax^2 + bxy + cy^2 + dx + ey + f = 0.$$

There are two possibilities, namely: first F is not prime, then the (degenerated) conic $F(x, y) = 0$ is a union of two distinct straight lines. Second, F is prime, in this case, we assume char $K \neq 2$. Then by an affine transformation of $K[X, Y]$, $F(X, Y)$ can be brought into one of the forms $Y^2 - X$, $aX^2 + bY^2 - 1$, $a, b \in K^\times$. These are called p a r a b o l a, e l l i p s e or h y p e r b o l a according as $aX^2 + bY^2$ is prime or not prime. Further, the defining polynomial of a hyperbola can be transformed into $XY - 1$. Note that a polynomial $aX^2 + bY^2 - 1$, $a, b \in K^\times$, is always prime and, if it has at least one zero [2]), then it has infinitely many zeros (see Exercise 2.B.7 (2)) and hence is a defining polynomial of a hyperbola or an ellipse.

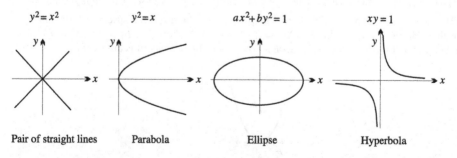

$y^2 = x^2$	$y^2 = x$	$ax^2 + by^2 = 1$	$xy = 1$
Pair of straight lines	Parabola	Ellipse	Hyperbola

---

[1]) The discovery of conic sections is attributed to Menaechmus (350 B. C.). They were intensively investigated by Apollonius of Perga (225 B. C.).
[2]) Depending on the ground field K, it can be very difficult to decide whether such a polynomial has a zero or not.

Now, let $F \in K[X, Y]$ be of degree 3 and assume that the homogeneous component F_3 of F splits into linear factors over K. Furthermore, if char $K \neq 2$, then by an affine tranformation of $K[X, Y]$, F can be brought (up to a non-zero constant) into one of the following polynomials: $Y^2 - (aX^3 + bX^2 + cX + d)$, $XY - (aX^3 + bX^2 + cX + d)$, $eY - (aX^3 + bX^2 + cX + d)$, or $XY^2 + eY - (aX^3 + bX^2 + cX + d)$. This was already done by Newton. The proof is elementary but consists of several cases. First one brings F_3 into one of the following polynomials: αX^3; $\alpha X(Y^2 - \beta X^2)$, $\alpha \in K^\times, \beta \in K$.

We give pictures of some real irreducible cubics:

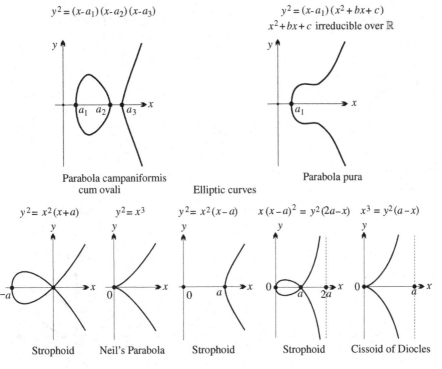

$$y^2 = (x-a_1)(x-a_2)(x-a_3)$$

$$y^2 = (x-a_1)(x^2+bx+c)$$
$$x^2+bx+c \text{ irreducible over } \mathbb{R}$$

Parabola campaniformis
cum ovali   Elliptic curves   Parabola pura

$$y^2 = x^2(x+a) \qquad y^2 = x^3 \qquad y^2 = x^2(x-a) \qquad x(x-a)^2 = y^2(2a-x) \qquad x^3 = y^2(a-x)$$

Strophoid   Neil's Parabola   Strophoid   Strophoid   Cissoid of Diocles

At the end we give some examples of real irreducible plane curves of higher degrees:

$$(x-a)^2(x^2+y^2) = b^2x^2$$

$b<a$ $b>a$ $b=a$

Conchoids of Nicomedes

$$(y^2+x^2)^2 - 2(x^2-y^2) + 1 - r^2$$

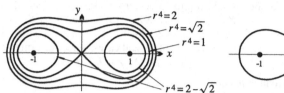

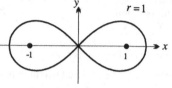

Cassini's curves   Lemniscate of Bernoulli

$$(x^2+y^2 - a z)^2 = b^2(x^2+y^2)$$

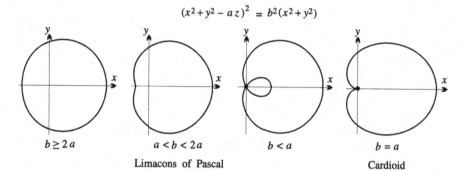

$b \geq 2a$ $a < b < 2a$ $b < a$ $b = a$

Limacons of Pascal   Cardioid

$$(x^2+y^2)^2 + 18a^2(x^2+y^2)$$
$$= 8ax(x^2-3y^2) + 27a^4$$

$$(1 - y^2 - x^2)^3 = 27 y^2 x^2$$

$$(y^2+x^2)^3 = 4 y^2 x^2$$

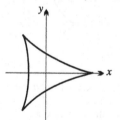

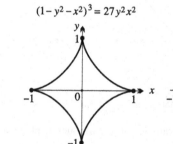

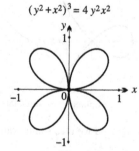

Steiner's quadric   Astroid   Quadrifolium

$$x^{2n}+y^{2n} = 1 \quad (n \in \mathbb{N}^*)$$

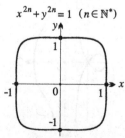

Fermat's curves

2.B.7. Exercise Let K be an infinite field.

(1) An algebraic plane curve $C \subseteq K^2$ over K of degree $d \in \mathbb{N}^*$ intersects with an arbitrary straight line in at most d points. Which straight lines intersect C in exactly d points? How many are they? (One has a better answer to this question if one works in a *projective plane* instead of affine plane. See also Chapter 5.)

(2) Assume that char $K \neq 2$ and let $F := aX^2 + bY^2 - 1, a, b \in K^\times$. If $V_K(F) \neq \emptyset$, then $V_K(F)$ is infinite. (**Hint:** Intersect the straight lines passing through a point $P \in V_K(F)$ with $V_K(F)$.)

(3) The subsets $\{ (x, y) \in \mathbb{R}^2 \mid y = e^x \}$ and $\{ (x, y) \in \mathbb{R}^2 \mid y = \sin x \}$ are dense in the Zariski topology of $\mathbb{R}^2 = \mathbb{R}\text{-Spec}\,\mathbb{R}[X, Y]$.

For a better understanding of the map $\mathfrak{a} \mapsto V_K(\mathfrak{a})$ from the set of ideals in $K[X_1, \ldots, X_n]$ to the set of affine algebraic sets in K^n, we introduce a map $\mathfrak{I}_K$ in the opposite direction. For this to *every* subset $E \subseteq K^n$, we associate the ideal

$$\mathfrak{I}_K(E) := \{ F \in K[X_1, \ldots, X_n] \mid F(a) = 0 \text{ for all } a \in E \} = \bigcap_{a \in E} \mathfrak{m}_a \,,$$

which is even a radical ideal in $K[X_1, \ldots, X_n]$. For an affine algebraic set $V \subseteq K^n$, the ideal $\mathfrak{I}_K(V)$ is called the **ideal of** V. The proofs of the following rules are simple verifications.

2.B.8. Proposition (1) *For a family $E_i, i \in I$, of subsets of K^n, $\mathfrak{I}_K(\bigcup_{i \in I} E_i) = \bigcap_{i \in I} \mathfrak{I}_K(E_i)$ and $\mathfrak{I}_K(\bigcap_{i \in I} E_i) \supseteq \sum_{i \in I} \mathfrak{I}_K(E_i)$.*

(2) $\mathfrak{I}_K(D) \subseteq \mathfrak{I}_K(E)$ *for subsets $D, E \subseteq K^n$ with $D \supseteq E$.*

(3) $\mathfrak{I}_K(\emptyset) = K[X_1, \ldots, X_n]$ *and if K is infinite then $\mathfrak{I}_K(K^n) = 0$.*

(4) $\mathfrak{I}_K(V_K(F_j, j \in J)) \supseteq \sum_{j \in J} F_j K[X_1, \ldots, X_n]$ *for a family $F_j, j \in J$, in $K[X_1, \ldots, X_n]$. In particular, $\mathfrak{a} \subseteq \mathfrak{I}_K(V_K(\mathfrak{a}))$ for an ideal $\mathfrak{a} \subseteq K[X_1, \ldots, X_n]$.*

(5) $V_K(\mathfrak{I}_K(E)) \supseteq E$ *for a subset $E \subseteq K^n$.*

The inclusion in (5) of the above proposition can be improved:

2.B.9. Proposition *Let E be a subset of K^n. Then $V_K(\mathfrak{I}_K(E)) = \overline{E}$ (where $\overline{E}$ the closure of E in K^n with respect to the Zariski topology). In particular, $\mathfrak{I}_K(V_K(\mathfrak{I}_K(E))) = \mathfrak{I}_K(E)$ and $V_K(\mathfrak{I}_K(V_K(\mathfrak{a}))) = V_K(\mathfrak{a})$ where $\mathfrak{a}$ is an ideal in $K[X_1, \ldots, X_n]$.*

PROOF. By (5) of 2.B.8 $\overline{E} \subseteq V_K(\mathfrak{I}_K(E))$. For the other inclusion, if $V_K(\mathfrak{b})$ is a closed subset containing E, where $\mathfrak{b}$ is an ideal in $K[X_1, \ldots, X_n]$, then $\mathfrak{b} \subseteq \bigcap_{a \in E} \mathfrak{m}_a = \mathfrak{I}_K(E)$ and hence $V_K(\mathfrak{b}) \supseteq V_K(\mathfrak{I}_K(E))$ by (5) of 2.B.2.   •

It is immediate from the last proposition that the map $\mathfrak{I}_K$ is injective on the set of affine algebraic subsets of K^n. Generally it is rather difficult to describe the image of the map $\mathfrak{I}_K$ in the set of all radical ideals of $K[X_1, \ldots, X_n]$. However, for an algebraically closed field we have a complete answer:

2.B.10. Theorem *Let K be an algebraically closed field. Then the map $\mathfrak{J}_K$ is a bijection from the set of affine algebraic subsets in K^n onto the set of radical ideals in $K[X_1, \ldots, X_n]$. In fact, the inverse is the map V_K.*

This theorem is an immediate consequence of the famous geometric version of Hilbert's Nullstellensatz.

2.B.11. Hilbert's Nullstellensatz (geometric version) *Let K be an algebraically closed field and let $\mathfrak{a} \subseteq K[X_1, \ldots, X_n]$ be an ideal. Then $\mathfrak{J}_K(V_K(\mathfrak{a})) = \sqrt{\mathfrak{a}}$.*

PROOF. Since K is algebraically closed, every maximal ideal $\mathfrak{m}$ in $K[X_1, \ldots, X_n]$ is of the form $\mathfrak{m}_a$ for some $a \in K^n$ by 1.F.8. Therefore by definition

$$\mathfrak{J}_K(V_K(\mathfrak{a})) = \bigcap_{a \in V_K(\mathfrak{a})} \mathfrak{m}_a = \bigcap_{\mathfrak{a} \subseteq \mathfrak{m}_a} \mathfrak{m}_a = \bigcap_{\substack{\mathfrak{a} \subseteq \mathfrak{m} \\ \mathfrak{m} \in \mathrm{Spm}\, K[X_1, \ldots, X_n]}} \mathfrak{m} .$$

Thus, in the residue class algebra $R := K[X_1, \ldots, X_n]/\mathfrak{a}$, the ideal $\mathfrak{J}_K(V_K(\mathfrak{a}))/\mathfrak{a}$ is the Jacobson radical

$$\mathfrak{m}_R = \bigcap_{\mathfrak{m} \in \mathrm{Spm}\, R} \mathfrak{m}$$

of R. But $\sqrt{\mathfrak{a}}/\mathfrak{a}$ is the nilradical $\mathfrak{n}_R$ of R. Now, the assertion is a special case of the following theorem.   ●

2.B.12. Theorem *Let R be an algebra of finite type over any field K. Then $\mathfrak{n}_R = \mathfrak{m}_R$.*

PROOF. Obviously, $\mathfrak{n}_R \subseteq \mathfrak{m}_R$. Conversely, let $f \notin \mathfrak{n}_R$. Now we use Rabinowitch's trick and consider the algebra $R_f = R[1/f]$ which is also of finite type over K. Since $f \notin \mathfrak{n}_R$, $R_f \neq 0$ and so there exists a maximal ideal $\mathfrak{m} \in \mathrm{Spm}\, R_f$. Then $\iota^{-1}(\mathfrak{m}) \in \mathrm{Spm}\, R$ by 1.F.9, where $\iota : R \to R_f$ is the canonical homomorphism. But $f \notin \iota^{-1}(\mathfrak{m})$ and hence $f \notin \mathfrak{m}_R$.   ●

2.B.13. Remark Let K be an arbitrary field, $\mathfrak{a} \subseteq K[X_1, \ldots, X_n]$ be an ideal and let $R := K[X_1, \ldots, X_n]/\mathfrak{a}$. The ideal $\mathfrak{J}_K(V_K(\mathfrak{a}))/\mathfrak{a}$ in R is the intersection $\bigcap_{\xi \in K\text{-Spec}\, R} \mathfrak{m}_\xi$ of the maximal ideals $\mathfrak{m}_\xi$ in R corresponding to the points $\xi \in K$-Spec R and therefore an invariant of the K-algebra R, called the K - r a d i c a l $\mathfrak{r} = \mathfrak{r}_R$ of R. The equality $\mathfrak{J}_K(V_K(\mathfrak{a})) = \sqrt{\mathfrak{a}}$ is equivalent with the condition that the nilradical of R and the K-radical of R coincide. Therefore the equality $\mathfrak{J}_K(V_K(\mathfrak{a})) = \sqrt{\mathfrak{a}}$ implies the equality $\mathfrak{J}_K(V_K(\mathfrak{b})) = \sqrt{\mathfrak{b}}$ for any ideal $\mathfrak{b}$ in a polynomial algebra $K[Y_1, \ldots, Y_m]$ with $R \cong K[Y_1, \ldots, Y_m]/\mathfrak{b}$.

2.B.14. Exercise (1) Let $\mathfrak{m}, \mathfrak{m}_1, \ldots, \mathfrak{m}_r$ be maximal ideals in $K[X_1, \ldots, X_n]$, where K is a field.

a) Show that $V_K(\mathfrak{m})$ contains at most one point. Moreover, $V_K(\mathfrak{m}) \neq \emptyset$ if and only if $\mathfrak{m} \in K$-Spec $K[X_1, \ldots, X_n]$.

b) There exist $F_i \in K[X_1, \ldots, X_i]$, $1 \leq i \leq n$, such that $\mathfrak{m}$ is generated by $F_1, \ldots, F_n$. (**Hint:** For each $1 \leq i \leq n$, $\mathfrak{m} \cap K[X_1, \ldots, X_i]$ is a maximal ideal in $K[X_1, \ldots, X_i]$ by 1.F.9. Use induction on n.)

c) More generally, there exist $F_i \in K[X_1, \ldots, X_i]$, $1 \le i \le n$, such that the ideal $\mathfrak{a} :=$ $\mathfrak{m}_1 \cap \cdots \cap \mathfrak{m}_r$ is generated by $F_1, \ldots, F_n$. In particular, the ideal $\mathfrak{I}_K(\{P_1, \ldots, P_r\})$ of a finite subset $\{P_1, \ldots, P_r\} \subseteq K^n$ is generated by n polynomials. By Chinese remainder theorem $K[X_1, \ldots, X_n]/\mathfrak{a} \cong \prod_{\rho=1}^{r} K[X_1, \ldots, X_n]/\mathfrak{m}_\rho$ and $K[X_1, \ldots, X_n]/\mathfrak{I}_K(\{P_1, \ldots, P_r\}) \cong$ K^r (as K-algebras).

(2) Let K be a non-algebraically closed field.

a) There exists a polynomial $F_n \in K[X_1, \ldots, X_n]$ such that $V_K(F_n) = \{(0, \ldots, 0)\}$. (**Hint:** Take $F_n = F_2(F_{n-1}, X_n)$ for $n \ge 3$.)

b) Any affine algebraic set in K^n is of the form $V_K(F)$ for some $F \in K[X_1, \ldots, X_n]$.

(3) (**A f f i n e   m o n o m i a l   c u r v e s**) Let K be an infinite field and let $m_1, \ldots, m_n$ be positive integers with $\mathrm{GCD}(m_1, \ldots, m_n) = 1$. Let $\gamma : K \to K^n$ be the curve defined by $t \mapsto (t^{m_1}, \ldots, t^{m_n})$.

a) Show that γ is injective and the image $\mathrm{Im}\,\gamma$ is an affine algebraic set. This is called the **a f f i n e   m o n o m i a l   c u r v e** defined by the sequence $m_1, \ldots, m_n$. The defining ideal $\mathfrak{I}_K(\mathrm{Im}\,\gamma)$ is the kernel $\mathfrak{a}$ of the K-algebra homomorphism $K[X_1, \ldots, X_n] \to K[T]$ defined by $X_i \mapsto T^{m_i}$, $i = 1, \ldots, n$, and so the coordinate K-algebra of $\mathrm{Im}\,\gamma$ is $R_\gamma :=$ $K[X_1, \ldots, X_n]/\mathfrak{a} \xrightarrow{\sim} K[T^{m_1}, \ldots, T^{m_n}] = K[M] \subseteq K[T]$ where $M = \mathbb{N}m_1 + \cdots + \mathbb{N}m_n$ is the numerical monoid generated by the elements $m_1, \ldots, m_n$.

b) The quotient field of R_γ is the rational function field $K(T)$. (This means, by definition, *affine monomial curves are rational curves*.) $K[T]$ is the normalization of R_γ, i.e. $K[T]$ is the integral closure of R_γ in $K(T)$. Again by definition, *the affine line $K = K$-Spec $K[T]$ is the normalization of the curve* $\mathrm{Im}\,\gamma$ (and $\gamma : K \to \mathrm{Im}\,\gamma$ is the normalization map).

c) $R_\gamma \simeq K[T]$ if and only if $m_i = 1$ for some i. In this case find a (minimal) set of generators for the ideal $\mathfrak{a}$.

d) If $n = 2$ then the ideal $\mathfrak{a}$ is generated by $X_1^{m_2} - X_2^{m_1}$.

e) Let $n = 3$ and $m_1 := 2m$, $m_2 := 2m + 1$, $m_3 := 2m + 2$, $m \in \mathbb{N}^*$. Then the ideal $\mathfrak{a}$ is generated by two binomials. (**Hint:** $X_2^2 - X_1 X_3$ and $X_3^m - X_1^{m+1}$ generate $\mathfrak{a}$.)

f) Let $n = 3$ and $m_1 := 2m + 1$, $m_2 := 2m + 2$, $m_3 := 2m + 3$, $m \in \mathbb{N}^*$. Then the ideal $\mathfrak{a}$ is generated by three binomials and can not be generated by two polynomials. (**Hint:** $X_2^2 - X_1 X_3$, $X_1^{m+2} - X_2 X_3^m$ and $X_3^{m+1} - X_1^{m+1} X_2$ generate $\mathfrak{a}$. – If $\mathfrak{a}$ is generated by $n - 1$ polynomials then we say that the curve $\mathrm{Im}\,\gamma$ is an (**i d e a l - t h e o r e t i c**)   **c o m p l e t e i n t e r s e c t i o n**. In this case $\mathfrak{a}$ is generated by $n - 1$ binomials. If there exist $n - 1$ polynomials $F_1, \ldots, F_{n-1} \in K[X_1, \ldots, X_n]$ such that $\mathfrak{a} = \sqrt{(F_1, \ldots, F_{n-1})}$, then we say that the curve $\mathrm{Im}\,\gamma$ is a **s e t - t h e o r e t i c c o m p l e t e i n t e r s e c t i o n**. J. Herzog has proved that the ideal $\mathfrak{a}$ of an affine monomial space curve ($n = 3$) is always generated by three binomials and using the explicit form of these generators, he proved that affine monomial space curves are set-theoretic complete intersections. (In the example f) try to find two polynomials F and G such that $\mathfrak{a} = \sqrt{(F, G)}$.) For general n this is still an open question.)

(4) Let $F \in K[X, Y]$ be a prime polynomial over an infinite field K. Then $\mathfrak{I}_K(V_K(F)) = (F)$ if and only if $V_K(F)$ has infinitely many points. (**Hint:** By Noether's normalization the K-algebra $R := K[X, Y]/(F)$ is a finite extension of a polynomial algebra $K[Z]$ in one variable. Let $\mathfrak{r}$ denote the K-radical $\mathfrak{I}_K(V_K(F))/(F)$ of R. If $V_K(F)$ is finite then $R/\mathfrak{r}$ is a finite K-algebra, i.e. $\mathfrak{r} \ne 0$. If $\mathfrak{r} \ne 0$ then $\mathfrak{r} \cap K[Z] \ne 0$ and $R/\mathfrak{r}$ is a finite K-algebra which implies that $V_K(F) = K$-Spec$(R/\mathfrak{r})$ is finite.) Of course, if K is algebraically

closed and F is prime polynomial in $K[X, Y]$, then obviously $V_K(F)$ is infinite and hence $\mathfrak{I}_K(V_K(F)) = (F)$. This is a special case of the general Nullstellensatz 2.B.11.

For an arbitrary K-algebra R, the a f f i n e   a l g e b r a i c   s u b s e t s (and hence the Zariski topology) can be defined on its K-spectrum

$$K\text{-Spec } R = \text{Hom}_{K-\text{alg}}(R, K) = \{\, \mathfrak{m} \in \text{Spm } R \mid R/\mathfrak{m} = K \,\}$$

in an obvious way. For $f \in R$ we put

$$V_K(f) := \{\xi \in K\text{-Spec } R \mid \xi(f) = 0\} = \{\xi \in K\text{-Spec } R \mid f(\xi) = 0\}$$
$$= \{\mathfrak{m} \in K\text{-Spec } R \subseteq \text{Spm } R \mid f \in \mathfrak{m}\}$$

and, for an arbitrary family f_j, $j \in J$, of elements in R,

$$V_K(f_j, j \in J) := \bigcap_{j \in J} V_K(f_j).$$

These affine algebraic subsets of K-Spec R satisfy the same properties of 2.B.2 and are the closed sets of the Zariski topology of K-Spec R. The open subsets are the complements $D_K(f_j, j \in J)$.

For a K-algebra R and a subset $E \subseteq K$-Spec R, we associate the radical ideal $\mathfrak{I}_K(E) := \{\, f \in R \mid f(\xi) = 0 \text{ for all } \xi \in E \,\} = \bigcap_{\xi \in E} \mathfrak{m}_\xi$ in R. These radical ideals satisfy the same properties of 2.B.8 and 2.B.9. Moreover, if R is a K-algebra of finite type over an *algebraically closed* field K, then the map $V_K :$ $\mathfrak{a} \mapsto V_K(\mathfrak{a})$ is bijective from the set of radical ideals in R onto the set of affine algebraic subsets in K-Spec R and the map $\mathfrak{I}_K : E \mapsto \mathfrak{I}_K(E)$ is its inverse. The assignments $R \rightsquigarrow K$-Spec R, $\varphi \rightsquigarrow \varphi^*$ define not only a contravariant functor from the catagory of K-algebras to the category of sets, but even to the category of topological spaces. For a K-algebra homomorphism $\varphi : R \to S$ the continuity of $\varphi^* : K$-Spec $S \to K$-Spec R, $\xi \mapsto \xi\varphi$, is immediate from the following more precise proposition which is easy to prove:

2.B.15. Proposition *Let* $\varphi : R \to S$ *be a K-algebra homomorphism. For any ideal* $\mathfrak{a}$ *in* R, $(\varphi^*)^{-1}(V_K(\mathfrak{a})) = V_K(\mathfrak{a}S)$, *where* $\mathfrak{a}S := \varphi(\mathfrak{a})S$ *is the extended ideal in* S.

We note the following three special cases which are used very often. Let R be an arbitrary K- algebra.

(1) For a surjective K-algebra homomorphism $\varphi : R \to R'$, the continuous map $\varphi^* : K$-Spec $R' \to K$-Spec R is a closed embedding with image $V_K(\text{Ker }\varphi)$. In particular, *for any ideal* $\mathfrak{a}$ *in* R *the map* $\pi^* : K\text{-Spec}(R/\mathfrak{a}) \to K$-Spec R, *where* $\pi : R \to R/\mathfrak{a}$ *is the canonical projection, identifies* $K\text{-Spec}(R/\mathfrak{a})$ *with the affine algebraic subset* $V_K(\mathfrak{a})$ *in* K-Spec R.

(2) *The canonical projection* $\pi : R \to R/\mathfrak{n}_R$, *where* $\mathfrak{n}_R$ *is the nilradical of* R, *induces a homeomorphism* $\pi^* : K\text{-Spec}(R/\mathfrak{n}_R) \xrightarrow{\sim} K$-Spec R. *This is a special*

case of (1) since $V_K(\mathfrak{n}_R) = K\text{-Spec }R$. The K-algebra $R_{\text{red}} := R/\mathfrak{n}_R$ is called the reduction of R. If $\mathfrak{n}_R = 0$, i.e. if $R = R_{\text{red}}$, then R is called reduced.

(3) *For $f \in R$, the canonical K-algebra homomorphism $\iota_f : R \to R_f = R[1/f]$ induces an open embedding $\iota_f^* : K\text{-Spec }R_f \to K\text{-Spec }R$ with image $D_K(f)$.* We use ι_f^* to identify $K\text{-Spec }R_f$ with the distinguished open subset $D_K(f)$ in $K\text{-Spec }R$. – More generally, for an arbitrary multiplicatively closed subset $T \subseteq R$, the canonical homomorphism $\iota_T : R \to T^{-1}R$ induces a homeomorphism from $K\text{-Spec }T^{-1}R$ to the subspace $\text{im}(\iota_T^*) = \{\mathfrak{m} \in K\text{-Spec }R \mid \mathfrak{m} \cap T = \emptyset\}$ of $K\text{-Spec }R$.

2.B.16. Example Let $\varphi : R \to S$ be an integral K-algebra homomorphism of K-algebras of finite type. Then the map $\varphi^* : K\text{-Spec }S \to K\text{-Spec }R$ is, in general, not closed. For example, for the natural inclusion $\iota : \mathbb{R}[X] \to \mathbb{R}[X, Y]/(Y^2 - X)$, the image $\iota^*(\mathbb{R}\text{-Spec }\mathbb{R}[X, Y]/(Y^2 - X)) = \{a \in \mathbb{R} \mid a \geq 0\}$ is not closed in the Zariski topology of $\mathbb{R}\text{-Spec }\mathbb{R}[X]$.

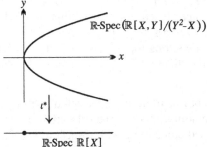

However, if K is algebraically closed the map φ^* is a closed map. This is immediate from the following assertion:

Let K be an algebraically closed field and let $\varphi : R \to S$ be an integral K-algebra homomorphism of K-algebras of finite type. Then $\varphi^*(K\text{-Spec }S) = V_K(\text{Ker }\varphi)$.

Replacing R by $R/\text{Ker }\varphi$ we may assume that φ is injective. Then for $\xi \in K\text{-Spec }R$ the maximal ideal $\mathfrak{m}_\xi$ is a φ-preimage of a maximal ideal $\mathfrak{n} \in \text{Spm }S$ (this is proved later in Lemma 3.B.9). Now by Hilbert's Nullstellensatz $\mathfrak{n} \in K\text{-Spec }S$.

2.B.17. Example Let R, S be K-algebras and let $\iota_R : R \to R \otimes_K S$, $\iota_S : S \to R \otimes_K S$ be the canonical maps. Then the continuous map

$$K\text{-Spec}(R \otimes_K S) \to (K\text{-Spec }R) \times (K\text{-Spec }S) \,, \quad \xi \mapsto (\iota_R^*(\xi), \iota_S^*(\xi))$$

is a bijection, but not a homeomorphism in general, i.e. in general, the Zariski topology on $K\text{-Spec}(R \otimes_K S)$ is stronger (= bigger) than the product of the Zariski topologies on $K\text{-Spec }R$ and $K\text{-Spec }S$. On the other hand, it is not too far from it, as is indicated by the following result:

If $Z \subseteq K\text{-Spec }R$ and $W \subseteq K\text{-Spec }S$ are dense subsets then $Z \times W$ is dense in the Zariski topology of $K\text{-Spec}(R \otimes_K S)$.

PROOF. Let $f = \sum_{i \in I} g_i \otimes h_i \in R \otimes_K S$ be an element which induces the non-zero function $f(\xi, \eta) = \sum_{i \in I} g_i(\xi) \otimes h_i(\eta)$ on $(K\text{-Spec }R) \times (K\text{-Spec }S)$. We need to show that this function is not identically zero on $Z \times W$. We may assume that the functions

$\eta \mapsto h_i(\eta)$, $i \in I$, on K-Spec S, are linearly independent over K. Since Z is dense in K-Spec R, there exists $\xi_0 \in Z$ such that $g_i(\xi_0) \neq 0$ for at least one $i \in I$. Now, the function $\eta \mapsto \sum_{i \in I} g_i(\xi_0) h_i(\eta)$ is a non-zero function on K-Spec S and since W is dense in K-Spec S, there exists $\eta_0 \in W$ such that $\sum_{i \in I} g_i(\xi_0) h_i(\eta_0) = f(\xi_0, \eta_0) \neq 0$.   •

One proves the equality $\mathfrak{r}_{R \otimes_K S} = \mathfrak{r}_R \otimes_K S + R \otimes_K \mathfrak{r}_S$ for the K-radicals of $R \otimes_K S$, R and S similarly.

2.C. Strong Topology

In this section the ground field is $\mathbb{R}$ or $\mathbb{C}$. The symbol $\mathbb{K}$ denotes one of these fields. Besides the Zariski topology, the affine space $\mathbb{K}^n$, $n \in \mathbb{N}$, has the usual topology which is, for $n > 0$, stronger than the Zariski topology, since every affine algebraic subset $V_K(F_j, j \in J)$, $F_j \in K[X_1, \ldots X_n]$ is the intersection of the closed sets $\{F_j = 0\}$ in $\mathbb{K}^n$ and hence closed in the usual topology. Therefore the usual topology $\mathbb{K}^n$ is called the s t r o n g t o p o l o g y on $\mathbb{K}^n$. The strong topology on $\mathbb{K}^n$ is the n-fold product of the strong topology on $\mathbb{K}$. However, the Zariski topology on $\mathbb{K}^n$, $n > 1$, is stronger than the n-fold product of the Zariski topology on $\mathbb{K}$ (cf. Example 2.B.17).

Now, let R be a $\mathbb{K}$-algebra of finite type and let $\alpha : \mathbb{K}[X_1, \ldots, X_n] \to R$ be a surjective $\mathbb{K}$-algebra homomorphism. Then the map $\alpha^* : \mathbb{K}$-Spec $R \to \mathbb{K}^n$ is injective and even a closed embedding with respect to the Zariski topologies. The topology on $\mathbb{K}$-Spec R induced from the strong topology of $\mathbb{K}^n$ via the injective map α^* is called the s t r o n g t o p o l o g y on $\mathbb{K}$-Spec R. Then by definition α^* is also a closed embedding with respect to the strong topologies. We have to verify that this strong topology on $\mathbb{K}$-Spec R is actually independent of the representation α of R. Let $\beta : \mathbb{K}[Y_1, \ldots, Y_m] \to R$ be another surjective $\mathbb{K}$-algebra homomorphism. Then there are $\mathbb{K}$-algebra homomorphisms $F : \mathbb{K}[X_1, \ldots, X_n] \to \mathbb{K}[Y_1, \ldots, Y_m]$ and $G : \mathbb{K}[Y_1, \ldots, Y_m] \to \mathbb{K}[X_1, \ldots, X_n]$ such that $\alpha = \beta \circ F$ and $\beta = \alpha \circ G$. Therefore $\alpha^* = F^* \circ \beta^*$ and $\beta^* = G^* \circ \alpha^*$ and hence the polynomial maps $G^* : \mathbb{K}^n \to \mathbb{K}^m$, $F^* : \mathbb{K}^m \to \mathbb{K}^n$ induce continuous maps $\mathbb{K}$-Spec $R = V_{\mathbb{K}}(\mathrm{Ker}\, \alpha) \to V_{\mathbb{K}}(\mathrm{Ker}\, \beta) = \mathbb{K}$-Spec R, $V_{\mathbb{K}}(\mathrm{Ker}\, \beta) \to V_{\mathbb{K}}(\mathrm{Ker}\, \alpha)$ which are inverses of each other and hence homeomorphisms.

If $\varphi : R \to S$ is a $\mathbb{K}$-algebra homomorphism of $\mathbb{K}$-algebras of finite type, then the map $\varphi^* : \mathbb{K}$-Spec $S \to \mathbb{K}$-Spec R is continuous with respect to the strong topologies on $\mathbb{K}$-Spec R and $\mathbb{K}$-Spec S. In fact, it is the restriction of a polynomial map $\mathbb{K}^m \to \mathbb{K}^n$ for some $m, n \in \mathbb{N}$. If we use these strong topologies then the assignments $R \rightsquigarrow \mathbb{K}$-Spec R, $\varphi \rightsquigarrow \varphi^*$ define a contravariant functor from the catagory of $\mathbb{K}$-algebras of finite type to the category of Hausdorff topological spaces.[3]

---

[3]) This functor can be easily extended to the category of all $\mathbb{K}$-algebras in a natural way.

2.C.1. Example Let R be a $\mathbb{K}$-algebra of finite type and let $f \in R$. Then the canonical $\mathbb{K}$-algebra homomorphism $\iota_f : R \to R_f$ induces an open embedding with respect to the strong topologies. In fact, if $\varphi : \mathbb{K}[X_1, \ldots, X_n] \to R$ is a surjective $\mathbb{K}$-algebra homomorphism then the map $\varphi' : \mathbb{K}[X_1, \ldots, X_n, X] \to R_f$, $X_i \mapsto \varphi(X_i)/1$, $X \mapsto 1/f$, is a surjective $\mathbb{K}$-algebra homomorphism and $\operatorname{Ker} \varphi'$ is generated by $\operatorname{Ker} \varphi$ and $FX - 1$, where F is an arbitrary lift of f in $\mathbb{K}[X_1, \ldots, X_n]$, i.e. $R_f \cong \mathbb{K}[X_1, \ldots, X_n]/(\operatorname{Ker} \varphi, FX - 1)$ and so $\mathbb{K}$-Spec R_f is homeomorphic to $\mathbb{K}$-Spec $\mathbb{K}[X_1, \ldots, X_n]/(\operatorname{Ker} \varphi, FX - 1)$. Under these identifications, $\mathbb{K}$-Spec $R_f = \{ (x, y) \in \mathbb{K}^n \times \mathbb{K} \mid x \in \mathbb{K}$-Spec R, $F(x) \neq 0$, $y = 1/F(x) \}$ and the map $\mathbb{K}$-Spec $R_f \to \mathbb{K}$-Spec R is the projection $(x, y) \mapsto x$, which is a homeomorphism onto the open subset $D_{\mathbb{K}}(f) = \{ x \in \mathbb{K}$-Spec $R \mid f(x) \neq 0 \}$ with inverse $x \mapsto (x, 1/f(x))$. For example, if $\mathbb{K} = \mathbb{R}$, $R := \mathbb{R}[X]$ and $f := X$, then $\mathbb{R}$-Spec $R = \mathbb{R}$, $R_f = \mathbb{R}[X, Y]/(fY - 1) = \mathbb{R}[X, Y]/(XY - 1)$,

$$\mathbb{R}\text{-Spec } R_f = \{ (x, y) \in \mathbb{R}^2 \mid y = 1/x,\ x \neq 0 \}$$

and the map $\mathbb{R}$-Spec $R_f \to \mathbb{R}$-Spec R is the projection from the standard hyperbola onto the x-axis.

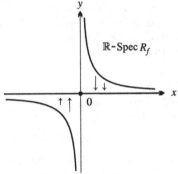

If $\mathbb{K} = \mathbb{R}$, $R := \mathbb{R}[X]$ and $f := (X^2 + 1)(X^2 - 1)$ then $\mathbb{R}$-Spec $R = \mathbb{R}$,

$$R_f = \mathbb{R}[X, Y]/(fY - 1) = \mathbb{R}[X, Y]/((X^2 + 1)(X^2 - 1)Y - 1),$$

$$\mathbb{R}\text{-Spec } R_f = \{ (x, y) \in \mathbb{R}^2 \mid y = \frac{1}{(x^2 + 1)(x^2 - 1)},\ x \neq 0 \}$$

and the map $\mathbb{R}$-Spec $R_f \to \mathbb{R}$-Spec R is again the projection onto the x-axis.

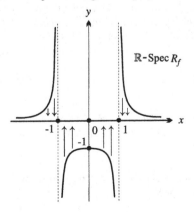

2.C.2. Example (The maximal spectrum of an $\mathbb{R}$-algebra of finite type)
Let R be an $\mathbb{R}$-algebra of finite type. The $\mathbb{C}$-algebra $R_{(\mathbb{C})} := \mathbb{C} \otimes_{\mathbb{R}} R = R \oplus Ri$ is
called the **complexification** of R. The usual complex conjugation $\kappa : \mathbb{C} \to \mathbb{C}$
induces the **conjugation** $\kappa_R := \kappa \otimes \mathrm{id}_R : R_{(\mathbb{C})} \to R_{(\mathbb{C})}$ which is the R-involution
$x + yi \mapsto x - yi$, $x, y \in R$, of $R_{(\mathbb{C})}$ with R as algebra of invariants. Further, $\kappa = \kappa_R$
is $\mathbb{C}$-antilinear, i.e. $\kappa(\alpha z) = \overline{\alpha}\kappa(z)$ for all $\alpha \in \mathbb{C}$ and $z \in R_{(\mathbb{C})}$. By functoriality the
involution $\kappa = \kappa_R$ of the R-algebra $R_{(\mathbb{C})}$ induces the involution $\sigma = \sigma_R$ of the maximal
spectrum Spm $R_{(\mathbb{C})}$, $\mathfrak{m} \mapsto \sigma(\mathfrak{m}) := \kappa^{-1}(\mathfrak{m}) = \kappa(\mathfrak{m})$, which we call the **conjugation**
of Spm $R_{(\mathbb{C})}$. By Hilbert's Nullstellensatz, Spm $R_{(\mathbb{C})}$ can be identified with the $\mathbb{C}$-spectrum
$\mathbb{C}\text{-Spec } R_{(\mathbb{C})} = \mathrm{Hom}_{\mathbb{C}-\mathrm{alg}}(R_{(\mathbb{C})}, \mathbb{C}) = \mathrm{Hom}_{\mathbb{R}-\mathrm{alg}}(R, \mathbb{C})$ (see 2.A.2). Under this identification, the involution σ of Spm $R_{(\mathbb{C})}$ is nothing else but the involution $\xi \mapsto \sigma(\xi) = \kappa \circ \xi \circ \kappa_R$
of $\mathrm{Hom}_{\mathbb{C}-\mathrm{alg}}(R_{(\mathbb{C})}, \mathbb{C})$ or the involution $\eta \mapsto \sigma(\eta) = \kappa \circ \eta$ of $\mathrm{Hom}_{\mathbb{R}-\mathrm{alg}}(R, \mathbb{C})$, and the $\mathbb{R}$-
spectrum $\mathbb{R}\text{-Spec } R$ of R is its invariant set, i.e. $\mathbb{R}\text{-Spec } R = \mathrm{Fix}\,\sigma = \{\xi \in \mathbb{C}\text{-Spec } R_{(\mathbb{C})} \mid$
$\sigma(\xi) = \xi\}$. For $\eta \in \mathrm{Hom}_{\mathbb{R}-\mathrm{alg}}(R, \mathbb{C})$ we have

$$\eta \in \mathrm{Fix}\,\sigma \Leftrightarrow \mathrm{im}\,\eta = \mathbb{R} \Leftrightarrow R/\mathrm{Ker}\,\eta = \mathbb{R} \text{ and } \eta \notin \mathrm{Fix}\,\sigma \Leftrightarrow \mathrm{im}\,\eta = \mathbb{C} \Leftrightarrow R/\mathrm{Ker}\,\eta \cong \mathbb{C}.$$

Therefore, *the map* $\mathrm{Hom}_{\mathbb{R}-\mathrm{alg}}(R, \mathbb{C}) \to \mathrm{Spm}\,R$, $\eta \mapsto \mathrm{Ker}\,\eta$ *induces a bijection from the*
orbit space $\mathbb{C}\text{-Spec } R_{(\mathbb{C})}/\{\mathrm{id}, \sigma\}$ *onto the maximal spectrum* Spm R *of* R. The singleton
orbits correspond to the real points of Spm R. The orbits with two points of $\mathbb{C}\text{-Spec } R_{(\mathbb{C})}$
correspond to the **complex points** of Spm R, i.e. to the maximal ideals $\mathfrak{m}$ in R with
$R/\mathfrak{m} \cong \mathbb{C}$.

In the special case $R := \mathbb{R}[X_1, \ldots, X_n]$, $n \geq 1$, we have $R_{(\mathbb{C})} = \mathbb{C}[X_1, \ldots, X_n]$ and the
conjugation of $\mathbb{C}\text{-Spec } R_{(\mathbb{C})} = \mathbb{C}^n$ is the involution $(z_1, \ldots, z_n) \mapsto (\overline{z}_1, \ldots, \overline{z}_n)$. If we
identify $\mathbb{C}^n = \mathbb{R}^n \times i\mathbb{R}^n \cong \mathbb{R}^n \times \mathbb{R}^n$ then

$$\mathrm{Spm}\,\mathbb{R}[X_1, \ldots, X_n] = \mathbb{C}^n/\{\mathrm{id}_{\mathbb{C}^n}, \sigma\} = \mathbb{R}^n \times (\mathbb{R}^n/\{\pm\mathrm{id}_{\mathbb{R}^n}\}).$$

To describe the quotient $\mathbb{R}^n/\{\pm\mathrm{id}\}$, we identify $\mathbb{R}^n$ with the cone over the $(n-1)$-sphere
$C(S^{n-1}) := \mathbb{R}_+ \times S^{n-1}/\{0\} \times S^{n-1}$.[4]) Then $\mathbb{R}^n/\{\pm\mathrm{id}\}$ is identified with the cone $C(\mathbb{P}^{n-1}) = $
$\mathbb{R}_+ \times \mathbb{P}^{n-1}/\{0\} \times \mathbb{P}^{n-1}$ over the real projective space $\mathbb{P}^{n-1} = \mathbb{P}^{n-1}(\mathbb{R})$ of dimension $n-1$.
Therefore

$$\mathrm{Spm}\,\mathbb{R}[X_1, \ldots, X_n] \cong \mathbb{R}^n \times C(\mathbb{P}^{n-1})$$

and $\mathbb{R}\text{-Spec }\mathbb{R}[X_1, \ldots, X_n] = \mathbb{R}^n$ corresponds to $\mathbb{R}^n \times \{*\}$, where $*$ is the vertex of the cone
$C(\mathbb{P}^{n-1})$. With this identification, the space of complex points of Spm $\mathbb{R}[X_1, \ldots, X_n]$ cor-
responds to $\mathbb{R}^n \times \mathbb{R}_+^\times \times \mathbb{P}^{n-1}$. For $n \neq 2$ the points of $\mathbb{R}\text{-Spec }\mathbb{R}[X_1, \ldots, X_n]$ are topological
singularities of Spm $\mathbb{R}[X_1, \ldots, X_n]$, i.e., for such a point there does not exist an open neigh-
bourhood which is homeomorphic to an open set in $\mathbb{R}^{2n}$. (This can be checked using simple
topological methods as in Example 2.C.4.) For $n = 1$ the space Spm $\mathbb{R}[X]$ is the closed
upper half plane $\overline{\mathbb{H}} = \mathbb{H} \uplus \mathbb{R} = \{z \in \mathbb{C} : \mathrm{Im}\,z \geq 0\}$ with the real line $\mathbb{R} = \mathbb{R}\text{-Spec }\mathbb{R}[X]$
as boundary. For an arbitrary n, describe the maximal ideals in $\mathbb{R}[X_1, \ldots, X_n]$ which
represent a point of Spm $\mathbb{R}[X_1, \ldots, X_n]$ but not a point of $\mathbb{R}\text{-Spec }\mathbb{R}[X_1, \ldots, X_n] = \mathbb{R}^n$.

2.C.3. Exercise For the following $\mathbb{R}$-algebras R of (Krull-)dimension 1, describe the $\mathbb{R}$-
spectrum $\mathbb{R}\text{-Spec } R$ and the maximal spectrum Spm R. Which of these algebras are reduced,
integral, normal, or factorial?

---

[4]) For a topological space X and a subset $Y \subseteq X$ we denote by X/Y the space obtained
from X by identifying the points of Y.

$$\mathbb{R}[X], \quad \mathbb{R}[X,Y]/(X^2 - Y^2), \quad \mathbb{R}[X,Y]/(X^2 + Y^2),$$

$$\mathbb{R}[X,Y]/(X^2 - Y^2 - 1), \quad \mathbb{R}[X,Y]/(X^2 + Y^2 - 1), \quad \mathbb{R}[X,Y]/(X^2 + Y^2 + 1).$$

(The real circle algebra $C := \mathbb{R}[X,Y]/(X^2 + Y^2 - 1)$ is not factorial, the maximal ideals correponding to the real points need two generators. For instance, if the maximal ideal $\mathfrak{m}_{(1,0)} \subseteq C$ is generated by f, then f induces an analytic function on the real circle with $V_\mathbb{R}(f) = \{(1,0)\}$ and $(1,0)$ is a simple zero of f. This contradicts the intermediate value theorem. The maximal ideals corresponding to the complex points are principal. – The maximal spectrum of the (factorial) algebra $\mathbb{R}[X,Y]/(X^2 + Y^2 + 1)$ contains only complex points and can be identified with the punctured real projective plane, i.e. the Möbius strip. For more examples see Example 2.B.6, cf. also Exercise 6.E.22.)

2.C.4. Example (S m o o t h (= r e g u l a r) and s i n g u l a r p o i n t s) The strong topology allows to use analytic methods to study affine algebraic sets over $\mathbb{K}$ and one might try to translate (if possible) the analytic concepts and results into an algebraic framework. In particular, the concept of an (analytically) smooth point is defined for an arbitrary subset $X \subseteq \mathbb{K}^n$ at a point $P \in X$. Let us recall this definition: $P \in X$ is called an (a n a l y t i c a l l y) s m o o t h or an (a n a l y t i c a l l y) r e g u l a r point of X in $\mathbb{K}^n$ if one of the following equivalent conditions holds:

(1) There exists an open neighbourhood U of P in $\mathbb{K}^n$ and, for some $k \le n$, an analytic submersion $F : U \to \mathbb{K}^k$ with $F(P) = 0$ and $U \cap X = F^{-1}(0)$.

(1') There exists an open neighbourhood U of P in $\mathbb{K}^n$ and, for some $k \le n$, $\mathbb{K}$-analytic functions $F_i : U \to \mathbb{K}, i = 1, \dots, k$, with $F_i(P) = 0$ and such that the Jacobian

$$\frac{\partial(F_1, \dots, F_k)}{\partial(X_1, \dots, X_n)} := \left(\frac{\partial F_i}{\partial X_j}\right)_{1 \le i \le k, 1 \le j \le n}$$

has maximal rank k on U and $U \cap X = \bigcap_{i=1}^k F_i^{-1}(0)$.

(2) There exists an open neighbourhood U of P in $\mathbb{K}^n$ and, for some $m \in \mathbb{N}$, a closed analytic embedding $G : V \to U$ of an open neighbourhood V of 0 in $\mathbb{K}^m$ such that $G(V) = X \cap U$.

The condition (1') which is called the J a c o b i a n c r i t e r i o n for smoothness, is just a reformulation of the condition (1) and the equivalence of (1) and (2) follows from the implicit function theorem. The integers k and m in the definition are related by $k + m = n$ and are called the $\mathbb{K}$-d i m e n s i o n of X at P and the $\mathbb{K}$-c o d i m e n s i o n of X at P respectively. They are denoted by $\dim_P X$ and $\text{codim}_P X$. The set X_{reg} (or reg X) of X of regular points is by definition an open subset of X and a locally closed analytic submanifold of $\mathbb{K}^n$. A point $P \in X$ is called a s i n g u l a r point of X if it is not regular and X_{sing} (or sing X) denotes the set of singular points of X.

Now let us consider an affine algebraic set $X = V_\mathbb{K}(\mathfrak{a}) \subseteq \mathbb{K}^n$, where $\mathfrak{a}$ is an ideal in $\mathbb{K}[X_1, \dots, X_n]$. Then the subsets X_{reg} and X_{sing} of $X = \mathbb{K}\text{-Spec } R$ are independent of the representation of $R := \mathbb{K}[X_1, \dots, X_n]/\mathfrak{a}$. For, if $R = \mathbb{K}[Y_1, \dots, Y_m]/\mathfrak{b}$ is another representation of R then the homeomorphisms $V_\mathbb{K}(\mathfrak{a}) \leftrightarrow V_\mathbb{K}(\mathfrak{b})$ are induced by polynomial maps $G^* : \mathbb{K}^n \to \mathbb{K}^m$, $F^* : \mathbb{K}^m \to \mathbb{K}^n$ (see the beginning of this section) and therefore analytic maps of the ambient spaces $\mathbb{K}^n$ and $\mathbb{K}^m$ respectively. Further, the sets X_{reg} and X_{sing} depend only on the ideal $\mathfrak{I}_\mathbb{K}(V_\mathbb{K}(\mathfrak{a}))$, i.e. on the $\mathbb{K}$-algebra $\mathbb{K}[X_1, \dots, X_n]/\mathfrak{I}_\mathbb{K}(V_\mathbb{K}(\mathfrak{a})) = R/\mathfrak{r}$ where $\mathfrak{r}$ is the $\mathbb{K}$-radical of R (see Remark 2.B.13). In the case $\mathbb{K} = \mathbb{C}$, $\mathfrak{r} = \mathfrak{n}_R$ by Hilbert's Nullstellensatz 2.B.11.

Let $F_1, \ldots, F_k$, $k \leq n$, be polynomials in $\mathbb{K}[X_1, \ldots, X_n]$ and $P \in V_{\mathbb{K}}(F_1, \ldots, F_k)$. If the Jacobian $\partial(F_1, \ldots, F_k)/\partial(X_1, \ldots, X_n)$ has maximal rank k at P then by definiton P is a smooth point of $V_{\mathbb{K}}(F_1, \ldots, F_k)$ of dimension $n - k$. Note that the set of points $Q \in \mathbb{K}^m$ such that the Jacobian $\partial(F_1, \ldots, F_k)/\partial(X_1, \ldots, X_n)$ has maximal rank k is an open set in $\mathbb{K}^n$ even in the Zariski topology. More generally, for a point P of an arbitrary affine algebraic set $V \subseteq \mathbb{K}^n$, if there exist polynomials $F_1, \ldots, F_k$, $k \leq n$, such that the Jacobian $\partial(F_1, \ldots, F_k)/\partial(X_1, \ldots, X_n)$ has maximal rank k at P and $V \cap U = V_{\mathbb{K}}(F_1, \ldots, F_k) \cap U$ for some open neighbourhood U of P in $\mathbb{K}^n$, then P is a smooth point of V.

In the complex case, for a smooth point $P \in V$, there always exist polynomials $F_1, \ldots, F_k \in \mathfrak{I}_{\mathbb{C}}(V)$ as above. The proof of this is not obvious. One needs some results from complex analytic geometry which will not be developed here. However, in the real case, this is not true in general. For example, for the prime polynomial $F := Y(X^2 + Y^2) - (X^4 + Y^4) \in \mathbb{R}[X, Y]$, all points of the real curve $V_{\mathbb{R}}(F) \subseteq \mathbb{R}^2$ are smooth. For a point $P \in V_{\mathbb{R}}(F)$, $P \neq (0, 0)$, the Jacobian $(\partial F/\partial X, \partial F/\partial Y)$ has rank 1 at P and hence P is smooth by the Jacobian criterion. However, $\mathfrak{I}_{\mathbb{R}}(V_{\mathbb{R}}(F)) = (F)$ by the Exercise 2.B.6 and so there is no polynomial $G \in \mathfrak{I}_{\mathbb{R}}(V_{\mathbb{R}}(F))$ for which the Jacobian $(\partial G/\partial X, \partial G/\partial Y)$ has rank 1 at $(0, 0)$.

$$V_{\mathbb{R}}(Y(Y^2 + X^2) - (Y^4 + X^4))$$

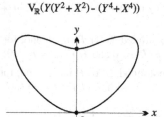

But the point $(0, 0)$ is also analytically smooth, since the (rational) map $\mathbb{R} \to \mathbb{R}^2$, $t \mapsto \frac{t^2 + 1}{t^4 + 1}(t, t^2)$, is a real-analytic embedding with image $V_{\mathbb{R}}(F) \setminus \{(0, 1)\}$. (By the way, this rational embedding shows that the real curve $V_{\mathbb{R}}(F)$ and also the complex curve $V_{\mathbb{C}}(F)$ are rational.)

2.C.5. Example (Quadrics and regular double points) In this example we study non-degenerate quadratic forms in $\mathbb{K}[X_1, \ldots, X_n]$. They describe the regular double points. First, let us recall the following famous classification theorem for quadratic forms over $\mathbb{K}$:

(1) *Up to a linear automorphism of* $\mathbb{R}[X_1, \ldots, X_n]$, *every non-degenerate real quadratic form* $F \in \mathbb{R}[X_1, \ldots, X_n]$ *is of the form*

$$F_{p,q} := X_1^2 + \cdots + X_p^2 - X_{p+1}^2 - \cdots - X_n^2,$$

where p and q are non-negative integers with $p + q = n$ *and which are uniquely determined by* F. (This is the Sylvester's law of inertia. The integers p and q are called the index of inertia and the Morse index of F, respectively. The difference $p - q$ is called the signature.)

(2) *Up to a linear automorphism of* $\mathbb{C}[X_1, \ldots, X_n]$, *every non-degenerate complex quadratic form* $F \in \mathbb{C}[X_1, \ldots, X_n]$ *is of the form*

$$F_n := X_1^2 + \cdots + X_n^2.$$

The reason why these quadratic forms are so important is apparent from the analytic Morse lemma. Before we state this, let us give the definition of a regular singular point of an analytic function f defined in an open neighbourhood of a point $P \in \mathbb{K}^n$, $n \geq 1$: The point P is called a r e g u l a r   s i n g u l a r   p o i n t   of f if f is singular at P, i.e. $\dfrac{\partial f}{\partial X_i}(P) = 0$ for all $i = 1, \ldots, n$, and if the H e s s i a n

$$\left(\frac{\partial^2 f}{\partial X_i \partial X_j} \right)_{1 \leq i, j \leq n}$$

is non-degenerate at P. In this case the a n a l y t i c   M o r s e   l e m m a holds:

Let P be a regular singular point of an analytic function f in an open neighbourhood of P in $\mathbb{K}^n$, $n \geq 1$ with $f(P) = 0$. Then there exist analytic coordinate functions $y_1, \ldots, y_n$ in P with $y_i(P) = 0$ and $f = F_{p,q}(y)$ if $\mathbb{K} = \mathbb{R}$ or $f = F_n(y)$ if $\mathbb{K} = \mathbb{C}$.

The type (p, q) of the quadratic form $F_{p,q}$ in the Morse lemma is of course the type of the quadratic form defined by the Hessian of f at P. In particular, for $n \geq 2$, the point P is an isolated zero of f if and only if $\mathbb{K} = \mathbb{R}$ and the quadratic form defined by the Hessian is definite.

Now we consider the case $\mathbb{K} = \mathbb{R}$ and the quadratic form $F_{p,q}$ with $p \geq q \geq 1$. For convenience we write:

$$F_{p,q} = X_1^2 + \ldots + X_p^2 - Y_1^2 - \ldots - Y_q^2 \in \mathbb{R}[X_1, \ldots, X_p, Y_1, \ldots, Y_q].$$

The zero set

$$X_{p,q} := V_{\mathbb{R}}(F_{p,q}) = \{ (x, y) \in \mathbb{R}^p \times \mathbb{R}^q \mid \|x\|^2 = x_1^2 + \cdots + x_p^2 = y_1^2 + \cdots + y_q^2 = \|y\|^2 \}$$

is called the light cone of the quadratic form $F_{p,q}$. Since $\partial F_{p,q}/\partial X_i = 2X_i$ for $i = 1, \ldots, p$ and $\partial F_{p,q}/\partial Y_j = -2Y_j$ for $j = 1, \ldots, q$, by the Jacobian criterion every point of $X_{p,q} \setminus \{0\}$ is a smooth point. *The vertex $0 \in X_{p,q}$ is even a* t o p o l o g i c a l   s i n g u l a r i t y *of $X_{p,q}$,* i. e. there does not exist any open neighbourhood of 0 in $X_{p,q}$ which is homeomorphic to an open subset of $\mathbb{R}^m$ for some $m \in \mathbb{N}$.[5])

PROOF. Note that $X_{p,q} \setminus \{0\}$ is as a real-analytic manifold isomorphic to the cylinder $\mathbb{R}_+^\times \times S^{p-1} \times S^{q-1}$ over $S^{p-1} \times S^{q-1}$ and $X_{p,q}$ itself is homeomorphic to the cone

$$\mathbb{R}_+ \times S^{p-1} \times S^{q-1}/\{0\} \times S^{p-1} \times S^{q-1}$$

over $S^{p-1} \times S^{q-1}$. The subsets

$$U_\varepsilon := [0, \varepsilon) \times S^{p-1} \times S^{q-1}/\{0\} \times S^{p-1} \times S^{q-1}, \ \varepsilon > 0,$$

form a fundamental system of open neighbourhoods of 0 in $X_{p,q}$. Let us first consider the simple case $q = 1$. Then for every open neighbourhood $U \subseteq X_{p,1}$ of 0, the punctured neighbourhood $U \setminus \{0\}$ has at least two connected components and for $p = 1$ even four connected components. Therefore no open neighbourhood of 0 in $X_{p,1}$ can be homoemorphic to an open ball in $\mathbb{R}^{n-1} = \mathbb{R}^p$.

Now assume $q \geq 2$ and that there exists an open neighbourhood B of 0 in $X_{p,q}$ which is homeomorphic to an open ball $B^{n-1} \subseteq \mathbb{R}^{n-1}$. Then there exists an $\varepsilon > 0$ such that $U_\varepsilon \subseteq B$. This inclusion induces canonical group homomorphisms $\pi(U_\varepsilon \setminus \{0\}) \to \pi(B \setminus \{0\})$

---

[5]) We note that this integer m must be $n - 1$, since every such neighbourhood contains smooth points of $X_{p,q}$ which are of dimension $n - 1$.

and $H_i(U_\varepsilon \setminus \{0\}) \to H_i(B \setminus \{0\})$, $i \geq 0$, of the fundamental and homology groups, respectively. These homomorphisms are injective, since there exists a homeomorphism $\varphi : X_{p,q} \setminus \{0\} \xrightarrow{\sim} U_\varepsilon \setminus \{0\}$ and since the canonical inclusion $U_\varepsilon \setminus \{0\} \hookrightarrow X_{p,q} \setminus \{0\}$ induces isomorphisms on the fundamental and homology groups. (To prove this use the composition

$$U_\varepsilon \setminus \{0\} \hookrightarrow B \setminus \{0\} \xrightarrow{\varphi \,|\, (B \setminus \{0\})} X_{p,q} \setminus \{0\} \,.)$$

For $q = 2$ this is a contradiction, since $\pi(U_\varepsilon \setminus \{0\}) = \pi(S^{p-1} \times S^1) \neq 0$ but $n \geq 4$ and $\pi(B \setminus \{0\}) = \pi(S^{n-2}) = 0$. For arbitrary $q \geq 2$, we use homology groups to get a contradiction: $\mathbb{Z} = H_{q-1}(S^{q-1}) \subseteq H_{q-1}(U_\varepsilon \setminus \{0\})$, but $0 < q - 1 < n - 2$ and hence $H_{q-1}(B \setminus \{0\}) = H_{q-1}(S^{n-2}) = 0$. $\quad\bullet$

Now for $n \geq 2$, consider the complex case $\mathbb{K} = \mathbb{C}$ and the complex cone

$$Z_n := \{ z \in \mathbb{C}^n \mid F_n(z) = 0 \} \,.$$

Let $\langle -, - \rangle$ denote the canonical *bilinear* form on $\mathbb{C}^n$. Then for $z = x + iy$, $x, y \in \mathbb{R}^n$, we get $F_n(z) = \langle z, z \rangle = \langle x, x \rangle - \langle y, y \rangle + 2i\langle x, y \rangle$. Therefore $z \in Z_n$ if and only if $\|x\|^2 = \|y\|^2$ and $\langle x, y \rangle = 0$, i.e. $x \perp y$. Hence $Z_n \setminus \{0\}$ is as a real-analytic manifold diffeomorphic to the cylinder $\mathbb{R}_+^\times \times \mathrm{St}_{\mathbb{R}}(2, n)$ over the Stiefel manifold

$$\mathrm{St}_{\mathbb{R}}(2, n) := \{ (x, y) \in \mathbb{R}^n \times \mathbb{R}^n \mid \|x\| = \|y\| = 1 \text{ and } x \perp y \}$$

of the pairs of orthonormal vectors in $\mathbb{R}^n$ and Z_n itself is homeomorphic to the cone $\mathbb{R}_+ \times \mathrm{St}_{\mathbb{R}}(2, n)/\{0\} \times \mathrm{St}_{\mathbb{R}}(2, n)$. The sets

$$U_\varepsilon := [0, \varepsilon) \times \mathrm{St}_{\mathbb{R}}(2, n)/\{0\} \times \mathrm{St}_{\mathbb{R}}(2, n) \,, \quad \varepsilon > 0 \,,$$

form a fundamental system of open neighbourhoods of 0 in Z_n. Similar to the real case, from this description, we conclude: *The vertex $0 \in Z_n$ is a topological singularity of Z_n.*

PROOF. Suppose that there exists an open neighbourhood B of 0 in Z_n which is homeomorphic to an open ball $B^{2(n-1)} \subseteq \mathbb{R}^{2(n-1)}$. Then there exists an $\varepsilon > 0$ such that $U_\varepsilon \subseteq B$. As in the real case, this inclusion induces injective group homomorphisms $\pi(U_\varepsilon \setminus \{0\}) = \pi(\mathrm{St}_{\mathbb{R}}(2, n)) \to \pi(B \setminus \{0\}) = \pi(S^{2n-3})$ and $H_i(U_\varepsilon \setminus \{0\}) = H_i(\mathrm{St}_{\mathbb{R}}(2, n)) \to H_i(B \setminus \{0\}) = H_i(S^{2n-3})$, $i \geq 0$, of the fundamental and homology groups, respectively. We show that this is not possible.

For $n = 2$, $\mathrm{St}_{\mathbb{R}}(2, 2) = O_2(\mathbb{R})$ has two connected components, but S^1 is connected.

For $n = 3$, $\mathrm{St}_{\mathbb{R}}(2, 3) = SO_3(\mathbb{R})$ has fundamental group $\mathbb{Z}_2$, but S^3 is simply connected. In fact, it is quite easy to give the two-sheeted universal covering of the punctured cone $Z_3 \setminus \{0\}$: For this we change coordinates and replace Z_3 by the quadric

$$Q := \{ (z, u, v) \in \mathbb{C}^3 \mid z^2 = uv \} \,.$$

Then $\mathbb{C}^2 \setminus \{0\} \to Q \setminus \{0\}$, $(w_1, w_2) \mapsto (w_1 w_2, w_1^2, w_2^2)$ is the required universal covering (which, of course, is closely related to the spin covering of the special orthogonal group $SO_3(\mathbb{R})$).

If $n \geq 4$, then $H_{n-1}(\mathrm{St}_{\mathbb{R}}(2, n)) \cong \mathbb{Z} \neq 0$, but $H_{n-1}(S^{2n-3}) = 0$. For even $n = 2m \geq 4$ we have a simpler argument: The canonical projection $p : (x, y) \mapsto x$ from $\mathrm{St}_{\mathbb{R}}(2, 2m)$ onto S^{2m-1} has the section $(x_1, \ldots, x_{2m}) \mapsto ((x_1, \ldots, x_{2m}), (-x_2, x_1, \ldots, -x_{2m}, x_{2m-1}))$ which induces an inclusion $\mathbb{Z} = H_{2m-1}(S^{2m-1}) \hookrightarrow H_{2m-1}(\mathrm{St}_{\mathbb{R}}(2, 2m - 1))$, however $H_{2m-1}(S^{4m-3}) = 0$. $\quad\bullet$

We remark that, *for $n \geq 4$, the Stiefel manifold* $\mathrm{St}_{\mathbb{R}}(2, n)$ *and hence the punctured cone* $Z_n \setminus \{0\}$ *is simply connected*. This follows from the fact that the fibre space $p : \mathrm{St}_{\mathbb{R}}(2, n) \to S^{n-1}$ is trivial over $S^{n-1} \setminus \{x\}$ for every $x \in S^{n-1}$. Therefore $p^{-1}(S^{n-1} \setminus \{x\}) \cong \mathbb{R}^{n-1} \times S^{n-2}$ is simply connected for every $x \in S^{n-1}$, which immediately implies that $\mathrm{St}_{\mathbb{R}}(2, n) = p^{-1}(S^{n-1} \setminus \{x\}) \cup p^{-1}(S^{n-1} \setminus \{x'\})$, $x \neq x'$, is also simply connected.

2.C.6. Remark Even in the complex case a singularity of an affine algebraic set need not be a topological singularity. For example, the origin is a singular point of the Neil's parabola $V_{\mathbb{K}}(Y^2 - X^3)$, but it is not a topological singularity, since the map $\mathbb{K} \to V_{\mathbb{K}}(Y^2 - X^3)$, $t \mapsto (t^2, t^3)$, is a homeomorphism.

2.C.7. Exercise In Example 2.C.5, for the real cone $X_{p,q} = V_{\mathbb{R}}(F_{p,q})$, $p \geq q \geq 1$, we have the equality $\mathfrak{I}_{\mathbb{R}}(X_{p,q}) = (F_{p,q})$. For $p + q \geq 3$, this is a special case of the following assertion: *Let* $F = Y^d + F_{d-1}Y^{d-1} + \cdots + F_0 \in \mathbb{R}[X_1, \ldots, X_n, Y]$, $F_i \in \mathbb{R}[X_1, \ldots, X_n]$, $i = 0, \ldots, d-1$, *be a monic prime polynomial in* Y. *Then* $\mathfrak{I}_{\mathbb{R}}(V_{\mathbb{R}}(F)) = (F)$ *if (and only if)* $V_{\mathbb{R}}(F) \cap D_{\mathbb{R}}(\partial F / \partial Y) \neq \emptyset$. (This assertion will be generalized later, cf. Exercise 6.E.23.)

For further comments on the strong topology see Exercise 6.E.22.

CHAPTER 3: Prime Spectra and Dimension

3.A. The Prime Spectrum of a Commutative Ring

In the previous chapter, we have associated a geometric object to any (commutative) algebra over a field K, namely its K-spectrum. A similar geometric object exists for an arbitrary commutative ring. This construction is one of the most important achievements of modern algebraic geometry.

Let R be a (commutative) ring (with identity). The set of all prime ideals of R is denoted by

$$\operatorname{Spec} R$$

and is called the prime spectrum or just the spectrum of R. The maximal spectrum $\operatorname{Spm} R$ (i.e. the set of all maximal ideals of R) is a subset of $\operatorname{Spec} R$. Moreover, for a K-algebra R we have the inclusions $K\text{-Spec} R \subseteq \operatorname{Spm} R \subseteq \operatorname{Spec} R$.

First, let us introduce some useful and elegant notations which are inspired from the notations for the K-spectrum of a K-algebra.

Let $X := \operatorname{Spec} R$. A point $x \in X$ is a prime ideal in R and we shall denote it by $\mathfrak{p}_x$ if we want to consider it in this way. If $x \in \operatorname{Spm} R \subseteq X$ we denote it also by $\mathfrak{m}_x$ (but notice the convention after Definition 4.A.5). The residue field

$$\kappa(x) := R_{\mathfrak{p}_x}/\mathfrak{p}_x R_{\mathfrak{p}_x} = Q(R/\mathfrak{p}_x)$$

is called the field of the point x. For a K-algebra R the K-spectrum $K\text{-Spec} R$ is the subset $\{x \in X \mid \kappa(x) = K\} \subseteq X$. For $f \in R$ and $x \in X$, we denote the image of f in $\kappa(x)$ by $f(x)$ and call it the value of f at the point x. This extends the analogous notation for the K-spectrum. But now the function $x \mapsto f(x)$, $x \in X$, has, in general, values in different fields.

A point $x \in X$ is a zero of $f \in R$ if and only if $f \in \mathfrak{p}_x$. The function $x \mapsto f(x)$ is identically zero on X if and only if $f \in \bigcap_{x \in X} \mathfrak{p}_x = \mathfrak{n}_R = \sqrt{0}$. Therefore, if R is not reduced, i.e. if R has non-trivial nilpotent elements, then the function $x \mapsto f(x)$ can be identically zero without f being the zero element of R. The set of all elements $f \in R$ which vanish on $\operatorname{Spm} R$ is the Jacobson radical $\mathfrak{m}_R = \bigcap_{x \in \operatorname{Spm} X} \mathfrak{m}_x$ of R and, for a K-algebra R, the set of all $f \in R$ which vanish on $K\text{-Spec} R$ is the K-radical $\mathfrak{r}_R$ of R (see Remark 2.B.13).

Now we define a topology on $X = \operatorname{Spec} R$, similar to the Zariski topology on the K-spectrum of a K-algebra. For $f \in R$, we put

$$\mathrm{V}(f) := \{x \in X \mid f(x) = 0\} = \{\mathfrak{p} \in \operatorname{Spec} R \mid f \in \mathfrak{p}\}$$

and, for an arbitrary family $\mathcal{F} = (f_j)_{j \in J}$ of elements $f_j \in R$, we put

$$\mathrm{V}(\mathcal{F}) = \mathrm{V}(f_j, j \in J) := \bigcap_{j \in J} \mathrm{V}(f_j) = \{x \in X \mid f_j(x) = 0 \text{ for all } j \in J\},$$

in particular, for a subset $\mathcal{F} \subseteq R$, we have $V(\mathcal{F}) = \bigcap_{f \in \mathcal{F}} V(f)$. For an arbitrary family $\mathcal{F} = (f_j)_{j \in J}$, we have $V(\mathcal{F}) = V(\mathfrak{a})$ where $\mathfrak{a} := \sum_{j \in J} R f_j$ is the ideal in R generated by the f_j, $j \in J$. The subsets of the form $V(\mathfrak{a})$ of $X = \operatorname{Spec} R$ are called the (a f f i n e) a l g e b r a i c s e t s i n $\operatorname{Spec} R$. They form the closed sets of a topology on X which is the Z a r i s k i t o p o l o g y on X. This is immediate from the simple rules stated in the following proposition which are the same as those for the affine algebraic sets in the K-spectrum of a K-algebra, cf. Proposition 2.B.2.

3.A.1. Proposition *Let* $\mathcal{F}_i, i \in I$, *be a family of subsets of a ring R and let* $\mathfrak{a}_j$, $j \in J$, *be a family of ideals in R. Then*:

(1) $V(\bigcup_{i \in I} \mathcal{F}_i) = \bigcap_{i \in I} V(\mathcal{F}_i)$.

(2) $V(\sum_{j \in J} \mathfrak{a}_j) = \bigcap_{j \in J} V(\mathfrak{a}_j)$.

(3) $V(fg) = V(f) \cup V(g)$ *for* $f, g \in R$.

(4) $V(\mathfrak{a}\mathfrak{b}) = V(\mathfrak{a} \cap \mathfrak{b}) = V(\mathfrak{a}) \cup V(\mathfrak{b})$ *for ideals* $\mathfrak{a}, \mathfrak{b} \subseteq R$.

(5) $V(\mathcal{F}) \subseteq V(\mathcal{G})$ *for subsets* $\mathcal{F}, \mathcal{G} \subseteq R$ *with* $\mathcal{F} \supseteq \mathcal{G}$.

(6) $V(1) = \emptyset$ *and* $V(0) = \operatorname{Spec} R$.

The open sets in the Zariski topology of $X = \operatorname{Spec} R$ are the complements

$$X \setminus V(f_j, j \in J) =: D(f_j, j \in J) = \bigcup_{j \in J} D(f_j).$$

In particular, $D(f) = X \setminus V(f) = \{x \in X \mid f(x) \neq 0\}$ for every $f \in R$. These open subsets are called the d i s t i n g u i s h e d o p e n s u b s e t s in X. *These subsets form a basis for the Zariski topology on X.* The last proposition translates into:

3.A.2. Proposition *Let* $\mathcal{F}_i, i \in I$, *be a family of subsets of a ring R and let* $\mathfrak{a}_j$, $j \in J$, *be a family of ideals in R. Then*:

(1) $D(\bigcup_{i \in I} \mathcal{F}_i) = \bigcup_{i \in I} D(\mathcal{F}_i)$.

(2) $D(\sum_{j \in J} \mathfrak{a}_j) = \bigcup_{j \in J} D(\mathfrak{a}_j)$.

(3) $D(fg) = D(f) \cap D(g)$ *for* $f, g \in R$.

(4) $D(\mathfrak{a}\mathfrak{b}) = D(\mathfrak{a} \cap \mathfrak{b}) = D(\mathfrak{a}) \cap D(\mathfrak{b})$ *for ideals* $\mathfrak{a}, \mathfrak{b} \subseteq R$.

(5) $D(\mathcal{F}) \supseteq D(\mathcal{G})$ *for subsets* $\mathcal{F}, \mathcal{G} \subseteq R$ *with* $\mathcal{F} \supseteq \mathcal{G}$.

(6) $D(1) = \operatorname{Spec} R$ *and* $D(0) = \emptyset$.

The set of closed points of $X = \operatorname{Spec} R$ is the maximal spectrum $\operatorname{Spm} R$ of R.

Another important property of the Zariski topology is:

3.A.3. Proposition *For every ring R, the spectrum $\operatorname{Spec} R$ is quasi-compact with respect to the Zariski topology.*

PROOF. It is enough to prove that an open cover of the form $\{ D(f_j) \mid j \in J \}$ has a finite subcover. Since $\emptyset = \bigcap_{j \in J} V(f_j) = V(\sum_{j \in J} Rf_j)$, we have $\sum_{j \in J} Rf_j = R$ and hence $1 = \sum_{j \in I} a_j f_j$ for some finite subset $I \subseteq J$. Then $\emptyset = \bigcap_{j \in I} V(f_j)$ and Spec $R = \bigcup_{j \in I} D(f_j)$.   •

Let $\varphi : R \to S$ be a ring homomorphism. Then $\mathfrak{q} \mapsto \varphi^{-1}(\mathfrak{q})$ is a canonical map $\varphi^* : \operatorname{Spec} S \to \operatorname{Spec} R$. As in the case of the K-spectra of K- algebras, *the assignments* $R \rightsquigarrow \operatorname{Spec} R$, $\varphi \rightsquigarrow \varphi^*$ *define a contravariant functor from the category of rings to the category of topological spaces.* Proposition 2.B.15 and the three special cases mentioned after it hold for arbitrary rings, replacing K-spectra by spectra. In particular, we identify for an ideal $\mathfrak{a}$ in R the closed set $V(\mathfrak{a}) \subseteq \operatorname{Spec} R$ with $\operatorname{Spec}(R/\mathfrak{a})$, and for a multiplicatively closed subset T of R the spectrum $\operatorname{Spec} T^{-1}R$ with the subspace $\{\mathfrak{p} \in \operatorname{Spec} R \mid \mathfrak{p} \cap T = \emptyset\}$, especially for $f \in R$ the basic open set $D(f) \subseteq \operatorname{Spec} R$ with $\operatorname{Spec} R_f$. For an $x \in \operatorname{Spec} R$ with corresponding prime ideal $\mathfrak{p}$, the fibre $(\varphi^*)^{-1}(x)$ can be identified with the spectrum $\operatorname{Spec} S_\mathfrak{p}/\mathfrak{p}S_\mathfrak{p}$. In particular, *the fibre* $(\varphi^*)^{-1}(x) \neq \emptyset$ *if and only if* $\mathfrak{p}S_\mathfrak{p} \neq S_\mathfrak{p}$, i. e. $\varphi^{-1}(\mathfrak{p}S) = \mathfrak{p}$.

3.A.4. Remark The assignment $R \rightsquigarrow \operatorname{Spm} R$ cannot be extended to a functor on the category of rings, since, in general, the preimage of a maximal ideal under a ring homomorphism is not a maximal ideal. For this reason we have to consider the whole prime spectrum. However, by Theorem 1.F.9, *for a fixed field* K, *the assignments* $R \rightsquigarrow \operatorname{Spm} R$, $\varphi \rightsquigarrow \varphi^*$ *define a functor on the category of K-algebras of finite type.*

As in the case of K-spectra, for any subset $E \subseteq X = \operatorname{Spec} R$, we define the ideal

$$\mathfrak{I}(E) := \{ f \in R \mid f(x) = 0 \text{ for all } x \in E \} = \bigcap_{x \in E} \mathfrak{p}_x .$$

$\mathfrak{I}(E)$ is a radical ideal. We have Propositions 2.B.8 and 2.B.9 (replacing V_K by V and $\mathfrak{I}_K$ by $\mathfrak{I}$), in particular, $V(\mathfrak{I}(E)) = \overline{E}$. Furthermore, we have the f o r m a l Hilbert's Nullstellensatz:

3.A.5. Theorem *For an arbitrary ideal* $\mathfrak{a}$ *in a commutative ring* R, *we have* $\mathfrak{I}(V(\mathfrak{a})) = \sqrt{\mathfrak{a}}$.

PROOF. Replacing R by $R/\mathfrak{a}$, we may assume $\mathfrak{a} = 0$. Then the assertion is the equation $\bigcap_{x \in X} \mathfrak{p}_x = \mathfrak{n}_R = \sqrt{0}$ mentioned before Proposition 2.B.4.   •

We therefore have: *The maps*

$$E \longmapsto \mathfrak{I}(E), \quad \mathfrak{a} \longmapsto V(\mathfrak{a})$$

are inclusion reversing bijective maps between the set of closed subsets in $X = \operatorname{Spec} R$ *and the set of radical ideals in* R, *which are inverse to each other.*

We now discuss some elementary topological properties of the prime spectra.

For every $f \in R$ the basic open set $D(f) = \operatorname{Spec} R_f$ is quasi-compact by 3.A.3. But, in general, an arbitrary open subset in $X = \operatorname{Spec} R$ is not quasi-compact. This leads us to the following definition:

3.A.6. Definition A topological space X is called N o e t h e r i a n if any one of the following equivalent conditions holds:

(1) Every open subset of X is quasi-compact.

(2) The open subsets of X satisfy the ascending chain condition.

(3) The closed subsets of X satisfy the descending chain condition.

(4) Every non-empty family of open subsets of X contains a maximal element.

(5) Every non-empty family of closed subsets of X contains a minimal element.

The proof of the equivalence of these conditions is a simple exercise and is left to the reader. By the formal Hilbert's nullstellensatz, the spectrum Spec R is a Noetherian topological space if and only if the radical ideals in R satisfy the ascending chain condition, in particular:

3.A.7. Proposition *If R is a Noetherian ring then $X =$ Spec R is a Noetherian topological space.*

More generally: If $\mathfrak{a} = \sum_{i=1}^{n} Rf_i$ is a finitely generated ideal in a commutative ring R, then $D(\mathfrak{a}) = \bigcup_{i=1}^{n} D(f_i)$ is quasi-compact.

3.A.8. Example The ring $\mathbb{Q}[X_i : i \in \mathbb{N}]/(X_i^2 : i \in \mathbb{N})$ is not Noetherian, but its spectrum is singleton and hence Noetherian.

The spectrum Spec R is in general far from being Hausdorff, indeed the only closed points are the points corresponding to the maximal ideals of R (see also Exercise 3.A.20 (2)). Further, if R is an integral domain, then any non-empty open subset contains the point corresponding to the zero prime ideal, in particular, $U \cap V \neq \emptyset$ for arbitrary non-empty open subsets $U, V \subseteq$ Spec R. This last property means that Spec R is irreducible in the sense of the following general definition:

3.A.9. Definition A topological space X is called i r r e d u c i b l e if it satisfies any one of the following equivalent conditions:

(1) $X \neq \emptyset$, and $U \cap V \neq \emptyset$ for arbitrary non-empty open subsets $U, V \subseteq X$.

(2) $X \neq \emptyset$, and every non-empty open subset of X is dense in X.

(3) $X \neq \emptyset$, and every non-empty open subset of X is connected.

(4) $X \neq \emptyset$, and $A \cup B$ is a proper subset of X for arbitrary proper closed subsets $A, B \subset X$.

The proof of the equivalence of these conditions is simple and again left to the reader. It is clear from this definition that irreducibility is a stronger condition than connectedness. A Hausdorff topological space is irreducible if and only if it is singleton.

3.A.10. Proposition *Let X be a topological space and let Y be any subset of X. Then Y is irreducible if and only if $\overline{Y}$ is irreducible.*

PROOF. Suppose Y is irreducible. Let U, V be open subsets in X with $U \cap \overline{Y} \neq \emptyset \neq V \cap \overline{Y}$. Then $U \cap Y \neq \emptyset \neq V \cap Y$ (by definition of $\overline{Y}$). Therefore $U \cap V \cap Y \neq \emptyset$ and hence $U \cap V \cap \overline{Y} \neq \emptyset$. – Conversely, suppose $\overline{Y}$ is irreducible and let U, V be open subsets in X with $U \cap Y \neq \emptyset \neq V \cap Y$. Then $U \cap V \cap \overline{Y} \neq \emptyset$ and hence $U \cap V \cap Y \neq \emptyset$.   •

3.A.11. Proposition *The closed irreducible subsets of* Spec R *are the sets* $V(\mathfrak{p})$, $\mathfrak{p} \in$ Spec R.

PROOF. Since $V(\mathfrak{p}) = \overline{\{\mathfrak{p}\}}$ for $\mathfrak{p} \in$ Spec R, the closed set $V(\mathfrak{p})$ is irreducible by 3.A.10. Conversely, let $V(\mathfrak{a})$ be an irreducible closed subset of Spec R, where $\mathfrak{a}$ is a radical ideal in R. Then $V(\mathfrak{a}) \neq \emptyset$ and hence $\mathfrak{a} \neq R$. Now to see that $\mathfrak{a}$ is a prime ideal, let f, g be elements in R with $fg \in \mathfrak{a}$. Then $V(\mathfrak{a}) = V(\mathfrak{a} + Rf) \cup V(\mathfrak{a} + Rg)$ and hence $V(\mathfrak{a}) = V(\mathfrak{a} + Rf)$ or $V(\mathfrak{a}) = V(\mathfrak{a} + Rg)$, since $V(\mathfrak{a})$ is irreducible. Therefore $f \in \mathfrak{a}$ or $g \in \mathfrak{a}$.   •

For a point x in an arbitrary topological space X, the closure $\overline{\{x\}}$ is irreducible by 3.A.10. A point x is called a g e n e r i c   p o i n t of X if $\overline{\{x\}} = X$. If X has a generic point then X is irreducible. By 3.A.11 *every closed irreducible subset of* Spec R *has a unique generic point.*

3.A.12. Definition Let X be a topological space. A maximal irreducible subset of X is called an i r r e d u c i b l e   c o m p o n e n t of X.

By Proposition 3.A.10, every irreducible component of a topological space is closed. A simple application of the Zorn's lemma gives:

3.A.13. Proposition *Every irreducible subset of a topological space X is contained in an irreducible component of X.*

Since $\{x\} \subseteq X$ is irreducible for every $x \in X$, the last proposition implies that *a topological space X is the union of its irreducible components.* With Proposition 3.A.11 follows:

3.A.14. Corollary *The irreducible components of* Spec R *are precisely the closed sets* $V(\mathfrak{p})$, *where $\mathfrak{p}$ is a minimal prime ideal in R. Every prime ideal in R contains a minimal prime ideal of R.*

Of course, the supplement in the last corollary can be proved directly, also using Zorn's lemma.

3.A.15. Lemma *Let X be a Noetherian topological space. Then any closed set in X is a finite union of irreducible subsets.*

PROOF. Let $\mathcal{F}$ denote the collection of those closed sets of X which cannot be expressed as a union of finitely many irreducible subsets. Suppose $\mathcal{F} \neq \emptyset$. Since X is Noetherian, there exists a minimal element $V_0 \in \mathcal{F}$. Then V_0 cannot be irreducible. Let $V_0 = V_1 \cup V_2$ with closed proper subsets $V_1, V_2 \subset V_0$. By minimality of V_0, both V_1 and V_2 are finite unions of irreducible subsets. Hence V_0 is a finite union of irreducible subsets, a contradiction.   •

3.A.16. Theorem *Let X be a Noetherian topological space and V be a closed subset of X. Then $V = V_1 \cup \cdots \cup V_r$, where $V_1, \ldots, V_r$ are the irreducible components of V and $V_i \not\subseteq \bigcup_{j \neq i} V_j$ for every $1 \leq i \leq r$. In particular, X has only finitely many irreducible components.*

PROOF. By the previous lemma we may write $V = V_1 \cup \cdots \cup V_r$, where the V_i's are closed irreducible subsets of X. Let this representation be a minimal one, i. e., $V_i \not\subseteq \bigcup_{j \neq i} V_j$ for every $i = 1, \ldots, r$. Let W be any irreducible subset of V. Then $W = (W \cap V_1) \cup \cdots \cup (W \cap V_r)$ and hence $W = W \cap V_i$ for some i, i. e. $W \subseteq V_i$. This and the minimality of r implies that $V_1, \ldots, V_r$ are the irreducible components of V.   •

From 3.A.16 and 3.A.14 we have:

3.A.17. Corollary *Any Noetherian ring R has only finitely many minimal prime ideals $\mathfrak{p}_1, \ldots, \mathfrak{p}_r$. The irreducible components of $\mathrm{Spec}\, R$ are $\mathrm{V}(\mathfrak{p}_1), \ldots, \mathrm{V}(\mathfrak{p}_r)$.*

Since any irreducible topological space is connected, from 3.A.16 follows:

3.A.18. Corollary *A Noetherian topological space X has only finitely many connected components and every connected component is a union of some irreducible components of X. In particular, the connected components of X are (closed and) open.*

3.A.19. Remark Let X be a topological space with only finitely many irreducible components $X_1, \ldots, X_r$. Then the connected components of X can be constructed in the following way: Take the irreducible components as the vertices of a graph and connect two vertices by an edge if and only if the corresponding irreducible components intersect. *Then the connected components of X are the unions of the irreducible components corresponding to the vertices of the connected components of this graph.* (A corresponding result is true for any finite cover $X = X_1 \cup \cdots \cup X_r$ with closed and connected (or with open and connected) subspaces $X_1, \ldots, X_r$.)

3.A.20. Exercise Let R be a ring.

(1) Spm R is dense in Spec R if and only if $\mathfrak{n}_R = \mathfrak{m}_R$, where $\mathfrak{n}_R$ and $\mathfrak{m}_R$ denote the nilradical and the Jacobson radical of R, respectively.

(2) Spec R is a Hausdorff space if and only if every prime ideal is maximal, i. e. dim $R \leq 0$. (The implication "dim $R = 0 \Rightarrow$ Spec R is Hausdorff" is not obvious.) If Spec R is a Hausdorff space, then Spec R is compact and totally disconnected. (**Hint:** R is reduced with

dim $R \leq 0$ if and only if every principal (or every finitely generated) ideal is generated by an idempotent element.)

(3) Show that the functors $C_{\mathbb{Z}/\mathbb{Z}2} : X \rightsquigarrow C_{\mathbb{Z}/\mathbb{Z}2}(X) := \{ f : X \to \mathbb{Z}/\mathbb{Z}2 \mid f \text{ is continuous} \}$ and Spec $: R \rightsquigarrow$ Spec R ($= (\mathbb{Z}/\mathbb{Z}2)$-Spec R) are contravariant functors between the category of compact and totally disconnected Hausdorff spaces X and the category of boolean rings R (i.e. rings with $f^2 = f$ for all $f \in R$) such that Spec $\circ C_{\mathbb{Z}/\mathbb{Z}2}$ and $C_{\mathbb{Z}/\mathbb{Z}2} \circ$ Spec are naturally equivalent to the identity functors respectively (S t o n e).

(4) Any subspace of a Noetherian topological space is Noetherian. In particular, using 3.A.18, any Noetherian topological space is locally connected.

(5) Show the equivalence of the following two conditions for an element $f \in R$: a) $D(f)$ is dense in Spec R. b) (The residue class of) f is a non-zero divisor in $R_{\text{red}} = R/\mathfrak{n}_R$. (The set of zero divisors of a reduced ring is the union of its minimal prime ideals.) – In particular: If f is a non-zero divisor in R, then $D(f)$ is dense in Spec R. Give an example which shows that the converse is not true in general. (**Remark:** Elements in R fulfilling conditions a) and/or b) above are called a c t i v e. Non-zero divisors are active.)

(6) If R is Noetherian and if the open set $U \subseteq$ Spec R is dense in Spec R, then there exists $f \in R$ such that $D(f) \subseteq U$ and $D(f)$ is dense in Spec R. (Use 3.A.17 and 3.B.31 below.)

(7) If $\varphi : R \to S$ is a ring homomorphism then $\overline{\varphi^*(V(\mathfrak{b}))} = V(\varphi^{-1}(\mathfrak{b}))$ for every ideal $\mathfrak{b} \subseteq S$. (**Hint:** $\varphi^{-1}(\sqrt{\mathfrak{b}}) = \sqrt{\varphi^{-1}(\mathfrak{b})}$.)

3.A.21. Exercise Let K be a field and R be a K-algebra.

(1) K-Spec R is dense in Spec R if and only if the nilradical $\mathfrak{n}_R$ of R coincides with the K-radical $\mathfrak{r}_R = \bigcap_{\xi \in K\text{-Spec } R} \mathfrak{m}_\xi$ of R.

(2) The closed irreducible subsets in K-Spec R are precisely the sets $V_K(\mathfrak{p})$, where $\mathfrak{p} \in$ Spec R is a prime ideal with $\mathfrak{J}_K(V_K(\mathfrak{p})) = \mathfrak{p}$.

3.A.22. Exercise Let $K \subseteq L$ be a field extension, R a K-algebra and $R_{(L)} := L \otimes_K R$.

(1) The canonical inclusion

$$K\text{-Spec } R = \mathrm{Hom}_{K\text{-alg}}(R, K) \subseteq \mathrm{Hom}_{K\text{-alg}}(R, L) = \mathrm{Hom}_{L\text{-alg}}(R_{(L)}, L) = L\text{-Spec } R_{(L)}$$

is an embedding of topological spaces (with respect to their Zariski topologies). (**Hint:** Let a_i, $i \in I$, be a K-base of L. For $f = \sum_i a_i \otimes g_i \in R_{(L)}$, $g_i \in R$, the equality $V_L(f) \cap (K\text{-Spec } R) = \bigcap_i V_K(g_i)$ holds. For any $g \in R \subseteq R_{(L)}$ one has $V_K(g) = V_L(g) \cap (K\text{-Spec } R)$.)

(2) If K-Spec R is dense in Spec R, then K-Spec $R \subseteq L$- Spec $R_{(L)}$ is dense in Spec $R_{(L)}$. More precisely, if $\mathfrak{n}_R = \bigcap_{\xi \in K\text{-Spec } R} \mathfrak{m}_\xi$, then $L \otimes_K \mathfrak{n}_R = \bigcap_{\xi \in K\text{-Spec } R} L \otimes \mathfrak{m}_\xi = \mathfrak{n}_{R_{(L)}}$.

(3) If R is an integral domain and if K-Spec R is dense in Spec R, then L- Spec $R_{(L)}$ is dense in Spec $R_{(L)}$ and $R_{(L)}$ is also an integral domain. In particular, if $P \in K[X, Y]$ is a prime polynomial with infinitely many zeros in K^2, then P is also a prime polynomial in $L[X, Y]$. (Cf. Exercise 2.B.14 (4). – Show by (simple) examples that the last result is not true in general for polynomials in three variables.)

3.A.23. Exercise Let K be an algebraically closed field. If R and S are K-algebras which are integral domains, then $R \otimes_K S$ is also an integral domain. (**Hint:** One may assume that R and/or S are of finite type over K. In this case use Exercise 3.A.22 (3) and Hilbert's nullstellensatz 2.B.11.) In particular: If R is a K-algebra and an integral domain, then $R_{(L)} = L \otimes_K R$ is an integral domain for every field extension $K \subseteq L$.

3.A.24. Exercise For a module M over a commutative ring R the s u p p o r t of M is by definition the subset

$$\operatorname{Supp} M = \operatorname{Supp}_R M := \{ \mathfrak{p} \in \operatorname{Spec} R \; : \; M_\mathfrak{p} \neq 0 \}$$

of Spec R. Show that if M is a finite R-module with generators $x_1, \ldots, x_n$ and $\mathfrak{a}_i := \operatorname{Ann} x_i$, then the support

$$\operatorname{Supp} M = \bigcup_{i=1}^{n} \operatorname{Supp} R x_i = \bigcup_{i=1}^{n} V(\mathfrak{a}_i) = V\left(\bigcap_{i=1}^{n} \mathfrak{a}_i \right) = V(\operatorname{Ann} M)$$

is the closed subset of Spec R defined by the annihilator Ann M of M.

3.B. Dimension

Let C be a chain in an ordered set, i.e. a totally ordered subset. If C is finite, then Card $C - 1$ is called the l e n g t h of C. For example, a singleton is a chain of length 0. The empty chain has length -1.

3.B.1. Definition Let R be a commutative ring. The (K r u l l -) d i m e n s i o n dim R of R is the supremum (in $\mathbb{N} \cup \{-1, \infty\}$) of the lengths of finite chains of prime ideals in R (ordered by inclusion), that is,

$$\dim R := \operatorname{Sup} \{ r \mid \text{ there exists a chain } \mathfrak{p}_0 \subset \mathfrak{p}_1 \subset \cdots \subset \mathfrak{p}_r, \; \mathfrak{p}_i \in \operatorname{Spec} R \} .$$

The dimension of the zero ring is -1.

A (strictly ascending) chain $\mathfrak{p}_0 \subset \mathfrak{p}_1 \subset \cdots \subset \mathfrak{p}_r$ of prime ideals in R of length r gives rise to a (strictly descending) chain

$$V(\mathfrak{p}_0) \supset V(\mathfrak{p}_1) \supset \cdots \supset V(\mathfrak{p}_r)$$

of closed irreducible subsets of Spec R and conversely. This motivates the following definition:

3.B.2. Definition Let X be a topological space. The (K r u l l -) d i m e n s i o n dim X of X is the supremum (in $\mathbb{N} \cup \{-1, \infty\}$) of the lengths of finite chains of closed irreducible subsets of X (ordered by inclusion).

By definition, dim X is the supremum of the $r \in \mathbb{N}$ such that there exists a (strictly ascending) chain

$$X_0 \subset X_1 \subset \cdots \subset X_r$$

of closed irreducible subsets of X. dim X is the supremum of the dimensions dim Y where Y runs through the set of the irreducible components of X. If all these components have the same dimension then X is called p u r e - d i m e n s i o n a l. The dimension of the empty space is -1. If Y is a subspace of X then dim $Y \leq$ dim X. The dimension of a commutative ring R coincides with the dimension of the topological space Spec R. The ring R is called p u r e - d i m e n s i o n a l if Spec R is pure-dimensional, i.e. if all the rings $R/\mathfrak{p}$, $\mathfrak{p}$ a minimal prime ideal in R, have the same dimension.

3.B.3. Remark Every non-empty Hausdorff space has Krull-dimension 0, since the only irreducible subsets of a Hausdorff space are the singletons. The concept of dimension given in Definition 3.B.2 is suitable in Algebraic Geometry for the topological spaces which are constructed from the spaces $\operatorname{Spec} R$ with their Zariski topologies, R a commutative ring.

Let X be a topological space and Y be a closed irreducible subset of X. The codimension $\operatorname{codim} Y = \operatorname{codim}_X Y = \operatorname{codim}(Y, X)$ of Y in X is the supremum (in $\mathbb{N} \cup \{-1, \infty\}$) of the lengths of those finite non-empty chains of closed irreducible subsets of X which have Y as the smallest element. By definition, $\operatorname{codim}_X Y$ is the supremum of the $r \in \mathbb{N}$ such that there exists a chain

$$Y \subset X_1 \subset \cdots \subset X_r$$

with closed irreducible subsets $X_1, \ldots, X_r \subseteq X$.

The codimension of an arbitrary closed subset Y of a topological space X is the infimum of the codimensions of the irreducible components of Y. With this definition, the empty set has codimension ∞. If all the irreducible components of Y have the same codimension $m \in \mathbb{N}$, then we say that Y is purely m-codimensional in X. If Y is a non-empty closed subspace of X then $\dim Y + \operatorname{codim}(Y, X) \le \dim X$. An immediate consequence of the definitions is:

3.B.4. Proposition *Let R be a ring and $\mathfrak{a}$ be an ideal in R. Then:*

(1) $\dim R = \dim \operatorname{Spec} R = \sup \{\dim R/\mathfrak{p} \mid \mathfrak{p} \text{ is minimal in } R\}$.

(2) *If S is a multiplicatively closed subset of R then* $\dim S^{-1}R \le \dim R$.

(3) $\operatorname{codim} V(\mathfrak{p}) = \dim R_\mathfrak{p} = \dim \operatorname{Spec} R_\mathfrak{p}$ *for every* $\mathfrak{p} \in \operatorname{Spec} R$.

(4) $\dim V(\mathfrak{a}) = \dim R/\mathfrak{a}$ *and* $\operatorname{codim} V(\mathfrak{a}) = \inf \{\operatorname{codim} V(\mathfrak{p}) \mid \mathfrak{p} \text{ minimal in } V(\mathfrak{a})\}$.

For an ideal $\mathfrak{a} \subseteq R$ the codimension $\operatorname{codim} V(\mathfrak{a})$ is also called the codimension or the height of $\mathfrak{a}$ and is denoted by

$$\operatorname{codim} \mathfrak{a} \quad \text{or} \quad \operatorname{ht} \mathfrak{a}.$$

The height of a prime ideal $\mathfrak{p}$ in R is the supremum of the lengths of strictly increasing chains $\mathfrak{p}_0 \subset \cdots \subset \mathfrak{p}_n$ with $\mathfrak{p}_n = \mathfrak{p}$. The following consequences are immediate.

3.B.5. Proposition *Let R be a ring and let $\mathfrak{a}$ be an ideal in R.*

(1) $\operatorname{ht} \mathfrak{p} = \dim R_\mathfrak{p}$ *for every* $\mathfrak{p} \in \operatorname{Spec} R$.

(2) $\operatorname{ht} \mathfrak{a} = \inf \{\operatorname{ht} \mathfrak{p} \mid \mathfrak{p} \in V(\mathfrak{a})\}$.

(3) *If S is a multiplicatively closed subset of R and $\mathfrak{p} \in \operatorname{Spec} R$ with $\mathfrak{p} \cap S = \emptyset$ then* $\operatorname{ht} \mathfrak{p} = \operatorname{ht} S^{-1}\mathfrak{p}$.

(4) $\dim R = \sup \{\dim R_\mathfrak{m} \mid \mathfrak{m} \in \operatorname{Spm} R\} = \sup \{\operatorname{ht} \mathfrak{m} \mid \mathfrak{m} \in \operatorname{Spm} R\}$.

(5) $\operatorname{ht} \mathfrak{a} + \dim R/\mathfrak{a} \le \dim R$.

3.B.6. Example Let R be a ring. For every ideal $\mathfrak{a}$ contained in the nilradical $\mathfrak{n}_R$ of R, the topological spaces $\operatorname{Spec} R$ and $\operatorname{Spec} R/\mathfrak{a}$ are homeomorphic and hence $\dim R = \dim R/\mathfrak{a}$. In particular, for $\mathfrak{a} = \mathfrak{n}_R$ we get $\dim R = \dim R_{\mathrm{red}}$. Furthermore, $\dim R < 0$ if and only if $R = 0$, and $\dim R \leq 0$ if and only if every prime ideal in R is maximal. If $\dim R = 0$ and R is an integral domain, then R is a field. More general: If $\dim R \leq 0$, then $R = Q(R)$. (A minimal prime ideal $\mathfrak{p}$ in an arbitrary ring R contains only zero-divisors because of $\mathfrak{p}R_{\mathfrak{p}} = \mathfrak{n}_{R_{\mathfrak{p}}}$.) The converse is not true, *even if R is reduced.* For example: Let A be a reduced local ring with maximal ideal $\mathfrak{m}$ and with a subfield $K \subseteq A$ such that $K = A/\mathfrak{m}$. Then $R := K \oplus \mathfrak{m}^{(\mathbb{N})} \subseteq A^{\mathbb{N}}$ is also a reduced local ring and $\mathfrak{m}_R = \mathfrak{m}^{(\mathbb{N})}$ contains only zero-divisors, i.e. $R = Q(R)$. But, all projections $R \to A$, $(a_n) \mapsto a_m$, are surjective, hence $\dim R > 0$ if $\dim A > 0$. Show: If R is reduced with only a finite number of minimal prime ideals and if $R = Q(R)$, then $\dim R \leq 0$.

One has the following important result, which characterizes Noetherian rings of dimension ≤ 0:

3.B.7. Theorem *For a ring R the following conditions are equivalent*:

(1) R *is Noetherian and* $\dim R \leq 0$.

(2) R *is Artinian.*[1])

(3) R *is of finite length (as an R-module)*.

PROOF. For an Artinian ring R, we first show:

a) Every prime ideal in R is maximal. In particular, $\dim R \leq 0$.

b) R has only finitely many maximal ideals.

c) $\mathfrak{n}_R = \mathfrak{m}_R$ and there exists $n \in \mathbb{N}^*$ such that $\mathfrak{m}_R^n = 0$.

To prove a) let $\mathfrak{p}$ be a prime ideal in R. Then $B := A/\mathfrak{p}$ is an Artinian integral domain. Let $x \in B$, $x \neq 0$. By the descending chain condition, there exists $n \in \mathbb{N}$, such that $Bx^n = Bx^{n+1}$. Therefore, $x^n = yx^{n+1}$ for some $y \in B$ and hence $1 = yx$.

For the proof of b) let $\mathcal{F} := \{\mathfrak{m}_1 \cap \cdots \cap \mathfrak{m}_n \mid n \in \mathbb{N}, \mathfrak{m}_i \in \operatorname{Spm} R\}$. Then $\mathcal{F} \neq \emptyset$ and since R is Artinian, $\mathcal{F}$ has a minimal element, say $\mathfrak{a} := \mathfrak{m}_1 \cap \cdots \cap \mathfrak{m}_r$. We shall show that $\mathfrak{m}_1, \ldots, \mathfrak{m}_r$ are the only maximal ideals in R. Let $\mathfrak{m} \in \operatorname{Spm} R$. By minimality of $\mathfrak{a}$ we have $\mathfrak{m} \cap \mathfrak{a} = \mathfrak{a}$ and hence $\mathfrak{m} \supseteq \mathfrak{a} = \mathfrak{m}_1 \cap \cdots \cap \mathfrak{m}_r$. Therefore $\mathfrak{m} \supseteq \mathfrak{m}_i$ and hence $\mathfrak{m} = \mathfrak{m}_i$ for some i.

Finally we prove c): By a) and $\operatorname{Spec} R = \operatorname{Spm} R = \{\mathfrak{m}_1, \ldots, \mathfrak{m}_r\}$. Therefore $\mathfrak{n}_R = \mathfrak{m}_R = \bigcap_{i=1}^r \mathfrak{m}_i = \prod_{i=1}^r \mathfrak{m}_i$. Since R is Artinian, there exists $n \in \mathbb{N}$ such that $\mathfrak{m}_R^n = \mathfrak{m}_R^{n+1}$. It is enough to prove that $0 : \mathfrak{m}_R^n = R$. Suppose $\mathfrak{a} := 0 : \mathfrak{m}_R^n \subset R$ and let $\mathfrak{b}$ be a minimal element in the set of ideals of R strictly

---

[1]) A commutative ring R is said to be A r t i n i a n if any one of the following equivalent conditions holds: (1) Any descending chain of ideals in R is stationary. (2) Any non-empty collection of ideals in R has a minimal element.

containing $\mathfrak{a}$. Then $\mathfrak{b} = Rx + \mathfrak{a}$ for every $x \in \mathfrak{b} \setminus \mathfrak{a}$. By Nakayama's lemma[2]), $\mathfrak{m}_R x + \mathfrak{a} \neq \mathfrak{b}$. Therefore, by the minimality of $\mathfrak{b}$, $\mathfrak{m}_R x + \mathfrak{a} = \mathfrak{a}$ and hence $\mathfrak{m}_R x \subseteq \mathfrak{a}$ and $x \in \mathfrak{a} : \mathfrak{m}_R = \mathfrak{a}$, a contradiction.

(2) $\Rightarrow$ (1): Let R be Artinian. Then $\dim R \leq 0$ by a). Furthermore, by b) and c) we have $0 = \mathfrak{m}_1 \cdots \mathfrak{m}_s$ with maximal ideals $\mathfrak{m}_1, \ldots, \mathfrak{m}_s$ (not necessarily distinct). The quotient modules $\mathfrak{m}_1 \cdots \mathfrak{m}_{i-1}/\mathfrak{m}_1 \cdots \mathfrak{m}_i$ of the chain $R \supseteq \mathfrak{m}_1 \supseteq \mathfrak{m}_1 \mathfrak{m}_2 \supseteq \cdots \supseteq \mathfrak{m}_1 \cdots \mathfrak{m}_s = 0$ are artinian vector spaces over the fields $R/\mathfrak{m}_i$ and hence finite-dimensional and Noetherian, $i = 1, \ldots, s$. Therefore R is also Noetherian.

(1) $\Rightarrow$ (2): Suppose that R is Noetherian of dimension ≤ 0. Then every prime ideal in R is maximal and by 3.A.17 there are only finitely many, say $\mathfrak{m}_1, \ldots, \mathfrak{m}_r$. Since $\mathfrak{n}_R = \mathfrak{m}_R = \bigcap_{j=1}^{r} \mathfrak{m}_j = \prod_{j=1}^{r} \mathfrak{m}_j$ is a finitely generated ideal, there exists $n \in \mathbb{N}$ such that $\mathfrak{n}_R^n = \mathfrak{m}_R^n = 0$. By a similar argument as in the proof of "(2) $\Rightarrow$ (1)" using now the fact that a Noetherian vector space is finite-dimensional and hence Artinian, we conclude that R itself is Artinian.

Since R is of finite length if and only if R is Noetherian and Artinian, the equivalence of (3) with (1) and (2) is now clear. $\quad\bullet$

For the computation of dimensions the following theorem is important.

3.B.8. Theorem *Let $R \subseteq S$ be an integral extension of commutative rings. Then* $\dim R = \dim S$.

The following lemma contains more precise informations.

3.B.9. Lemma *Let $R \subseteq S$ be an integral extension of rings.*

(1) *Let $\mathfrak{q}$ be a prime ideal in S and $\mathfrak{p} := \mathfrak{q} \cap R$. Then $\mathfrak{q}$ is maximal if and only if $\mathfrak{p}$ is maximal.*

(2) *If $\mathfrak{q} \subseteq \mathfrak{q}'$ are prime ideals in S such that $\mathfrak{q} \cap R = \mathfrak{q}' \cap R$ then $\mathfrak{q} = \mathfrak{q}'$.*

(3) *If $\mathfrak{p}$ is a prime ideal in R then there exists a prime ideal $\mathfrak{q}$ in S with $\mathfrak{q} \cap R = \mathfrak{p}$, i. e. the map $\operatorname{Spec} S \to \operatorname{Spec} R$ is surjective.* (Lying-over theorem)

(4) *Let $\mathfrak{p} \subseteq \mathfrak{p}'$ be prime ideals in R and let $\mathfrak{q}$ be a prime ideal in S such that $\mathfrak{q} \cap R = \mathfrak{p}$. Then there exists a prime ideal $\mathfrak{q}'$ in S such that $\mathfrak{q}' \cap R = \mathfrak{p}'$ and $\mathfrak{q} \subseteq \mathfrak{q}'$.* (Going-up theorem)

---

[2]) For the convenience of the reader we recall the following Lemma of Nakayama: *Let R be a commutative ring and N be an R-submodule of the R-module M. Assume that M/N is finitely generated and $M = \mathfrak{m}_R M + N$. Then $M = N$.* For the *proof* we replace M by M/N and can, therefore, assume that $N = 0$. Assume $M \neq 0$ and let $x_1, \ldots, x_r, r \geq 1$, be a minimal system of generators of M. Then, by assumption, $x_1 = a_1 x_1 + \cdots + a_r x_r$ with $a_1, \ldots, a_r \in \mathfrak{m}_R$. It follows that $(1 - a_1)x_1 = a_2 x_2 + \cdots + a_r x_r$ and $x_1 \in R x_2 + \cdots + R x_r$, since $1 - a_1 \in 1 + \mathfrak{m}_R$ is a unit in R. This contradicts the minimality of the generating system $x_1, \ldots, x_r$.

PROOF of Lemma 3.B.9 : (1) follows from 1.F.5, because $R/\mathfrak{p} \subseteq S/\mathfrak{q}$ is an integral extension of integral domains.

(2) Let $\mathfrak{p} := \mathfrak{q} \cap R = \mathfrak{q}' \cap R$. Replacing R by $R_\mathfrak{p}$ and S by $S_\mathfrak{p}$, we may assume that R is local and $\mathfrak{p} = \mathfrak{m}_R$. Then (2) is immediate from (1).

(3) Again replacing R by $R_\mathfrak{p}$ and S by $S_\mathfrak{p}$, we may assume that R is local and $\mathfrak{p} = \mathfrak{m}_R$. It is enough to prove that $\mathfrak{m}_R S \neq S$. Suppose that $\mathfrak{m}_R S = S$. Then $1 = a_1 x_1 + \cdots + a_n x_n$ with $a_i \in \mathfrak{m}_R$ and $x_i \in S$ and $\mathfrak{m}_R R[x_1, \ldots, x_n] = R[x_1, \ldots, x_n]$ and hence by Nakayama's Lemma, $R[x_1, \ldots, x_n] = 0$, since the extension $R \subseteq R[x_1, \ldots, x_n]$ is finite. This is a contradiction.

(4) Replacing R by $R/\mathfrak{p}$ and S by $S/\mathfrak{q}$, we may assume that $\mathfrak{p} = \mathfrak{q} = 0$. Then (4) follows from (3).   ●

PROOF of Theorem 3.B.8 : If $\mathfrak{q}_0 \subset \cdots \subset \mathfrak{q}_r$ is a strictly ascending chain of prime ideals in S, then by 3.B.9(2) $\mathfrak{q}_0 \cap R \subset \cdots \subset \mathfrak{q}_r \cap R$ is a strictly ascending chain of prime ideals in R. Conversely, if $\mathfrak{p}_0 \subset \cdots \subset \mathfrak{p}_r$ is a strictly ascending chain of prime ideals in R, then by 3.B.9 (3) and (4) there exists a strictly ascending chain $\mathfrak{q}_0 \subset \cdots \subset \mathfrak{q}_r$ of prime ideals in S with $\mathfrak{p}_i = \mathfrak{q}_i \cap R$.   ●

3.B.10. Example (G o i n g - d o w n   t h e o r e m) Let $R \subseteq S$ be an integral extension of rings. By Lemma 3.B.9 (2) and (4) for every prime ideal $\mathfrak{p} \subseteq R$ and every prime ideal $\mathfrak{q} \subseteq S$ with $\mathfrak{q} \cap R = \mathfrak{p}$, $\mathfrak{q}$ is a minimal prime ideal over $\mathfrak{p}S$. The following lemma gives a partial converse of this remark:

3.B.11. Lemma *Let $R \subseteq S$ be an integral extension of integral domains and suppose that R is normal. If $\mathfrak{p} \subseteq R$ is a prime ideal in R and if $\mathfrak{q} \subseteq S$ is a prime ideal in S which is minimal over $\mathfrak{p}S$, then $\mathfrak{q} \cap R = \mathfrak{p}$.*

PROOF. (Cf. Remark to Theorem 6 in [14], Vol 1, Ch.V, § 3.) $\mathfrak{q}S_\mathfrak{q}$ is nilpotent modulo $\mathfrak{p}S_\mathfrak{q}$. Therefore, for $x \in \mathfrak{q} \cap R$, there exist $n \in \mathbb{N}^*$ and $y \in S \setminus \mathfrak{q}$ such that $z := yx^n \in \mathfrak{p}S$. Let $F = X^m + a_{m-1}X^{m-1} + \cdots + a_0 \in R[X]$ be the minimal polynomial of y (see Lemma 1.E.9). Then $G := X^m + a_{m-1}x^n X^{m-1} + \cdots + a_0 x^{mn}$ is the minimal polynomial of z. All its coefficients $a_{m-1}x^n, \ldots, a_0 x^{mn}$ belong to $\mathfrak{p}$. This follows from $z \in \mathfrak{p}S$ and the facts that, by Remark 1.E.3, there exists an integral equation $H(z) = z^d + b_{d-1}z^{d-1} + \cdots + b_0 = 0$ with $b_0, \ldots, b_{d-1} \in \mathfrak{p}$ and that G divides the polynomial $H = X^d + b_{d-1}X^{d-1} + \cdots + b_0$. Now, if $x \notin \mathfrak{p}$ then $a_{m-1}, \ldots, a_0 \in \mathfrak{p}$ and $y^m \in \mathfrak{p}S \subseteq \mathfrak{q}$, i. e. $y \in \mathfrak{q}$, a contradiction.   ●

An important consequence of the above lemma is the following theorem:

3.B.12. Going-down Theorem *Let $R \subseteq S$ be an integral extension of integral domains and suppose that R is normal. If $\mathfrak{p}_1 \supset \cdots \supset \mathfrak{p}_n$ is a chain of prime ideals in R and if $\mathfrak{q}_1 \supset \cdots \supset \mathfrak{q}_m$ is a chain of prime ideals in S, with $\mathfrak{q}_i \cap R = \mathfrak{p}_i$, $1 \leq i \leq m < n$, then the chain $\mathfrak{q}_1 \supset \cdots \supset \mathfrak{q}_m$ can be extended to a chain $\mathfrak{q}_1 \supset \cdots \supset \mathfrak{q}_n$ of prime ideals in S with $\mathfrak{q}_i \cap R = \mathfrak{p}_i$, $1 \leq i \leq n$.*

PROOF. We may assume that $m = 1$ and $n = 2$. By Lemma 3.B.11, one can take for $\mathfrak{q}_2$ any minimal prime ideal over $\mathfrak{p}_2 S$ contained in $\mathfrak{q}_1$.   ●

Show the following abstract Going-down theorem: *Let* $\varphi : R \to S$ *be a ring homomorphism,* $\mathfrak{p}, \mathfrak{p}' \in \operatorname{Spec} R$ *with* $\mathfrak{p} \supset \mathfrak{p}'$ *and* $\mathfrak{q} \in \operatorname{Spec} S$ *with* $\mathfrak{q} \cap R := \varphi^{-1}(\mathfrak{q}) = \mathfrak{p}$. *Then there exists a* $\mathfrak{q}' \in \operatorname{Spec} S$ *with* $\mathfrak{q} \supset \mathfrak{q}'$ *and* $\mathfrak{q}' \cap R = \mathfrak{p}'$ *if and only if* $\mathfrak{p}' S_{\mathfrak{q}} \cap R_{\mathfrak{p}} = \mathfrak{p}' R_{\mathfrak{p}}$. – *The last equality is fulfilled if* $S_{\mathfrak{q}}$ *is flat (and hence faithfully flat) over* $R_{\mathfrak{p}}$. *In particular, the Going-down theorem holds in general if* S *is flat over* R.

3.B.13. Example Let R be a ring. Then $\dim R[X_1, \ldots, X_n] \geq \dim R + n$. For the proof we may assume $n = 1$. But if $\mathfrak{p}_1 \subset \cdots \subset \mathfrak{p}_m$ is a strictly increasing chain of prime ideals in R then $\mathfrak{p}_1 R[X] \subset \cdots \subset \mathfrak{p}_m R[X] \subset \mathfrak{p}_m R[X] + X R[X]$ is a strictly increasing chain of prime ideals in $R[X]$. If R is Noetherian then the above inequality is an equality, see Exercise 3.B.41 below. In the special case when R is a field, we prove this directly:

3.B.14. Proposition *If K is a field then the polynomial ring* $K[X_1, \ldots, X_n]$ *has dimension* n.

PROOF. We use induction on n. The case $n = 0$ is trivial. Let $R := K[X_1, \ldots, X_n]$ and let $0 = \mathfrak{p}_0 \subset \mathfrak{p}_1 \subset \cdots \subset \mathfrak{p}_m$ be a chain of prime ideals in $\operatorname{Spec} R$. Let $\overline{R} := R/\mathfrak{p}_1$ and let x_i denote the image of X_i in $\overline{R}$. Then $x_1, \ldots, x_n$ are algebraically dependent over K and hence by Noether's Normalization Lemma 1.F.2 there exist $y_1, \ldots, y_r \in \overline{R}$ with $r < n$ which are algebraically independent over K, such that $\overline{R}$ is integral over $K[y_1, \ldots, y_r]$. Therefore by Theorem 3.B.8 and by the induction hypothesis, we have $\dim \overline{R} = \dim K[y_1, \ldots, y_r] = r$. Now, since $m - 1 \leq \dim \overline{R} = r < n$, we have $m \leq n$ and hence $\dim R \leq n$. $\bullet$

One simple consequence of 3.B.14 is:

3.B.15. Proposition *Let K be a field and let* $f \in K[X_1, \ldots, X_n]$ *be a non-constant polynomial. Then the ring* $R := K[X_1, \ldots, X_n]/(f)$ *has dimension* $n - 1$.

PROOF. By 1.F.1 we may assume that f is a monic polynomial in X_n. Then R is integral over $K[X_1, \ldots, X_{n-1}] \subseteq R$ and hence $\dim R = \dim K[X_1, \ldots, X_{n-1}] = n - 1$ by Theorems 3.B.8 and 3.B.14. $\bullet$

For algebras of finite type over a field K there is a classical characterization of the dimension which generalizes 3.B.14 and 3.B.15. To describe it we start with the following definition:

3.B.16. Definition Let $K \subseteq L$ be a field extension. A maximal algebraically independent subset of L over K is called a t r a n s c e n d e n c e b a s i s of L over K.

A simple application of Zorn's lemma shows: *Every field extension* $K \subseteq L$ *has transcendence bases.* Furthermore: *Any two transcendence bases have the same cardinality.* For infinite transcendence bases this is a purely set theoretic argument. In the finite case it follows from the following exchange lemma which is proved easily by induction on n: *Let $L \supseteq K$ be an extension of fields. Let $x_1, x_2, \ldots, x_m \in L$ be algebraically independent over K and let $y_1, y_2, \ldots, y_n$ be a transcendence basis of L over K. Then $m \leq n$ and there exist $n - m$ elements among $y_1, y_2, \ldots, y_n$ which together with $x_1, x_2, \ldots, x_m$ form a transcendence basis of L over K.*

The cardinality of a transcendence basis of L over K is called the t r a n s c e n d e n c e
d e g r e e of L over K and denoted by

$$\operatorname{trdeg}_K L .$$

If a subset S of L is a transcendence basis of L over K then L is algebraic over
$K(S)$ and conversely, if a subset S of L is algebraically independent over K and
if L is algebraic over $K(S)$ then S is a transcendence basis of L over K. If
$L = K(T)$ then there exists a transcendence basis S of L over K with $S \subseteq T$.
In particular, if L is finitely generated over K (as a field) then $\operatorname{trdeg}_K L$ is finite.

Let K be a field and let R be a K-algebra which is an integral domain. Then
the transcendence degree of the field of fractions $Q(R)$ over K is called the
t r a n s c e n d e n c e   d e g r e e   o f R o v e r K and denoted by $\operatorname{trdeg}_K R$. For
example, $\operatorname{trdeg}_K K[X_1, \ldots, X_n] = n$.

Now we can give the following classical description of the dimension for K-algebras
of finite type:

3.B.17. Theorem *Let K be a field and let R be a K-algebra of finite type which is
an integral domain. Then* $\dim R = \operatorname{trdeg}_K R$.

PROOF. By Noether's normalization lemma there exist elements $f_1, \ldots, f_m \in$
R which are algebraically independent over K and such that R is integral over
$K[f_1, \ldots, f_m]$. Therefore, by 3.B.8 and 3.B.14, $\dim R = \dim K[f_1, \ldots, f_m] =$
$m = \operatorname{trdeg}_K Q(R) = \operatorname{trdeg}_K R$.   •

3.B.18. Corollary *Let K be a field and let $R \subseteq S$ be an extension of integral
domains of finite type over K. Then $\dim R \leq \dim S$ and $\dim R = \dim S$ if and
only if the field extension $Q(R) \subseteq Q(S)$ is algebraic and hence finite.*

3.B.19. Corollary *Let K be a field, R be a K-algebra of finite type and let $f \in R$.
If f does not belong to any minimal prime ideal of R, then $\dim R = \dim R_f$.*

PROOF. The map $\mathfrak{p} \mapsto \mathfrak{p} R_f$ is a bijection from the set of all minimal prime ideals
of R onto the set of all minimal prime ideals of R_f, and for these prime ideals we
have $Q(R/\mathfrak{p}) = Q(R_f/\mathfrak{p} R_f)$.   •

3.B.20. Corollary *Let K be a field, R and R' be K-algebras of finite type. Then
the K-algebra $R \otimes_K R'$ is of finite type and $\dim(R \otimes_K R') = \dim R + \dim R'$.*

PROOF. Let $n := \dim R$, $m := \dim R'$. By Noether's normalization lemma, there
exist injective and finite K-algebra homomorphisms $\varphi : K[X_1, \ldots, X_n] \to R$ and
$\varphi' : K[Y_1, \ldots, Y_m] \to R'$. Then

$$\varphi \otimes \varphi' : K[X_1, \ldots, X_n] \otimes_K K[Y_1, \ldots, Y_m] \to R \otimes_K R'$$

is also injective and finite. Hence $\dim(R \otimes_K R') = n + m = \dim R + \dim R'$.   •

3.B.21. Corollary *Let* $K \subseteq L$ *be a field extension and let* R *be a* K-*algebra of finite type. Then the* L-*algebra* $S := L \otimes_K R$ *is of finite type and* $\dim S = \dim R$.

PROOF. By Noether's normalization lemma, there exists an injective and finite K-algebra homomorphism $\varphi : K[X_1, \ldots, X_n] \to R$, $n := \dim R$. Then the L-algebra homomorphism $L \otimes_K \varphi : L[X_1, \ldots, X_n] \to S$ is also injective and finite. Therefore $\dim S = n = \dim R$. •

The following theorem is a stronger version of 3.B.17:

3.B.22. Theorem *Let* K *be a field and let* R *be a* K-*algebra of finite type which is an integral domain. If* $0 = \mathfrak{p}_0 \subset \mathfrak{p}_1 \subset \cdots \subset \mathfrak{p}_n$ *is a maximal chain of prime ideals in* R, *then* $n = \dim R$.

PROOF. We use induction on n. If $n = 0$ then $0 = \mathfrak{p}_0$ is a maximal ideal, that is, R is a field and hence $\dim R = 0$. Now assume that $n > 0$. By Noether's normalization lemma there exist elements $f_1, \ldots, f_m \in R$ which are algebraically independent over K and such that R is integral over $R' := K[f_1, \ldots, f_m]$, in particular, $m = \dim R$. Let $\mathfrak{p}'_i := \mathfrak{p}_i \cap R'$, $1 \le i \le n$. Since $R' \cong K[X_1, \ldots, X_m]$ is a factorial domain, it is a normal domain, so we can apply the Going-down Theorem 3.B.12 to the ring extension $R' \subseteq R$ and conclude that $\mathfrak{p}'_1$ is a prime ideal of codimension 1 in R' and hence principal. Consider the integral extension $\overline{R'} := R'/\mathfrak{p}'_1 \subseteq \overline{R} := R/\mathfrak{p}_1$. Then by 3.B.8 and 3.B.15, $\dim \overline{R} = \dim \overline{R'} = m-1$. Since the chain $0 \subset \mathfrak{p}_2/\mathfrak{p}_1 \subset \cdots \subset \mathfrak{p}_n/\mathfrak{p}_1$ of prime ideals in $\overline{R}$ is also maximal, by induction hypothesis we have $\dim \overline{R} = n - 1$ and hence $m = \dim R = n$. •

3.B.23. Corollary *Let* K *be a field and let* R *be a* K-*algebra of finite type. If* $\mathfrak{p} \subseteq R$ *is a prime ideal in* R *and if* $\mathfrak{p} = \mathfrak{p}_0 \subset \mathfrak{p}_1 \subset \cdots \subset \mathfrak{p}_n$ *is maximal under all chains of prime ideals in* R *starting in* $\mathfrak{p}$, *then* $n = \dim R/\mathfrak{p} = \mathrm{trdeg}_K \kappa(\mathfrak{p})$, $\kappa(\mathfrak{p}) := R_\mathfrak{p}/\mathfrak{p}R_\mathfrak{p} = Q(R/\mathfrak{p})$.

3.B.24. Corollary *Let* K *be a field,* R *a* K-*algebra of finite type and let* $\mathfrak{p}$, $\mathfrak{q} \subseteq R$ *be prime ideals in* R *with* $\mathfrak{p} \subset \mathfrak{q}$. *Then all the maximal chains of prime ideals* $\mathfrak{p} = \mathfrak{p}_0 \subset \mathfrak{p}_1 \subset \cdots \subset \mathfrak{p}_r = \mathfrak{q}$ *starting with* $\mathfrak{p}$ *and ending with* $\mathfrak{q}$ *have the same length* $\dim R_\mathfrak{q}/\mathfrak{p}R_\mathfrak{q}$.

A ring with the property given in the last corollary is called c a t e n a r y. Thus, *algebras of finite type over a field are catenary.*

3.B.25. Corollary *Let* K *be a field and let* R *be a pure dimensional* K-*algebra of finite type. Then* $\mathrm{codim}\, \mathfrak{p} = \dim R - \dim R/\mathfrak{p}$ *for every* $\mathfrak{p} \in \mathrm{Spec}\, R$. *In particular,* $\mathrm{codim}\, \mathfrak{m} = \dim R$ *for every* $\mathfrak{m} \in \mathrm{Spm}\, R$.

A reformulation of the last corollary in geometric terms is:

3.B.26. Corollary *Let K be a field and let R be a pure dimensional K-algebra of finite type.   Then for every irreducible closed subset $V \subseteq X := \operatorname{Spec} R$, $\operatorname{codim}(V, X) + \dim V = \dim X$.*

3.B.27. Exercise Let K be a field and let R be a K-algebra of finite type. Let $\mathfrak{p} \subseteq R$ be a prime ideal in R and let $f \in R \setminus \mathfrak{p}$. Then $\mathfrak{p}$ is a maximal ideal in R if and only if $\mathfrak{p} R_f$ is a maximal ideal in R_f.

We end this lecture with one of the most important results for the dimension theory of general Noetherian rings and show at first that if R is a Noetherian ring and f is a non-unit in R then the codimension of $\mathrm{V}(f)$ in $\operatorname{Spec} R$ is at most 1. This is known as Krull's principal ideal theorem.

3.B.28. Krull's Principal Ideal Theorem *Let R be a Noetherian ring and let f be a non-unit in R. Then $\operatorname{ht} Rf = \operatorname{codim}(\mathrm{V}(f), \operatorname{Spec} R) \leq 1$.*

PROOF. Let $\mathfrak{p}$ be a minimal prime ideal containing f. Then we have to show that $\operatorname{codim} \mathfrak{p} = \dim R_{\mathfrak{p}} \leq 1$. Since $\mathfrak{p} R_{\mathfrak{p}}$ is the minimal prime ideal containing $f/1$ in $R_{\mathfrak{p}}$ and $\operatorname{codim} \mathfrak{p} = \operatorname{codim} \mathfrak{p} R_{\mathfrak{p}}$, we may assume that R is a local ring and that $\mathfrak{p} = \mathfrak{m}_R$ is the unique maximal ideal of R. If $\mathfrak{m}_R$ is a minimal prime ideal in R then $\operatorname{codim} \mathfrak{m}_R = 0$. Otherwise, by reducing modulo a minimal prime ideal of R, we may further assume that R is an integral domain and $f \neq 0$. Now, let $\mathfrak{q} \in \operatorname{Spec} R$ with $\mathfrak{q} \subset \mathfrak{m}_R$. Then $f \notin \mathfrak{q}$. Consider the s y m b o l i c p o w e r s $\mathfrak{q}^{(n)} := \mathfrak{q}^n R_{\mathfrak{q}} \cap R$ and the descending chain $\mathfrak{q}^{(1)} + Rf \supseteq \cdots \supseteq \mathfrak{q}^{(n)} + Rf \supseteq \cdots$ in R. The ring R/Rf is a Noetherian of dimension zero and hence Artinian by 3.B.7. Therefore there exists $n \in \mathbb{N}^*$ such that $\mathfrak{q}^{(n)} + Rf = \mathfrak{q}^{(n+1)} + Rf$. We claim that $\mathfrak{q}^{(n)} = \mathfrak{q}^{(n+1)} + \mathfrak{q}^{(n)} f$. To prove this, let $g = h + rf \in \mathfrak{q}^{(n)}$ with $h \in \mathfrak{q}^{(n+1)}$ and $r \in R$. Since $\mathfrak{m}_R$ is a minimal prime containing f, we have $f \notin \mathfrak{q}$ and $r = (g - h)/f \in \mathfrak{q}^n R_{\mathfrak{q}} \cap R = \mathfrak{q}^{(n)}$. By Nakayama's lemma $\mathfrak{q}^{(n)} = \mathfrak{q}^{(n+1)}$ and hence $\mathfrak{q}^n R_{\mathfrak{q}} = \mathfrak{q}^{n+1} R_{\mathfrak{q}}$. Again by the same lemma, $\mathfrak{q}^n R_{\mathfrak{q}} = 0$ and hence $\mathfrak{q} = 0$ (since R is an integral domain).   •

The following generalization of the principal ideal theorem is the announced main result:

3.B.29. Theorem *Let R be a Noetherian ring and let $f_1, \dots, f_r \in R$. Then, for every irreducible component Y of $\mathrm{V}(f_1, \dots, f_r)$, we have $\operatorname{codim}(Y, \operatorname{Spec} R) \leq r$. In particular, if $\mathrm{V}(f_1, \dots, f_r) \neq \emptyset$ then*

$$\operatorname{ht}(Rf_1 + \cdots + Rf_r) = \operatorname{codim}(\mathrm{V}(f_1, \dots, f_r), \operatorname{Spec} R) \leq r.$$

PROOF. We use induction on r. Let $\mathfrak{p}$ be a minimal prime ideal over $Rf_1 + \cdots + Rf_r$. By passing to the ring $R_{\mathfrak{p}}$, we may assume that R is local and $\mathfrak{p} = \mathfrak{m} = \mathfrak{m}_R$. Let $\mathfrak{p}_0 \subset \cdots \subset \mathfrak{p}_{s-1} \subset \mathfrak{p}_s = \mathfrak{m}$, $s \geq 1$, be a chain of prime ideals in R. Let i be the maximal index such that $\mathfrak{m}$ is not a minimal prime ideal of $\mathfrak{a} := \mathfrak{p}_{s-1} + Rf_1 + \cdots + Rf_{i-1}$, $1 \leq i \leq r$. Then $\mathfrak{m}$ is a minimal prime ideal of $\mathfrak{a} + Rf_i$ and by 3.B.28 the ideal $\mathfrak{m}/\mathfrak{a}$ has codimension 1 in $R/\mathfrak{a}$. Therefore, by replacing $\mathfrak{p}_{s-1}$ with any minimal prime ideal of $\mathfrak{a}$, we may assume that there is no prime ideal between $\mathfrak{p}_{s-1}$ and $\mathfrak{p}_s = \mathfrak{m}$ and $f_i \not\subseteq \mathfrak{p}_{s-1}$. Then $\mathfrak{m}$ is nilpotent modulo $\mathfrak{p}_{s-1} + Rf_i$ and hence there is $n \in \mathbb{N}^*$ such that for every $1 \le j \le r$, $j \neq i$, $f_j^n \in \mathfrak{p}_{s-1} + Rf_i$, i.e. $f_j^n = g_j + h_j f_i$ with $g_j \in \mathfrak{p}_{s-1}$ and $h_j \in R$. Let $\mathfrak{b} := \sum_{j \neq i} Rg_j$. Since $\mathfrak{m}$ is a minimal prime ideal of $Rf_i + \sum_{j \neq i} Rf_j^n = Rf_i + \mathfrak{b}$, again by 3.B.28 $\mathfrak{p}_{s-1}$ is a minimal prime ideal of $\mathfrak{b} = \sum_{j \neq i} Rg_j$. Now, by induction hypothesis, $s - 1 \le \operatorname{ht} \mathfrak{p}_{s-1} \le r - 1$. $\bullet$

The following proposition is a "converse" of 3.B.29:

3.B.30. Proposition *Let R be a Noetherian ring and let $\mathfrak{p} \subseteq R$ be a prime ideal of codimension r. Then there exist r elements $f_1, \dots, f_r \in \mathfrak{p}$ such that $\mathfrak{p}$ is a minimal prime ideal over $Rf_1 + \cdots + Rf_r$, i.e. such that $V(\mathfrak{p})$ is an irreducible component of $V(f_1, \dots, f_r)$.*

PROOF. We use induction on r and may assume $r > 0$. Let $\mathfrak{p}_1, \dots, \mathfrak{p}_m$ be the minimal prime ideals of R contained in $\mathfrak{p}$ and choose an element $f_1 \in \mathfrak{p} \setminus \bigcup_{j=1}^m \mathfrak{p}_j$ (cf. the following Lemma 3.B.31). Then $\operatorname{codim} \overline{\mathfrak{p}} < r$ for $\overline{\mathfrak{p}} := \mathfrak{p}/Rf_1 \subseteq \overline{R} := R/Rf_1$. By induction hypothesis, there are $r - 1$ residue classes $\overline{f}_2, \dots, \overline{f}_r \in \overline{R}$ such that $\overline{\mathfrak{p}}$ is a minimal prime ideal over $\overline{R} \overline{f}_2 + \cdots + \overline{R} \overline{f}_r$. Then $\mathfrak{p}$ is a minimal prime ideal over $Rf_1 + \cdots + Rf_r$. $\bullet$

3.B.31. Lemma *Let R be a commutative ring, $\mathfrak{a}$, $\mathfrak{b}$ arbitrary ideals and $\mathfrak{p}_1, \dots, \mathfrak{p}_m$ prime ideals in R with $\mathfrak{a} \not\subseteq \mathfrak{b}$, $\mathfrak{a} \not\subseteq \mathfrak{p}_j$, $j = 1, \dots, m$. Then $\mathfrak{a} \not\subseteq \mathfrak{b} \cup \mathfrak{p}_1 \cup \cdots \cup \mathfrak{p}_m$.*

PROOF. We use induction on m. For the induction step we may assume that none of the ideals $\mathfrak{b}, \mathfrak{p}_1, \dots, \mathfrak{p}_m, \mathfrak{p}_{m+1}$ is contained in another of them. By induction hypothesis there is an element $a \in \mathfrak{a}$, $a \notin \mathfrak{b} \cup \mathfrak{p}_1 \cup \cdots \cup \mathfrak{p}_m$. If $a \notin \mathfrak{p}_{m+1}$, then $a \notin \mathfrak{b} \cup \mathfrak{p}_1 \cup \cdots \cup \mathfrak{p}_{m+1}$. Otherwise choose an element $b \in \mathfrak{a} \cap \mathfrak{b} \cap \mathfrak{p}_1 \cap \cdots \cap \mathfrak{p}_m$, $b \notin \mathfrak{p}_{m+1}$. Then $a + b \in \mathfrak{a}$, $a + b \notin \mathfrak{b} \cup \mathfrak{p}_1 \cup \cdots \cup \mathfrak{p}_{m+1}$. $\bullet$

Proposition 3.B.30 is of particular interest for the case of Noetherian local rings. Let R be such a ring with maximal ideal $\mathfrak{m}_R$ and of dimension $d = \operatorname{codim} \mathfrak{m}_R$. Then d is the minimal number in $\mathbb{N}$ such that there exists a system of elements $f_1, \dots, f_d \in \mathfrak{m}_R$ with $\mathfrak{m}_R = \sqrt{Rf_1 + \cdots + Rf_d}$. Every such a system is called a **system of parameters** of R.

3.B.32. Exercise Let R be a Noetherian ring. Any non-empty set of closed *irreducible* subsets of $X := \operatorname{Spec} R$ contains a maximal element, i. e. the set of closed irreducible subsets of X is Noetherian (with respect to inclusion).

3.B.33. Exercise Let K be a field and let R be a K-algebra of finite type with $\dim R > 0$. Then $\operatorname{Spm} R = K$-$\operatorname{Spec} R$ if and only if K is algebraically closed.

3.B.34. Exercise An algebra R of finite type over a field K is finite if and only if $\dim R \leq 0$. Moreover, in this case

$$\text{Card}\,(K\text{-Spec}\,R) \leq \text{Card}\,(\text{Spm}\,R) \leq \text{Dim}_K R =: r\,.$$

Further, if the second inequality is an equality, then so is the first and R is isomorphic as a K-algebra to the product algebra K^r. (Recall that $\text{Dim}_K R$ denotes the dimension of R as a K-vector space.)

3.B.35. Exercise Let K be a field and let R be an affine domain over K (i. e. a K-algebra of finite type which is an integral domain) of dimension d. Then every irreducible closed subset of codimension 1 in $\text{Spec}\,R$ has dimension $d - 1$. Give an example of a Noetherian integral domain R such that this assertion is false for $\text{Spec}\,R$. (**Hint:** Look at $R := K[[X]][Y]$. More generally, describe explicitly $\text{Spec}\,A[Y]$ for a discrete valuation ring A.)

3.B.36. Exercise Let K be a field and let R be a K-algebra of finite type. Then $\dim R \geq d$ if and only if there exists an injective K-algebra homomorphism $K[X_1, \ldots, X_d] \to R$. (**Hint** : Reduce to the case of an integral domain R.) If S is another K-algebra of finite type with $R \subseteq S$, then $\dim R \leq \dim S$.

3.B.37. Exercise Let K be a field and $R = K[x_1, \ldots, x_n]$ a K-algebra $\neq 0$ of finite type with algebra generators $x_1, \ldots, x_n$.

(1) Let $I \subseteq \{1, \ldots, n\}$ and x_i, $i \in I$, algebraically independent over K. The following conditions are equivalent:

a) R is algebraic over $K[x_i : i \in I]$, i. e. every element of R is algebraic over $K[x_i : i \in I]$.

b) The elements x_j, $j \notin I$, are algebraic over $K[x_i : i \in I]$.

c) R_S is algebraic (and hence finite) over the quotient field $K(x_i : i \in I)$ of $K[x_i : i \in I]$, $S := K[x_i : i \in I] \setminus \{0\}$.

(2) Show that

$$\dim R = \max\,\{\,\text{Card}\,I : I \subseteq \{1, \ldots, n\}\,,\; x_i\,,\; i \in I\,, \text{ algebraically independent over } K\}\,.$$

(Note that not necessarily $\dim R = \text{Card}\,I$ if I is a maximal subset of $\{1, \ldots, n\}$ for which x_i, $i \in I$, are algebraically independent over K. But this is true if R is an integral domain.)

3.B.38. Exercise (L e m m a   o f   A b h y a n k a r) Let K be a field and $n \in \mathbb{N}^*$. Then there exists an injective K-algebra homomorphism $\varphi : K[[X_1, \ldots, X_n]] \to K[[X, Y]]$. (**Hint:** *There exists a sequence* $f_1 := X$, f_2, f_3, $\ldots$, f_n, $\ldots$ *of (over K) algebraically independent elements in* $K[[X]]$ (L e m m a   o f   S.   M a c L a n e   and   D. F. G.   S c h i l l i n g). Now define φ by $X_i \mapsto f_i Y$, $i = 1, 2, \ldots$.)

3.B.39. Exercise For a Noetherian integral domain R with quotient field K, the following statements are equivalent: (1) K is an R-subalgebra of an R-algebra of finite type. (2) K is an R-algebra of finite type. (3) $K = R_f$ for some $f \in R$, $f \neq 0$. (4) The generic point of $\text{Spec}\,R$ is open. (5) R is a semi-local ring of dimension ≤ 1. (6) $\text{Spec}\,R$ is finite.

3.B.40. Exercise Let R be a ring, $\mathbb{A}_R^n$ the "affine space" $\text{Spec}\,R[X_1, \ldots, X_n]$ and $F = \sum_{\nu \in \mathbb{N}^n} a_\nu X^\nu \in R[X_1, \ldots, X_n]$. Let $\pi : \mathbb{A}_R^n \to \text{Spec}\,R$ denote the canonical projection (i. e. the map corresponding to the canonical inclusion $R \hookrightarrow R[X_1, \ldots, X_n]$).

(1) Show that $\pi(\mathrm{D}(F)) = \mathrm{D}(\mathrm{C}(F))$, where $\mathrm{C}(F) := \sum_\nu Ra_\nu$ is the so called $\mathrm{c\,o\,n\,t\,e\,n\,t}$ of F. In particular, π is an open map.

(2) Show that $\pi(\mathrm{V}(F)) = \mathrm{V}(a_0) \cup \mathrm{D}(\mathrm{C}^*(F))$, where $\mathrm{C}^*(F) := \sum_{\nu \neq 0} Ra_\nu$. In particular, if $\dim R \geq 1$ and $n \geq 1$, then π is not a closed map.

3.B.41. Exercise If R is a non-zero Noetherian ring, then $\dim R[X_1, \ldots, X_n] = n + \dim R$. (**Hint:** Let $n = 1$ and let $\mathfrak{M}$ be a maximal ideal in $R[X]$ with $\mathfrak{p} := \mathfrak{M} \cap R$. Replacing R by $R_\mathfrak{p}$ one may assume that R is local and $\mathfrak{M} \cap R = \mathfrak{m} := \mathfrak{m}_R$ the maximal ideal of R. If $f_1, \ldots, f_d \in R$ is a system of parameters of R and if the residue class of the polynomial $F \in \mathfrak{M}$ generates the maximal ideal $\mathfrak{M}/\mathfrak{m}R[X]$ in $(R/\mathfrak{m})[X]$, then $f_1, \ldots, f_d, F$ is a system of parameters of $R[X]_\mathfrak{M}$.)

3.B.42. Exercise Let $R \hookrightarrow S$ be a ring extension.

(1) If S is integral over R, then $\operatorname{Spec} S \to \operatorname{Spec} R$ is a closed map.

(2) If R and S are of finite type over $\mathbb{K}$ ($= \mathbb{R}$ or $= \mathbb{C}$) and if S is integral over R, then $\mathbb{K}\text{-Spec}\, S \to \mathbb{K}\text{-Spec}\, R$ is a proper map with respect to the strong topologies on $\mathbb{K}\text{-Spec}\, S$ and $\mathbb{K}\text{-Spec}\, R$. Further, if $\mathbb{K} = \mathbb{R}$, then $\operatorname{Spm} S \to \operatorname{Spm} R$ is proper with respect to the strong topologies.

(3) The following conditions are equivalent: a) S is integral over R. b) For every $y \in S$, the element $1/y$ is a unit in $R[1/y] \subseteq S[1/y] := S_y$. c) The map $\operatorname{Spec} S[X] \to \operatorname{Spec} R[X]$ is closed. (**Hint:** Prove c) $\Rightarrow$ b) by using the following commutative diagrams.)

3.B.43. Exercise Let $\varphi : R \to S$ be a local homomorphism of Noetherian local rings (i. e. $\varphi(\mathfrak{m}_R) \subseteq \mathfrak{m}_S$). Show that $\dim S \leq \dim R + \dim S/\mathfrak{m}_R S$. Assume furthermore that R is an integral domain and that $\dim S = \dim R + \dim S/\mathfrak{m}_R S$, then show that φ is injective. (**Hint:** If $x_1, \ldots, x_r \in \mathfrak{m}_R$ is a system of parameters in R and if the residue classes of $y_1, \ldots, y_s \in \mathfrak{m}_S$ form a system of parameters in $S/\mathfrak{m}_R S$, then $\mathfrak{m}_S$ is minimal over the ideal in S generated by $\varphi(x_1), \ldots, \varphi(x_r), y_1, \ldots, y_s$ and hence $\dim S = \operatorname{ht} \mathfrak{m}_S \leq r + s$ by 3.B.29.)

3.B.44. Remark We add the following characterisation of local Artinian principal ideal rings: *For an Artinian local ring $(A, \mathfrak{m})$ the following statements are equivalent*: (1) *Every ideal in A is principal.* (2) $\mathfrak{m}$ *is principal.* (3) $\operatorname{Dim}_{A/\mathfrak{m}}(\mathfrak{m}/\mathfrak{m}^2) \leq 1$. (**Hint:** For the only non-trivial implication (3) $\Rightarrow$ (1) let $\mathfrak{a}$ be a non-zero ideal in A. By (3) and Nakayama's lemma $\mathfrak{m} = Aa$. Since A is Artinian, by 3.B.7, the ideal $Aa = \mathfrak{m} = \mathfrak{n}_A$ is nilpotent and there exists a non-negative integer r such that $\mathfrak{a} \subseteq Aa^r$ but $\mathfrak{a} \not\subseteq Aa^{r+1}$. Hence, there exists $y \in \mathfrak{a}$ such that $y = xa^r$ and $x \notin Aa$. Then $a^r \in \mathfrak{a}$ and $\mathfrak{a} = Aa^r$. – The proof shows that the powers of $\mathfrak{m} = Aa$ are the only ideals in A and, in particular, there are only finitely many ideals in A.) An arbitrary Artinian ring is a principal ideal ring if all its maximal ideals are principal.

CHAPTER 4: Schemes

4.A. Sheaves of Rings

Let R be a commutative ring. In general, the spectrum $X := \operatorname{Spec} R$ with the Zariski topology does not contain enough information about the structure of the ring R. For instance, for any field K, the space $\operatorname{Spec} K$ is just a singleton. Or, for a Noetherian integral domain R of dimension 1, the closed sets of $X = \operatorname{Spec} R$ are X itself and the finite subsets of $\operatorname{Spm} R \subseteq \operatorname{Spec} R$. So in this case, the cardinality of $\operatorname{Spec} R$ is the only invariant we can derive from the topological space $\operatorname{Spec} R$. Therefore we endow $\operatorname{Spec} R$ with an additional structure, namely its structure sheaf, from which the ring R can be recovered.

4.A.1. Definition Let X be a topological space. A p r e s h e a f $\mathcal{O}$ o f r i n g s on X is given by the following data:

(1) To every open set $U \subseteq X$ is associated a ring $\mathcal{O}(U)$;

(2) to every inclusion $V \subseteq U$ of open subsets of X there is associated a ring homomorphism $\rho_V^U : \mathcal{O}(U) \to \mathcal{O}(V)$ subject to the conditions:

a) $\rho_U^U = \operatorname{id}_{\mathcal{O}(U)}$,

b) if $W \subseteq V \subseteq U$ are open subsets of X then $\rho_W^V \circ \rho_V^U = \rho_W^U$.

The maps ρ_V^U are called the r e s t r i c t i o n m a p s and for $f \in \mathcal{O}(U)$ we will frequently use the notation $f|V$ instead of $\rho_V^U(f)$. If $\mathcal{O}'$ is another presheaf of rings on X, a m o r p h i s m $\varphi : \mathcal{O} \to \mathcal{O}'$ is a family of ring homomorphisms $\varphi_U : \mathcal{O}(U) \to \mathcal{O}'(U)$, U open in X, such that $\varphi_U(f)|V = \varphi_V(f|V)$ for $V \subseteq U$ and $f \in \mathcal{O}(U)$.

For a sheaf one requires additional conditions on a presheaf:

4.A.2. Definition A presheaf $\mathcal{O}$ of rings on a topological space X is called a s h e a f o f r i n g s if it satisfies the following conditions:

(S1) For every open subset U of X and every open covering $U_i, i \in I$, of U, if $f, g \in \mathcal{O}(U)$ and $f|U_i = g|U_i$ for every $i \in I$, then $f = g$;

(S2) for every open subset U of X and every open covering $U_i, i \in I$, of U, if $f_i \in \mathcal{O}(U_i)$, $i \in I$, are elements with $f_i|U_i \cap U_j = f_j|U_i \cap U_j$ for every $i, j \in I$, then there exists an element $f \in \mathcal{O}(U)$ such that $f|U_i = f_i$ for every $i \in I$.

The above conditions (S1) and (S2) are known as S e r r e ' s c o n d i t i o n s. Note that condition (S1) guarantees that the element f in condition (S2) is uniquely determined. Considering the empty covering of the empty set, this uniqueness implies $\mathcal{O}(\emptyset) = 0$.

A pair $(X, \mathcal{O}_X)$ of a topological space X and a sheaf of rings $\mathcal{O}_X$ on X is called a r i n g e d s p a c e. The sheaf $\mathcal{O}_X$ is called the s t r u c t u r e s h e a f of the ringed space $(X, \mathcal{O}_X)$. One often denotes $(X, \mathcal{O}_X)$ by X, if there is no ambiguity about the structure sheaf. If one wants to emphasize only the topological structure then one writes $|X|$ for the underlying topological space. For an open subset $U \subseteq X$, the ring $\mathcal{O}_X(U)$ is also denoted by $\Gamma(U, \mathcal{O}_X)$ or in short by $\Gamma(U)$ and is called the r i n g o f s e c t i o n s on U.

4.A.3. Example The concept of a structure sheaf is well illustrated by the following examples: Let X be an arbitrary topological space. For any open set $U \subseteq X$, define $\mathcal{C}_X(U)$ as the ring of continuous real-valued functions on U. For the restriction maps ρ_V^U, $V \subseteq U$, take the usual restriction maps. Then trivially $(X, \mathcal{C}_X)$ is a ringed space. With this example in mind, one can consider Serre's conditions as an abstract characterization of properties which are local, such as the continuity of a function.

In a similar way, if X is an open subset of $\mathbb{R}^n$, one constructs the sheaf $\mathcal{C}_X^k$ of C^k-functions, $k \in \mathbb{N} \cup \{\infty\}$. In this case, $\mathcal{C}_X^k(U)$, U open in X, is the ring of real-valued C^k-functions on U. The real-valued analytic functions also define a sheaf $\mathcal{C}_X^\omega$ on X. In all these examples, one can replace the value field $\mathbb{R}$ by the field $\mathbb{C}$ of complex numbers.

Let X be again an arbitrary topological space. If one sets $\mathcal{O}(U) := \mathbb{R}$ and $\rho_V^U = \mathrm{id}_\mathbb{R}$ for all open sets U, V, $V \subseteq U \subseteq X$, one gets a presheaf on X. This is never a sheaf because $\mathcal{O}(\emptyset) \neq 0$. If one replaces $\mathcal{O}(\emptyset) = \mathbb{R}$ by $\mathcal{O}(\emptyset) = 0$, then one gets a sheaf if and only if X is empty or if X is an irreducible topological space. But, for every topological space X, the locally constant functions define a sheaf. (Cf. Exercise 4.A.11.)

Let $\mathcal{O}$ be a presheaf of rings on a topological space X and $X' \subseteq X$ be an open subspace. Then, the collection of rings $\mathcal{O}(U)$, U open in X', is a presheaf of rings on X'. It is called the r e s t r i c t i o n of $\mathcal{O}$ to X' and denoted by $\mathcal{O}|X'$. If $\mathcal{O}$ is a sheaf then $\mathcal{O}|X'$ is also a sheaf. For instance, in the above example the sheaf $\mathcal{C}_X^k$ of C^k-functions on an open set $X \subseteq \mathbb{R}^n$ is the restriction of the sheaf $\mathcal{C}_{\mathbb{R}^n}^k$ of C^k-functions on $\mathbb{R}^n$ to X.

For a point x in a topological space X, we denote by $\mathcal{U}(x)$ the set of open neighbourhoods of x in X. Let $\mathcal{O}$ be a presheaf (of rings) on X. For $U, V \in \mathcal{U}(x)$ and $f \in \Gamma(U), g \in \Gamma(V)$, we say that f and g define the same g e r m at x if there exists a neighbourhood $W \in \mathcal{U}(x)$, $W \subseteq U \cap V$, such that $f|W = g|W$. This defines an equivalence relation on the disjoint union $\biguplus_{U \in \mathcal{U}(x)} \mathcal{O}(U)$. The quotient set is called the s t a l k a t x and it is denoted by $\mathcal{O}_{X,x}$ or $\mathcal{O}_x$. It carries a natural ring structure such that the canonical mappings

$$\rho_x^U : \mathcal{O}(U) \to \mathcal{O}_x,$$

$U \in \mathcal{U}(x)$, are ring homomorphisms. Indeed, $\mathcal{O}_x$ (together with the homomorphisms ρ_x^U) is the direct limit

$$\mathcal{O}_x = \varinjlim_{U \in \mathcal{U}(x)} \mathcal{O}(U)$$

of the directed system of rings $\mathcal{O}(U)$, $U \in \mathcal{U}(x)$ (where $\mathcal{U}(x)$ is considered as a directed set with respect to anti-inclusion). [1])

4.A.4. Example For the ringed space $(\mathbb{R}^n, \mathcal{C}^\omega_{\mathbb{R}^n})$ and a point $x := (a_1, \ldots, a_n) \in \mathbb{R}^n$, the stalk $\mathcal{C}^\omega_{\mathbb{R}^n, x}$ can be identified with the ring $\mathbb{R}\langle\!\langle X_1 - a_1, \ldots, X_n - a_n \rangle\!\rangle$ of convergent power series centred in x.

In the last example all the stalks are local rings, i.e. (commutative) rings with just one maximal ideal. The same is true for the sheaves $\mathcal{C}^k_{\mathbb{R}^n}$. This is a typical and important property of ringed spaces.

4.A.5. Definition A ringed space $(X, \mathcal{O}_X)$ is called a l o c a l l y r i n g e d s p a c e if all its stalks $\mathcal{O}_x = \mathcal{O}_{X,x}$, $x \in X$, are local rings.

Let $(X, \mathcal{O}_X)$ be a locally ringed space. For $x \in X$, the maximal ideal and the residue field of the stalk $\mathcal{O}_x$ are denoted by $\mathfrak{m}_x$ and by

$$\kappa(x) = \mathcal{O}_x/\mathfrak{m}_x.$$

The residue class $[f_x] \in \kappa(x)$ of a germ $f_x \in \mathcal{O}_x$ is called the v a l u e of f_x. If $U \subseteq X$ is open, every section $f \in \Gamma(U, \mathcal{O}_X)$ defines on U the function

$$x \mapsto f(x) := [f_x] \in \kappa(x), \quad x \in U.$$

Note that this function has values in different fields. But, if all the residue fields can be identified with one and the same field K, then f defines a function $U \to K$ in the usual sense. For instance, if $\mathcal{O}_X = \mathcal{C}_X$ is the sheaf of continuous real-valued functions on X, then $\kappa(x) = \mathbb{R}$ for every $x \in X$, in a canonical way, and the function $U \to \mathbb{R}$ associated to a section $f \in \Gamma(U, \mathcal{C}_X)$ is just f itself. For an arbitrary locally ringed space $(X, \mathcal{O}_X)$ and a section $f \in \Gamma(X)$ we set

$$X_f = \mathrm{D}(f) := \{x \in X \mid f(x) \neq 0\} \text{ and } \mathrm{V}(f) := X \setminus X_f = \{x \in X \mid f(x) = 0\}.$$

4.A.6. Example To illustrate the concept of the value of a section at a point we prove the following two propositions.

4.A.7. Proposition *Let $(X, \mathcal{O}_X)$ be a locally ringed space and U be an open subset of X. A section $f \in \Gamma(U, \mathcal{O}_X)$ is a unit in $\Gamma(U, \mathcal{O}_X)$ if and only if $f(x) \neq 0$ for every $x \in X$, i.e. if and only if the germ $f_x = \rho^U_x(f)$ is a unit in $\mathcal{O}_x$ for every $x \in X$.*

---

[1]) Let us recall that a d i r e c t e d s e t is a (partially) ordered set I with the following property: For arbitrary $i, j \in I$ there exists an element $k \in I$ with $i \leq k$, $j \leq k$. A d i r e c t e d s y s t e m of rings over I is a family of rings $A_i, i \in I$, and ring homomorphisms $\varphi_{ji} : A_i \to A_j$ for the pairs $(i, j) \in I \times I$ with $i \leq j$ such that $\varphi_{ii} = \mathrm{id}_{A_i}$ for all i and $\varphi_{kj}\varphi_{ji} = \varphi_{ki}$ for all i, j, k with $i \leq j \leq k$. The d i r e c t l i m i t $A := \lim_{\substack{\longrightarrow \\ i \in I}} A_i$ and the canonical ring homomorphisms $\varphi_i : A_i \to A$ with $\varphi_i = \varphi_j\varphi_{ji}$ for $i \leq j$ (constructed similarly as above) have the following universal property: For any ring B and ring homomorphisms $\psi_i : A_i \to B, i \in I$, with $\psi_i = \psi_j\varphi_{ji}$ for $i \leq j$, there exists a unique ring homomorphism $\psi : A \to B$ with $\psi_i = \psi\varphi_i$ for all i. – Direct limits are defined analogously for other structures like sets, groups, modules *etc.*

PROOF. If $f \in \Gamma(U, \mathcal{O}_X)^\times$ then its homomorphic image f_x is a unit in $\mathcal{O}_x$ for every $x \in X$. Conversely, let f_x be a unit in $\mathcal{O}_x$ for every $x \in U$. Then by definition, for each $x \in X$, there exists an open neighbourhood U^x of x in U and a section $g^x \in \Gamma(U^x, \mathcal{O}_X)$ such that $(f|U^x)\,g^x = 1$. By Serre's conditions (S1) and (S2) in Definition 4.A.2 there exists a $g \in \Gamma(U, \mathcal{O}_X)$ such that $g|U^x = g^x$ and $fg = 1$ in $\Gamma(U, \mathcal{O}_X)$.   •

4.A.8. Proposition *Let* $(X, \mathcal{O}_X)$ *be a locally ringed space and let* $f \in \Gamma(X, \mathcal{O}_X)$. *Then,* $X_f = \mathrm{D}(f)$ *is open in* X *and its complement* $\mathrm{V}(f) = X \setminus X_f$ *is closed in* X. *Moreover,* $f \mid X_f$ *is a unit in* $\Gamma(X_f, \mathcal{O}_X)$.

PROOF. Let $x \in X_f$. Then the germ f_x is a unit in $\mathcal{O}_x$ and therefore there exists a neighbourhood $U \in \mathcal{U}(x)$ and a $g \in \Gamma(U, \mathcal{O}_X)$ such that $(f \mid U)\,g = 1$ in $\Gamma(U, \mathcal{O}_X)$. Hence $U \subseteq X_f$ and X_f is open. The last part follows from the previous proposition.   •

To motivate the definition of a morphism of ringed spaces consider topological spaces X, Y again endowed with the structure sheaves $\mathcal{C}_X$ and $\mathcal{C}_Y$ of continuous real-valued functions. A continuous map $F : X \to Y$ induces for every open subset $V \subseteq Y$ a ring homomorphism $\Theta_V : \Gamma(V, \mathcal{C}_Y) \to \Gamma(F^{-1}(V), \mathcal{C}_X)$ by $f \mapsto f \circ F$. If X and Y are open sets in $\mathbb{R}^n$ and $\mathbb{R}^m$, respectively, then a C^k-map $X \to Y$ can be characterized as a continuous map $F : X \to Y$ such that the homomorphisms Θ_V map the ring $\Gamma(V, \mathcal{C}_Y^k)$ of C^k-functions on $V \subseteq Y$ into the ring $\Gamma(F^{-1}(V), \mathcal{C}_X^k)$ of C^k-functions on $F^{-1}(V) \subseteq X$. (Cf. Example 4.A.10 below.) In the abstract set-up of ringed spaces a continuous map $F : X \to Y$ of topological spaces does not define in a natural way ring homomorphisms $\Theta_V : \Gamma(V, \mathcal{O}_Y) \to \Gamma(F^{-1}(V), \mathcal{O}_X)$. In this case these homomorphisms have to be part of the definition of a morphism.

4.A.9. Definition (Morphism of ringed spaces) Let $(X, \mathcal{O}_X)$ and $(Y, \mathcal{O}_Y)$ be ringed spaces. A morphism from $(X, \mathcal{O}_X)$ to $(Y, \mathcal{O}_Y)$ is a pair (F, Θ), where $F : X \to Y$ is a continuous map and Θ a family of ring homomorphisms $\Theta_V : \Gamma(V, \mathcal{O}_Y) \longrightarrow \Gamma(F^{-1}(V), \mathcal{O}_X)$, V open in Y, such that for all open subsets $V, W \subseteq Y$ with $W \subseteq V$ the diagrams

$$\begin{array}{ccc} \Gamma(V, \mathcal{O}_Y) & \xrightarrow{\;\Theta_V\;} & \Gamma(F^{-1}(V), \mathcal{O}_X) \\ {\scriptstyle \rho^V_W}\downarrow & & \downarrow{\scriptstyle \rho^{F^{-1}(V)}_{F^{-1}(W)}} \\ \Gamma(W, \mathcal{O}_Y) & \xrightarrow{\;\Theta_W\;} & \Gamma(F^{-1}(W), \mathcal{O}_X) \end{array}$$

are commutative. (F, Θ) is called an isomorphism if F is a homeomorphism and each Θ_V is a ring isomorphism.

A morphism of ringed spaces (F, Θ) will be denoted in short by F, when there is no scope of confusion. In this case the homomorphism Θ_V is also denoted by $\Theta_F(V)$. A morphism $F = (F, \Theta)$ of ringed spaces $(X, \mathcal{O}_X)$ and $(Y, \mathcal{O}_Y)$ induces, for every $x \in X$, a ring homomorphism

$$\Theta_x : \mathcal{O}_{Y, F(x)} \to \mathcal{O}_{X, x}$$

such that the diagrams

$$
\begin{array}{ccc}
\Gamma(V, \mathcal{O}_Y) & \xrightarrow{\;\Theta_V\;} & \Gamma(F^{-1}(V), \mathcal{O}_X) \\
\downarrow{\scriptstyle \rho_y^V} & & \downarrow{\scriptstyle \rho_x^{F^{-1}(V)}} \\
\mathcal{O}_{Y, F(x)} & \xrightarrow{\;\Theta_x\;} & \mathcal{O}_{X, x}
\end{array}
$$

are commutative, $x \in X$, $V \in \mathcal{U}(F(x))$.

A morphism (F, Θ) of locally ringed spaces $(X, \mathcal{O}_X)$ and $(Y, \mathcal{O}_Y)$ is said to be a morphism of locally ringed spaces if $\Theta_x : \mathcal{O}_{Y, F(x)} \to \mathcal{O}_{X, x}$ is a local homomorphism of local rings for every $x \in X$. Let us recall that a homomorphism $\varphi : R \to S$ of local rings R, S with maximal ideals $\mathfrak{m}_R$ and $\mathfrak{m}_S$ is said to be a local homomorphism if $\varphi(\mathfrak{m}_R) \subseteq \mathfrak{m}_S$ or equivalently if $\varphi^{-1}(\mathfrak{m}_S) = \mathfrak{m}_R$. Such a homomorphism induces a homomorphism $\overline{\varphi} : R/\mathfrak{m}_R \to S/\mathfrak{m}_S$ of the residue fields. So a morphism F of locally ringed spaces induces homomorphisms $\kappa(F(x)) \to \kappa(x)$, $x \in X$.

A morphism between two locally ringed spaces is always assumed to be a morphism of locally ringed spaces, unless otherwise stated.

For the sheaf $\mathcal{C}_X$ of real-valued continuous functions on a topological space X, the sets of sections $\Gamma(U, \mathcal{C}_X)$ are not only rings but even algebras over $\mathbb{R}$, and all the restriction maps ρ_V^U, $V \subseteq U$ open in X, are $\mathbb{R}$-algebra homomorphisms. In general, if all the rings of sections $\Gamma(U, \mathcal{O}_X)$ of a ringed space $(X, \mathcal{O}_X)$ are algebras over a fixed base ring A and if all the restriction maps ρ_V^U, $V \subseteq U$ open in X, are A-algebra homomorphisms, then we call $(X, \mathcal{O}_X)$ a ringed space over A. Accordingly, an A-morphism of ringed spaces $(X, \mathcal{O}_X)$ and $(Y, \mathcal{O}_Y)$ over A is a morphism (F, Θ) of ringed spaces such that all the homomorphisms $\Theta_V : \Gamma(V, \mathcal{O}_Y) \to \Gamma(F^{-1}(V), \mathcal{O}_X)$, V open in Y, are A-algebra homomorphisms.

4.A.10. Example Let $k \in \mathbb{N} \cup \{\infty\}$. A (real) n-dimensional C^k-manifold X can be defined as a ringed space $(X, \mathcal{O}_X)$ over $\mathbb{R}$ which is locally isomorphic to $(\mathbb{R}^n, \mathcal{C}_{\mathbb{R}^n}^k)$, i. e. for every $x \in X$ there exists $U \in \mathcal{U}(x)$ and an open subset U' in $\mathbb{R}^n$ such that $(U, \mathcal{O}_X|U)$ is $\mathbb{R}$-isomorphic to $(U', \mathcal{C}_{\mathbb{R}^n}^k|U')$. In a similar way one can define real-analytic and complex manifolds as ringed spaces over $\mathbb{R}$ and $\mathbb{C}$, respectively. One proves easily that *the morphisms of C^k-manifolds or analytic manifolds as ringed spaces over $\mathbb{R}$ or $\mathbb{C}$ are precisely those morphisms which are induced by the C^k-maps of C^k-manifolds and the analytic maps of analytic manifolds.*

We end this section with two important constructions.

To every presheaf $\mathcal{O}$ of rings on a topological space X can be associated, in a canonical way, a sheaf of rings $\mathcal{O}^{\#}$. For the construction of $\mathcal{O}^{\#}$ we use the stalks $\mathcal{O}_x = \varinjlim_{U \in \mathcal{U}(x)} \mathcal{O}(U)$ which are defined for presheaves in the same way as for sheaves. Then $\mathcal{O}^{\#}(U)$ is defined as a subring of $\prod_{x \in U} \mathcal{O}_x$. A tuple $(s^x)_{x \in U}$ belongs to $\mathcal{O}^{\#}(U)$ if and only if for every $x \in U$ there exists a $V \in \mathcal{U}(x)$,

$V \subseteq U$, and an element $f \in \mathcal{O}(V)$ with $s^y = f_y$ for each $y \in V$. The restriction maps are obvious. It is easy to prove that $\mathcal{O}^\#$ is a sheaf of rings and it is called the s h e a f i f i c a t i o n of $\mathcal{O}$. The canonically defined morphism $\iota : \mathcal{O} \to \mathcal{O}^\#$ of presheaves has the following universal property: If $\varphi : \mathcal{O} \to \mathcal{O}'$ is a morphism of presheaves of rings on X, where $\mathcal{O}'$ is even a sheaf, there exists a unique morphism $\varphi^\# : \mathcal{O}^\# \to \mathcal{O}'$ of sheaves such that $\varphi = \varphi^\# \circ \iota$. In particular, ι is an isomorphism if $\mathcal{O}$ is already a sheaf. In every case, ι induces isomorphisms $\mathcal{O}_x \xrightarrow{\sim} \mathcal{O}_x^\#$ of stalks for all $x \in X$.

To motivate the next construction let X be a ringed space with structure sheaf $\mathcal{O}_X$. Let $\mathcal{B}$ be a base of the topology of X. Then for every open set U of X, the restriction maps induce a canonical isomorphism

$$\Gamma(U, \mathcal{O}_X) \xrightarrow{\sim} \varprojlim_{U' \subseteq U, U' \in \mathcal{B}} \Gamma(U', \mathcal{O}_X)$$

given by $f \mapsto (f|U')_{U' \subseteq U, U' \in \mathcal{B}}$. This follows immediately from Serre's conditions (S1) and (S2). So the structure sheaf $\mathcal{O}_X$ is completely determined by the rings $\mathcal{O}_X(U)$ and the restriction maps ρ_V^U, $U, V \in \mathcal{B}$, $V \subseteq U$. [2])

Conversely, assume that there is given a family of rings $\mathcal{O}(U)$, $U \in \mathcal{B}$, with restriction maps ρ_V^U, $U, V \in \mathcal{B}$, $V \subseteq U$, which fulfills all the conditions for a sheaf as far as the open sets of the base $\mathcal{B}$ are involved. Then *define*

$$\mathcal{O}(U) := \varprojlim_{U' \subseteq U, U' \in \mathcal{B}} \mathcal{O}(U')$$

for an arbitrary open set $U \subseteq X$. For $U \in \mathcal{B}$ these rings can be identified with the given rings $\mathcal{O}(U)$ and for open sets $U, V \subseteq X$, $V \subseteq U$, we have canonical restriction maps $\rho_V^U : \mathcal{O}(U) \to \mathcal{O}(V)$. It is easy to check that *the data* $(\mathcal{O}(U), \rho_V^U)$ *define a sheaf of rings*.

We will use the last construction quite often. The first example is given in the following section.

Presheaves and sheaves of rings are particular cases of presheaves and sheaves of sets. A p r e s h e a f $\mathcal{A}$ of sets on a topological space X is a collection of sets $\mathcal{A}(U)$, U open in X, with restriction maps $\rho_V^U : \mathcal{A}(U) \to \mathcal{A}(V)$ for open sets $V \subseteq U \subseteq X$ such that $\rho_W^U = \rho_V^U \circ \rho_W^V$ for all open sets $W \subseteq V \subseteq U \subseteq X$. A presheaf $\mathcal{A}$ of sets is called a s h e a f if it satisfies the conditions (S1) and (S2)

---

[2]) Recall that for an ordered index set I, a family A_i, $i \in I$, of rings and ring homomorphisms $\varphi_{ij} : A_j \to A_i$ for $i \leq j$ with $\varphi_{ii} = \mathrm{id}_{A_i}$ and $\varphi_{ij}\varphi_{jk} = \varphi_{ik}$ for $i \leq j \leq k$, the p r o j e c t i v e   l i m i t

$$A := \varprojlim_{i \in I} A_i = \{ (a_i)_{i \in I} \in \textstyle\prod_i A_i : \varphi_{ij}(a_j) = a_i \text{ for } i \leq j \}$$

and the canonical projections $\varphi_i : A \to A_i$ have the following universal property: For any ring B and ring homomorphisms $\psi_i : B \to A_i$ with $\psi_i = \varphi_{ij}\psi_j$, there is a unique ring homomorphism $\psi : B \to A$ with $\psi_i = \varphi_i \psi$ for all i. – This definition naturally extends to other structures like sets, groups, modules *etc.*

of Definition 4.A.2. If $\mathcal{A}$ is a sheaf, then $\mathcal{A}(\emptyset)$ is a singleton. The s t a l k s $\mathcal{A}_x$, $x \in X$, and the s h e a f i f i c a t i o n $\mathcal{A}^{\#}$ are defined according to sheaves of rings. A m o r p h i s m $f : \mathcal{A} \to \mathcal{B}$ of presheaves $\mathcal{A}$, $\mathcal{B}$ of sets (defined over the same space X) is a collection of mappings $f_U : \mathcal{A}(U) \to \mathcal{B}(U)$ with $\rho_V^U \circ f_U = f_V \circ \rho_V^U$ for all open sets $U, V \subseteq X$ with $V \in X$. Such a morphism f induces mappings $f_x : \mathcal{A}_x \to \mathcal{B}_x$ of the stalks.

In an analogous fashion one defines presheaves and sheaves of other structures, e.g. groups, modules *etc.* If $(X, \mathcal{O}_X)$ is a ringed space, an $\mathcal{O}_X$-m o d u l e is a sheaf $\mathcal{F}$ over X such that for every open set $U \subseteq X$ the set $\mathcal{F}(U) = \Gamma(U, \mathcal{F})$ is an $\mathcal{O}_X(U)$-module and for every inclusion $V \subseteq U$ of open sets the restriction map $\rho_V^U : \mathcal{F}(U) \to \mathcal{F}(V)$ is compatible with the scalar multiplications. For two $\mathcal{O}_X$-modules $\mathcal{F}$ and $\mathcal{G}$, a homomorphism $f : \mathcal{F} \to \mathcal{G}$ is a morphsim of sheaves such that the mappings $f_U : \mathcal{F}(U) \to \mathcal{G}(U)$ are $\mathcal{O}(U)$-linear for all U open in X. The set of all these homomorphisms is denoted by $\text{Hom}_{\mathcal{O}_X}(\mathcal{F}, \mathcal{G})$. The presheaf $U \mapsto \text{Hom}_{\mathcal{O}|U}(\mathcal{F}|U, \mathcal{G}|U)$ is obviously a sheaf and hence an $\mathcal{O}_X$-module. This $\mathcal{O}_X$-module is denoted by $\mathcal{H}om_{\mathcal{O}_X}(\mathcal{F}, \mathcal{G})$. The dual $\mathcal{H}om_{\mathcal{O}_X}(\mathcal{F}, \mathcal{O}_X)$ is usually denoted by $\mathcal{F}^*$. For a point $x \in X$, there is a canonical $\mathcal{O}_x$-homomorphism $\mathcal{H}om_{\mathcal{O}_X}(\mathcal{F}, \mathcal{G})_x \to \text{Hom}_{\mathcal{O}_x}(\mathcal{F}_x, \mathcal{G}_x)$ which is, in general, not an isomorphism. The tensor products $\mathcal{F}(U) \otimes_{\mathcal{O}_X(U)} \mathcal{G}(U)$, $U \subseteq X$ open, define a presheaf of $\mathcal{O}_X(U)$-modules (in general this is not a sheaf), its sheafication is called the tensor product of $\mathcal{F}$ and $\mathcal{G}$ and is denoted by $\mathcal{F} \otimes_{\mathcal{O}_X} \mathcal{G}$. The canonical $\mathcal{O}_x$-homomorphism $(\mathcal{F} \otimes_{\mathcal{O}_X} \mathcal{G})_x \to \mathcal{F}_x \otimes_{\mathcal{O}_x} \mathcal{G}_x$ is always an isomorphism.

4.A.11. Exercise (1) Let X be a topological space and A be an arbitrary set. The c o n s t a n t p r e s h e a f on X corresponding to A associates to every non-empty open set $U \subseteq X$ the set A and to the empty set a singleton, and all the restriction maps for the non-empty open subsets are the identity of A. The sheafification of this presheaf is the sheaf of locally constant functions with values in A and is usually denoted by A_X. It is called the c o n s t a n t s h e a f on X corresponding to A. If A is a group or a ring, then A_X is a sheaf of groups or a sheaf of rings.

(2) Now assume that $X \neq \emptyset$ and that A contains at least two elements. Then show that the constant presheaf on X corresponding to A is already a sheaf, if and only if X is an irreducible topological space.

4.A.12. Exercise Let X be a topological space with the discrete topology and let $\mathcal{A}$ be a sheaf over X. Show that for every point $x \in X$, $\Gamma(\{x\}, \mathcal{A}) = \mathcal{A}_x$ and that for every open subset $U \subseteq X$, $\Gamma(U, \mathcal{A}) \cong \prod_{x \in U} \mathcal{A}_x$. More generally, for an arbitrary topological space X and for a family U_i, $i \in I$, of pairwise disjoint open sets $U_i \subseteq X$ show that $\Gamma(\biguplus_{i \in I} U_i, \mathcal{A}) = \prod_{i \in I} \Gamma(U_i, \mathcal{A})$.

4.A.13. Exercise Let X be a topological space and $\mathcal{A}$ a sheaf over X. Show that for every open set $U \subseteq X$ the canonical map $s \mapsto (s_x)_{x \in U}$ is an injective map $\Gamma(U, \mathcal{A}) \to \prod_{x \in U} \mathcal{A}_x$.

4.A.14. Exercise (D i r e c t i m a g e s) Let $F : X \to Y$ be a continuous map of topological spaces and let $\mathcal{A}$ be a presheaf on X. Show that through $\mathcal{B}(V) := \mathcal{A}(F^{-1}(V))$, $\rho_W^V := \rho_{F^{-1}(W)}^{F^{-1}(V)}$, $W \subseteq V \subseteq Y$ open, a presheaf $\mathcal{B}$ on Y is defined. This is denoted by $F_*\mathcal{A}$ and

is called the d i r e c t i m a g e of $\mathcal{A}$ with respect to F. If $\mathcal{A}$ is a sheaf, so also is $F_*\mathcal{A}$.
– If $(X, \mathcal{O}_X)$ and $(Y, \mathcal{O}_Y)$ are ringed spaces, a morphism $(F, \Theta) : X \to Y$ is given by a continuous map $F : X \to Y$ and a morphism $\Theta : \mathcal{O}_Y \to F_*\mathcal{O}_X$ of sheaves of rings (over Y).

4.A.15. Exercise Let $(X, \mathcal{O}_X)$ be a ringed space and $f \in \Gamma(X, \mathcal{O}_X) = \mathcal{O}_X(X)$ a global section. Show that:

(1) f is a unit in $\Gamma(X, \mathcal{O}_X)$, if and only if for each point $x \in X$ the germ f_x is a unit in $\mathcal{O}_{X,x}$. Through $\mathcal{O}_X^\times(U) := (\mathcal{O}_X(U))^\times$, $U \subseteq X$ open, a sheaf $\mathcal{O}_X^\times$ of groups on X is defined. (It is called the s h e a f o f u n i t s of $\mathcal{O}_X$.)

(2) If X is quasi-compact, then f is nilpotent in $\Gamma(X, \mathcal{O}_X)$, if and only if for each point $x \in X$ the germ f_x is nilpotent in $\mathcal{O}_{X,x}$. Give an example to demonstrate that this does not hold in general for an arbitrary X. (**Hint:** Use Exercise 4.A.12.)

4.A.16. Exercise Let X be a topological space and $\varphi : \mathcal{A} \to \mathcal{B}$ be a morphism of sheaves over X.

(1) Show that $\varphi_U : \mathcal{A}(U) \longrightarrow \mathcal{B}(U)$ is injective (resp. bijective) for all open sets $U \subseteq X$ if and only if $\varphi_x : \mathcal{A}_x \to \mathcal{B}_x$ is injective (resp. bijective) for all $x \in X$.

(2) Let $X := \mathbb{C}$ and $\mathcal{A} = \mathcal{B} := \mathcal{C}_\mathbb{C}^\times$ be the sheaf (of groups) of nowhere-zero, continuous, complex-valued functions on $\mathbb{C}$. Let $m \in \mathbb{N}^*$, $m \geq 2$, and $\varphi : \mathcal{C}_\mathbb{C}^\times \to \mathcal{C}_\mathbb{C}^\times$ be the morphism sending a function f to its m-th power f^m. Show that φ_z is surjective for all $z \in \mathbb{C}$, but there are open sets $U \subseteq \mathbb{C}$ for which φ_U is *not* surjective. (Indeed φ_U is surjective if and only if all connected components of U are simply connected.) A similar example is given by $\mathcal{A} := \mathcal{C}_\mathbb{C}$, $\mathcal{B} := \mathcal{C}_\mathbb{C}^\times$ and the exponential morphism $\exp : \mathcal{C}_\mathbb{C} \to \mathcal{C}_\mathbb{C}^\times$ which sends a function f to $\exp f$.

4.A.17. Exercise Let $(X, \mathcal{O}_X)$ be a locally ringed space and let $r \in \mathbb{N}$. Then the following conditions are equivalent: (1) X has exactly r connected components. (2) $\Gamma(X, \mathcal{O}_X)$ has exactly 2^r idempotent elements. (**Hint:** If a topological space X has infinitely many connected components, then (by induction) for every $t \in \mathbb{N}^*$ there is a decomposition $X = X_1 \uplus \cdots \uplus X_t$ of X as a disjoint union of non-empty open subsets X_i.) In particular, X is connected if and only if $0 \neq 1$ are the only idempotent elements in $\Gamma(X, \mathcal{O}_X)$. (This criterion for connectedness is used very often.)

4.B. Schemes

Now for a commutative ring R we will construct a structure sheaf $\mathcal{O}_X$ on the topological space $X = \operatorname{Spec} R$. We will denote this sheaf of rings by $\widetilde{R}$. With this structure sheaf $(X, \mathcal{O}_X) = (\operatorname{Spec} R, \widetilde{R})$ will be a locally ringed space from which the original ring R can be recovered. By the construction described in the previous section it is enough to define the rings of sections only for the basic open sets $D(f)$, $f \in R$.

Let $f \in R$. Then $D(f)$ as a subspace of $\operatorname{Spec} R$ can be identified with $\operatorname{Spec} R_f$. Therefore it is natural to associate to $D(f)$ the ring R_f as the ring of sections over $D(f)$, i. e.

$$\Gamma(D(f), \widetilde{R}) := R_f .$$

But the element f is not uniquely determined by the open set $D(f)$. For $f, g \in R$ we have $D(f) = D(g)$ if and only if there are $s, t \in \mathbb{N}$ such that $f|g^s$ and $g|f^t$ or, equivalently, the saturations $\operatorname{sat} S_f$ and $\operatorname{sat} S_g$ of the multiplicatively closed sets generated by f and g respectively coincide.[3] If we identify R_f with $(\operatorname{sat} S_f)^{-1}R$, then there is no ambiguity in the definition $\Gamma(D(f), \tilde{R}) = R_f$ for the ring of sections over $D(f)$. More generally, $D(g) \subseteq D(f)$ is equivalent to $\operatorname{sat} S_f \subseteq \operatorname{sat} S_g$ which gives a canonical restriction map

$$\rho_{D(g)}^{D(f)} : R_f \to R_g$$

from $\Gamma(D(f), \tilde{R})$ to $\Gamma(D(g), \tilde{R})$. Obviously these assignments define a presheaf $\tilde{R}$ of rings on $\operatorname{Spec} R$.

4.B.1. Theorem $\tilde{R}$ is a sheaf of rings with $\tilde{R}_x = R_{\mathfrak{p}_x}$ for all $x \in X = \operatorname{Spec} R$.

PROOF. We have to verify the conditions (S1) and (S2) with respect to the canonical basis $D(f)$, $f \in R$, of the Zariski topology on X. Because of $\tilde{R}|D(f) = \tilde{R}_f$, it is enough to prove these conditions for $X = D(1)$ instead of considering an arbitrary $D(f)$.

(S1) Let $D(f_i)$, $i \in I$, be an open covering of X and let $g \in R$. Then $g|D(f_i) = 0$ implies that $g/1 = 0 \in R_{f_i}$ and hence $f_i^{k_i} g = 0$ for some $k_i \in \mathbb{N}$. Now, $\bigcup_{i \in I} D(f_i) = X$ implies $\sum_{i \in I} Rf_i = R$ and hence $\sum_{i \in I} Rf_i^{k_i} = R$. Therefore, $1 = \sum_{i \in I} a_i f_i^{k_i}$ with $a_i \in R$ (and $a_i = 0$ for almost all $i \in I$) and $g = 1 \cdot g = \sum_{i \in I} a_i f_i^{k_i} g = 0$.

(S2) Let $D(f_i)$, $i \in I$, be an open cover of X and $g_i \in R_{f_i}$. Since X is quasi-compact (and condition (S1) is already verified), we may assume that I is finite. Then there exists $k \in \mathbb{N}$ and $h_i \in R$ such that $g_i = h_i/f_i^k$ for every $i \in I$. By assumption $g_i \,|\, D(f_i) \cap D(f_j) = g_i \,|\, D(f_i f_j) = g_j \,|\, D(f_i f_j) = g_j \,|\, D(f_i) \cap D(f_j)$. Hence $f_i^l f_j^l h_i f_j^k = f_i^l f_j^l h_j f_i^k$ for some $l \in \mathbb{N}$. We have $\sum_{j \in I} Rf_j^{k+l} = R$ and $1 = \sum_{j \in I} a_j f_j^{k+l}$ with $a_j \in R$. Set $g = \sum_{j \in I} a_j f_j^l h_j$. Then $g \,|\, D(f_i) = g_i$ for every $i \in I$, because of $g f_i^{k+l} = \sum_{j \in I} a_j f_j^l h_j f_i^{k+l} = \sum_{j \in I} a_j f_i^l f_j^{k+l} h_i = f_i^l h_i$ and hence $g/1 = f_i^l h_i/f_i^{k+l} = h_i/f_i^k = g_i$ in R_{f_i}.

The proof that the canonical homomorphism $\tilde{R}_x = \varinjlim_{f \notin \mathfrak{p}_x} R_f \to R_{\mathfrak{p}_x}$ is an isomorphism is straightforward. ●

4.B.2. Definition (1) An a f f i n e s c h e m e is a locally ringed space $(X, \mathcal{O}_X)$ which is isomorphic to $(\operatorname{Spec} R, \tilde{R})$ for some commutative ring R.

---

[3] For an arbitrary multiplicatively closed subset S of a ring R the s a t u r a t i o n sat S of S is the multiplicatively closed set comprising the divisors of the elements of S. It is the biggest multiplicatively closed subset S' of R containing S for which the canonical homomorphism $S^{-1}R \to S'^{-1}R$ is bijective or the preimage of the units in $S^{-1}R$ with respect to the canonical homomorphism $R \to S^{-1}R$.

(2) A s c h e m e is a locally ringed space $(X, \mathcal{O}_X)$ such that there exists an open covering $U_i, i \in I$, of X and $(U_i, \mathcal{O}_X | U_i)$ is an affine scheme for every $i \in I$.

(3) A m o r p h i s m of schemes is a morphism of the locally ringed spaces.

As in the previous section, we often denote a ringed space $(X, \mathcal{O}_X)$ by X, if no confusion is likely.

The d i m e n s i o n of a scheme $X = (X, \mathcal{O}_X)$ is (by definition) the dimension of the topological space X in the sense of Definition 3.B.2. (See also Exercise 4.B.8 (3).)

If X is an affine scheme, then necessarily $(X, \mathcal{O}_X) \cong (\operatorname{Spec} R, \widetilde{R})$ with $R := \Gamma(X, \mathcal{O}_X)$.

For a scheme $X = (X, \mathcal{O}_X)$ an open set $U \subseteq X$ is called an a f f i n e   o p e n s e t, if $(U, \mathcal{O}_X | U)$ is an affine scheme. If $(X, \mathcal{O}_X) = (\operatorname{Spec} R, \widetilde{R})$ is affine, then the standard open sets $\mathrm{D}(f)$, $f \in R$, are affine open sets in X, because of $(\mathrm{D}(f), \widetilde{R} | \mathrm{D}(f)) = (\mathrm{D}(f), \widetilde{R}_f)$. This implies that for an arbitrary scheme $(X, \mathcal{O}_X)$ the set of affine open sets $U \subseteq X$ form a basis of the topology of X, and that for any open subset $U \subseteq X$ the locally ringed space $(U, \mathcal{O}_X | U)$ is again a scheme called an o p e n   s u b s c h e m e of X. A scheme which is isomorphic to an open subscheme of an affine scheme is called q u a s i - a f f i n e.

4.B.3. Example (1) Let A be a ring. Then the affine scheme $\operatorname{Spec} A[T_1, \ldots, T_n]$ defined by the polynomial algebra $A[T_1, \ldots, T_n]$ is denoted by $\mathbb{A}_A^n$. For a field K the affine scheme $\mathbb{A}_K^n = \operatorname{Spec} K[T_1, \ldots, T_n]$ is called the n - d i m e n s i o n a l   a f f i n e   s p a c e over K. Recall that $K^n = K\text{-}\operatorname{Spec} K[T_1, \ldots, T_n] \subseteq \operatorname{Spm} K[T_1, \ldots, T_n] \subseteq \mathbb{A}_K^n$.

(2) Let $X_1 = X_2 = \mathbb{A}_K^1$ be the affine line over a field K. Let P be the zero point of $\mathbb{A}_K^1$, i.e. the point corresponding to the maximal ideal (T) of $K[T]$ and $U := \mathbb{A}_K^1 \setminus \{P\}$. We construct the topological space $X = X_1 \uplus X_2 / \sim$, where the equivalence relation $\sim$ identifies each point $Q \in X_1$ with the corresponding point $Q \in X_2$ if and only if $Q \neq P$. A subset $V \subseteq X$ is open if and only if $\iota_1^{-1}(V)$ and $\iota_2^{-1}(V)$ are open in X_1 and X_2 respectively, where $\iota_1 : X_1 \to X$ and $\iota_2 : X_2 \to X$ are the canonical inclusion maps. They are even open inclusions and X carries a structure sheaf $\mathcal{O}_X$ for which $\mathcal{O}_X | X_i = \mathcal{O}_{X_i}, i = 1, 2$. So $(X, \mathcal{O}_X)$ is a scheme called the a f f i n e   l i n e $\mathbb{A}_K^1$ w i t h   t h e   p o i n t P d o u b l e d. By Theorem 4.B.11 below this is not affine.

4.B.4. Exercise (S u m s   o f   s c h e m e s) (1) Let $X_i = (X_i, \mathcal{O}_{X_i})$, $i \in I$, be a family of non-empty schemes and $X := \uplus_{i \in I} X_i$ be the disjoint union of the topological spaces $X_i, i \in I$, (for which the canonical inclusions $X_i \to X$ are open). Then $\mathcal{O}_X(U) := \prod_{i \in I} \Gamma(U \cap X_i, \mathcal{O}_{X_i})$, $U \subseteq X$ open, is the unique sheaf of rings on X with $\mathcal{O}_X | X_i = \mathcal{O}_{X_i}$ for all $i \in I$. $(X, \mathcal{O}_X)$ is a scheme called the s u m of the schemes $X_i, i \in I$, and denoted by $\uplus_{i \in I} X_i$ or by $\coprod_{i \in I} X_i$.

(2) Let $R = R_1 \times \cdots \times R_s$ be a product of commutative rings $R_1, \ldots, R_s$. Then the affine scheme $(\operatorname{Spec} R, \widetilde{R})$ is canonically isomorphic to the sum of the affine schemes $(\operatorname{Spec} R_i, \widetilde{R}_i)$, $i = 1, \ldots, s$.

(3) A sum $\uplus_{i \in I} X_i$ of non-empty schemes $X_i = (X_i, \mathcal{O}_{X_i})$, $i \in I$, is affine if and only if I is finite and all the summands X_i are affine. (**Hint:** Note that the topological space X of an affine scheme $(X, \mathcal{O}_X)$ is always quasi-compact.)

4.B.5. Exercise Derive from Exercise 4.B.4 (3) the following result: Let R be a (commutative) ring and let $r \in \mathbb{N}$. Then the following conditions are equivalent: (1) Spec R has exactly r connected components. (2) R has exactly 2^r idempotent elements. (3) There exist rings $R_1, \dots, R_r$ such that each R_i has exactly two idempotent elements and $R \simeq R_1 \times \cdots \times R_r$ (as rings). (4) R_{red} has exactly 2^r idempotent elements. (For any ring A the reduction A_{red} is the quotient $A/\mathfrak{n}_A$ of A by its nilradical $\mathfrak{n}_A$. Note that Spec A = Spec A_{red}. See Exercise 4.A.17.)

4.B.6. Exercise (R e d u c e d   s c h e m e s) A scheme $(X, \mathcal{O}_X)$ is called r e d u c e d, if for every open set $U \subseteq X$ the ring $\mathcal{O}_X(U)$ has no non-zero nilpotent elements.

(1) Show that an affine scheme (Spec $R, \widetilde{R}$) is reduced if and only if R is reduced (i.e. R has no non-zero nilpotent elements).

(2) Show that $(X, \mathcal{O}_X)$ is reduced if and only if for every $x \in X$ the local ring $\mathcal{O}_{X,x}$ is reduced.

(3) Let $(X, \mathcal{O}_X)$ be a scheme. Let $(\mathcal{O}_X)_{\mathrm{red}}$ be the sheaf associated to the presheaf which assigns to an open set $U \subseteq X$ the reduction $\mathcal{O}_X(U)_{\mathrm{red}}$ with the natural restriction maps. Show that $(X, (\mathcal{O}_X)_{\mathrm{red}})$ is a scheme, called the r e d u c t i o n X_{red} of X. If $(X, \mathcal{O}_X) = ($Spec $R, \widetilde{R})$, then $(X, (\mathcal{O}_X)_{\mathrm{red}}) = ($Spec $R_{\mathrm{red}}, \widetilde{R_{\mathrm{red}}})$. (In general $\mathcal{O}_{X_{\mathrm{red}}} = \mathcal{O}_X/\mathcal{N}_X$, where $\mathcal{N}_X \subseteq \mathcal{O}_X$ is the ideal sheaf of locally nilpotent sections with $\Gamma(U, \mathcal{N}_X) = \mathfrak{n}_{\Gamma(U, \mathcal{O}_X)}$, the nilradical of $\Gamma(U, \mathcal{O}_X)$, for every open affine subset $U \subseteq X$. – See Definition 4.E.6 and Example 4.E.7 for general closed embeddings and closed subschemes.)

(4) Show that there is a canonical morphism of schemes $X_{\mathrm{red}} \to X$ which is the identity on the underlying topological spaces.

(5) Every morphism $(F, \Theta) : X \to Y$ of schemes induces a morphism $(F_{\mathrm{red}}, \Theta_{\mathrm{red}}) : X_{\mathrm{red}} \to Y_{\mathrm{red}}$ with $F_{\mathrm{red}} = F$ as continuous maps between topological spaces. It is uniquely determined by the condition that the diagram

$$
\begin{array}{ccc}
X & \xrightarrow{(F, \Theta)} & Y \\
\uparrow & & \uparrow \\
X_{\mathrm{red}} & \xrightarrow{(F_{\mathrm{red}}, \Theta_{\mathrm{red}})} & Y_{\mathrm{red}}
\end{array}
$$

is commutative, where the vertical morphisms are the canonical ones as in part (4).

4.B.7. Exercise (I n t e g r a l   and   i r r e d u c i b l e   s c h e m e s) A scheme $(X, \mathcal{O}_X)$ is called i n t e g r a l, if it is non-empty and if for every non-empty open set $U \subseteq X$, the ring $\Gamma(U, \mathcal{O}_X)$ is an integral domain. $(X, \mathcal{O}_X)$ is called i r r e d u c i b l e, if X is an irreducible topological space. Show that:

(1) An affine scheme $X = $ Spec R is integral if and only if R is an integral domain.

(2) For a scheme X (not necessarily affine) the following conditions are equivalent: a) X is integral. b) X is reduced and irreducible. c) $X \neq \emptyset$ and for every non-empty open affine subset $U \subseteq X$ the ring $\Gamma(U, \mathcal{O}_X)$ is an integral domain.

(3) A scheme X is irreducible if and only if its reduction X_{red} is integral. (Cf. Exercise 4.B.6.)

4.B.8. Exercise If X is a topological space and if Z is an irreducible closed subset of X, then a g e n e r i c   p o i n t for Z is a point $z \in Z$ such that $Z = \overline{\{z\}}$ (i.e. the closure of $\{z\}$).

(1) If X is a scheme, show that every irreducible closed subset of X has a unique generic point. (**Hint:** Cf. Proposition 3.A.11 for the affine case.) For $x, y \in X$ with $x \in \overline{\{y\}}$, i.e. with $\overline{\{x\}} \subseteq \overline{\{y\}}$, we call y a g e n e r a l i z a t i o n of x and x a s p e c i a l i z a t i o n of y. Show that the set of generalizations of x is the intersection of all neighbourhoods of x and can be identified (as a topological space) with $\operatorname{Spec} \mathcal{O}_x$. Furthermore, $\overline{\{x\}}$ is an irreducible component of X if and only if $\dim \mathcal{O}_x = 0$, i.e. x has no generalizations besides x.

(2) If X is an integral scheme, show that the local ring $\mathcal{O}_z$ of the generic point z of X is a field. It is called the (r a t i o n a l) f u n c t i o n f i e l d of X and often denoted by $\mathcal{R}(X)$. Show that $\mathcal{R}(X)$ is the quotient field of $\Gamma(U, \mathcal{O}_X)$ for *every* non-empty open affine subset $U \subseteq X$. (See Exercise 6.E.13 for generalizations.)

(3) The dimension of a scheme X is the supremum of the (Krull-)dimensions of its stalks $\mathcal{O}_x, x \in X$. More precisely, for $x \in X$ the codimension of the closed irreducible set $\overline{\{x\}}$ coincides with the (Krull-)dimension of $\mathcal{O}_x$.

4.B.9. Theorem *Let X be a locally ringed space and let Y be an affine scheme. Then the canonical map*

$$\varphi : \operatorname{Hom}_{\mathrm{loc}}(X, Y) \longrightarrow \operatorname{Hom}(\Gamma(Y, \mathcal{O}_Y), \Gamma(X, \mathcal{O}_X)),$$

$$(F, \Theta) \longmapsto \big(\Theta_Y : \Gamma(Y, \mathcal{O}_Y) \to \Gamma(X, \mathcal{O}_X) \big)$$

is a bijection between the set of morphisms of the locally ringed spaces X, Y and the set of ring homomorphisms from $\Gamma(Y, \mathcal{O}_Y)$ to $\Gamma(X, \mathcal{O}_X)$.

PROOF. For $R := \Gamma(Y, \mathcal{O}_Y)$ we define $\psi : \operatorname{Hom}(R, \Gamma(X, \mathcal{O}_X)) \to \operatorname{Hom}_{\mathrm{loc}}(X, Y)$ in the following way: Let $h : R \to \Gamma(X, \mathcal{O}_X)$ be a given ring homomorphism. $h_x : R \to \kappa(x)$ denotes for $x \in X$ the following composition map:

$$R \xrightarrow{\; h \;} \Gamma(X, \mathcal{O}_X) \longrightarrow \mathcal{O}_{X,x} \longrightarrow \kappa(x).$$

Then $\psi(h)(x) := \operatorname{Ker} h_x \in \operatorname{Spec} R = Y$. One has $\psi(h)^{-1}(Y_f) = X_{h(f)}$ for $f \in R$ (where $Y_f = \mathrm{D}(f)$). Hence the map $\psi(h) : X \to Y$ is continuous. Furthermore, there is a unique ring homomorphism $h_f : \Gamma(Y_f, \mathcal{O}_Y) \to \Gamma(X_{h(f)}, \mathcal{O}_X)$ that makes the following diagram commutative:

$$
\begin{array}{ccc}
R = \Gamma(Y, \mathcal{O}_Y) & \xrightarrow{\;\; h \;\;} & \Gamma(X, \mathcal{O}_X) \\[2pt]
{\scriptstyle \rho^Y_{Y_f}} \downarrow & & \downarrow {\scriptstyle \rho^X_{X_{h(f)}}} \\[2pt]
R_f = \Gamma(Y_f, \mathcal{O}_Y) & \xrightarrow[\;\; h_f \;\;]{} & \Gamma(X_{h(f)}, \mathcal{O}_X)
\end{array}
$$

The existence of h_f follows from Proposition 4.A.8 and the universal property of $R \to R_f$. It is easy to check that $\psi(h) : X \to Y$ and the family $h_f, f \in R$, define an element in $\operatorname{Hom}_{\mathrm{loc}}(X, Y)$ and that φ and ψ are inverse to each other. ●

4.B.10. Corollary *Let X, Y be affine schemes with $R := \Gamma(Y, \mathcal{O}_Y)$ and $S := \Gamma(X, \mathcal{O}_X)$. Then the canonical map $\operatorname{Hom}(X, Y) \to \operatorname{Hom}(R, S)$ is a bijection.*

Let **Rings** denote the category of commutative rings and **Affine** denote the category of affine schemes. Then Spec : **Rings** $\to$ **Affine**, $R \rightsquigarrow (\operatorname{Spec} R, \widetilde{R})$, and Γ : **Affine** $\to$ **Rings**, $X \rightsquigarrow \Gamma(X, \mathcal{O}_X)$, are contravariant functors. Clearly, $\Gamma \circ \operatorname{Spec} = \operatorname{id}$ and $\operatorname{Spec} \circ \Gamma$ is naturally equivalent to id.

4.B.11. Corollary *Let X be a locally ringed space with $R := \Gamma(X, \mathcal{O}_X)$. Then X is an affine scheme if and only if the morphism $X \to \operatorname{Spec} R$ belonging to the identity map of R is an isomorphism of ringed spaces.*

4.B.12. Exercise Let $(X, \mathcal{O}_X)$ be a locally ringed space. There is exactly one morphism $F : X \to \operatorname{Spec} \mathbb{Z}$ of locally ringed spaces. For a prime number p the fibre $F^{-1}(\mathbb{Z}p)$ is the set of points $x \in X$ with $\operatorname{char} \kappa(x) = p$ and the fibre $F^{-1}((0))$ is the set of points $x \in X$ with $\operatorname{char} \kappa(x) = 0$. For $X := \operatorname{Spec} \mathbb{Z}[X_1, \ldots, X_m]$, for example, one has $F^{-1}(\mathbb{Z}p) \cong \operatorname{Spec}(\mathbb{Z}/\mathbb{Z}p)[X_1, \ldots, X_m]$ and $F^{-1}(0) \cong \operatorname{Spec} \mathbb{Q}[X_1, \ldots, X_m]$. Identify $\operatorname{Spm} \mathbb{Z}[X_1, \ldots, X_m] \subseteq X$. (Cf. the remarks at the end of Section 1.F.) List all the elements of $\operatorname{Spec} \mathbb{Z}[X_1]$ and $\operatorname{Spm} \mathbb{Z}[X_1]$.

4.B.13. Exercise Let R be a factorial domain. The only non-empty closed subsets in $\operatorname{Spec} R$ of pure codimension 1 are the hypersurfaces $V(f)$, $0 \neq f \in R \setminus R^{\times}$. The only affine open subsets in $\operatorname{Spec} R$ are the basic open subsets $D(f)$, $f \in R$. (**Hint:** For a non-empty open set $U = D(\mathfrak{a}) \subseteq \operatorname{Spec} R$ one has

$$\Gamma(U, \widetilde{R}) = \{a/b \mid a, b \in R, b \neq 0, \operatorname{GCD}(a, b) = 1, V(b) \subseteq V(\mathfrak{a})\}.)$$

In particular: Let K be a field. For $n \geq 2$ the only non-trivial affine open sets in $\mathbb{A}_K^n := \operatorname{Spec} K[X_1, \ldots, X_n]$ are the complements $D(f) = \operatorname{Spec} K[X_1, \ldots, X_n]_f$ of the hypersurfaces $V(f)$, f a non-constant polynomial in $K[X_1, \ldots, X_n]$, and for any closed point $x \in \mathbb{A}_K^n$ the punctured affine space $\mathbb{A}_K^n \setminus \{x\}$ is quasi-affine but not affine.

4.B.14. Exercise Let R be a ring and $\mathfrak{a}$ an ideal in R. The open set $U = D(\mathfrak{a}) \subseteq \operatorname{Spec} R$ is affine if and only if the ideal $\mathfrak{a}$ generates in $S := \Gamma(U, \widetilde{R})$ the unit ideal. (**Hint:** If $S = \mathfrak{a}S$, then the canonical morphism $(U, \widetilde{R} \mid U) \to (\operatorname{Spec} S, \widetilde{S})$ is an isomorphism.)

4.B.15. Exercise Let K be a field and R be the structure algebra of the three-dimensional quadric $\operatorname{Spec} R = V(UX - VY) \subseteq \mathbb{A}_K^4$, that is, $R = K[U, V, X, Y]/(UX - VY) = K[u, v, x, y]$. Show that:

(1) $\pi := x + u$ is a prime element in R and $\mathfrak{p} := (x, y)$ and $\mathfrak{q} := (u, v)$ are prime ideals of codimension 1 in R which are not principal (and so R is a non-factorial integral domain).

(2) The complement $D(\mathfrak{p})$ of the hypersurface $V(\mathfrak{p}) \subseteq \operatorname{Spec} R$ is not affine. In particular, $D(\mathfrak{p})$ is not a basic open subset. (**Hint:** Note that $V(\mathfrak{q}) = \operatorname{Spec} R/(u, v) \cong \mathbb{A}_K^2$. If $D(\mathfrak{p})$ were affine, then the closed subset $D(\mathfrak{p}) \cap V(\mathfrak{q})$ of $D(\mathfrak{p})$ would also be affine. But $D(\mathfrak{p}) \cap V(\mathfrak{q}) \cong \mathbb{A}_K^2 \setminus \{(0, 0)\}$ is not affine by Exercise 4.B.13.)

(3) $U := D(\pi) \cap D(\mathfrak{p}) = D(\pi \mathfrak{p})$ is an affine open subset of $\operatorname{Spec} R$ and not a basic open subset. (This is a classical example of a non-basic affine open subset.) (**Hint:** Let $t := v/x = u/y \in Q(R)$. Then $t, 1/\pi \in S := \Gamma(U, \widetilde{R})$ and $S = \pi \mathfrak{p}S$, because of $1 = (1/\pi)x + (t/\pi)y$. Now use Exercise 4.B.14. – If U were a basic open set, then $D(\mathfrak{p})$ too would be so. But $D(\mathfrak{p})$ is not affine.)

Let A be a ring. In view of Theorem 4.B.9, a scheme $(X, \mathcal{O}_X)$ together with a distinguished morphism $X \to \operatorname{Spec} A$ is equivalent to an A-algebra structure of $\Gamma(X, \mathcal{O}_X)$. If such a structure is given, we call $(X, \mathcal{O}_X)$ an A - s c h e m e or a s c h e m e o v e r A or o v e r $\operatorname{Spec} A$. A simple example of an A-scheme is the affine scheme $\mathbb{A}_A^n$. More generally, we define:

4.B.16. Definition Let S be a fixed scheme. Then a scheme X is called a s c h e m e o v e r S or an S - s c h e m e, if a s t r u c t u r e m o r p h i s m $\varphi : X \to S$ is given. If X and Y are two S-schemes with structure morphisms $\varphi : X \to S$ and $\psi : Y \to S$, then a morphism $F : X \to Y$ is called an S - m o r p h i s m, if $\varphi = \psi \circ F$.

By this definition a morphism $F : X \to Y$ of A-schemes is an A-morphism, if and only if the induced morphism $\Gamma(Y, \mathcal{O}_Y) \to \Gamma(X, \mathcal{O}_X)$ is an A-algebra homomorphism. In this case for any open set $V \subseteq Y$ the homomorphism $\Gamma(V, \mathcal{O}_Y) \to \Gamma(F^{-1}(V), \mathcal{O}_X)$ is an A-algebra homomorphism.

A scheme which is a singleton as a topological space is called a f a t p o i n t. These are exactly the affine schemes associated to the local rings R of dimension 0, i. e. commutative rings with exactly one prime ideal. In the Noetherian case these are the Artinian local rings. A reduced fat point is the spectrum of a field. Such a scheme is called a p o i n t. If x is a point of an arbitrary scheme X, then there is a canonical embedding from the point $\operatorname{Spec} \kappa(x)$ into X which has the point $\{x\}$ as the image and which induces the canonical projection $\mathcal{O}_{X,x} \to \kappa(x)$ of the stalks. For an arbitrary point $Z := \operatorname{Spec} K$, K a field, a morphism $Z \to X$ is given by a point $x \in X$ *and* a field homomorphism $\kappa(x) \to K$. By definition two morphisms $\varphi, \psi : X \to Y$ of schemes c o i n c i d e (s c h e m e - t h e o r e t i c a l l y) a t a p o i n t $x \in X$, denoted by

$$\varphi(x) \equiv \psi(x),$$

if $\varphi(x) = \psi(x) =: y$ *and* if the induced morphisms $\operatorname{Spec} \kappa(x) \to \operatorname{Spec} \kappa(y)$ of schemes coincide, i. e. if the induced field homomorphisms $_\varphi\overline{\Theta}_x, {}_\psi\overline{\Theta}_x : \kappa(y) \to \kappa(x)$ coincide. Note that the condition $\varphi(x) \equiv \psi(x)$ is equivalent to the condition that the compositions $f \circ \iota_x$ and $g \circ \iota_x$ coincide, where $\iota_x : \operatorname{Spec} \kappa(x) \to X$ is the canonical inclusion mentioned above.

For a K-scheme X, K a field, the K-morphisms $\operatorname{Spec} K \to X$ are in a bijective correspondence with the set of points $x \in X$ for which the residue field $\kappa(x)$ coincides with K. Such a point is called a K - r a t i o n a l p o i n t of X and the set of K-rational points of X is usually denoted by $X(K)$. *For an affine K-scheme $X := \operatorname{Spec} R$, R a K-algebra, the set of K-rational points of X coincides with the K-spectrum K-$\operatorname{Spec} R$ of R introduced already in Chapter 2.*

4.B.17. Example A typical fat point is the n - f o l d p o i n t $\operatorname{Spec} K[T]/(T^n)$ over the field K, $n \in \mathbb{N}^*$. In particular, for $n = 2$ we obtain the d o u b l e p o i n t $\operatorname{Spec} K[\varepsilon]$ with $\varepsilon := \overline{T}$, $\varepsilon^2 = 0$. The algebra $K[\varepsilon]$ is also called the a l g e b r a o f d u a l n u m b e r s over K.

For a K-scheme X a K-morphism $\operatorname{Spec} K[\varepsilon] \to X$ is given by a K-morphism $\operatorname{Spec} K \to X$, i. e. a K-rational point $x \in X(K)$, and a K-algebra homomorphism $\xi : \mathcal{O}_x \to K[\varepsilon]$. If we define the K-linear map $\delta : \mathcal{O}_x \to K$ by the equation $\xi(f) = f(x) + \delta(f)\varepsilon$, $f \in \mathcal{O}_x$, then for $f, g \in \mathcal{O}_x$ we have

$$\xi(fg) = \xi(f)\xi(g) = (f(x) + \delta(f)\varepsilon)(g(x) + \delta(g)\varepsilon) = f(x)g(x) + (g(x)\delta(f) + f(x)\delta(g))\varepsilon\,.$$

Hence δ satisfies the p r o d u c t r u l e

$$\delta(fg) = g(x)\delta(f) + f(x)\delta(g)\,.$$

So δ is a K-derivation $\mathcal{O}_x \to K$. As in analysis such a derivation is called a t a n g e n t v e c t o r at the K-rational point x. *The set of K-morphisms* Spec $K[\varepsilon] \to X$ *with a fixed image point* $x \in X(K)$ *can be identified with the space of tangent vectors at x called the* t a n g e n t s p a c e at $x \in X(K)$. A K-derivation vanishes on the square $\mathfrak{m}_x^2$ of the maximal ideal $\mathfrak{m}_x \subseteq \mathcal{O}_x$. Therefore the tangent space at x is canonically isomorphic to the K-dual of the so called c o t a n g e n t s p a c e $\mathfrak{m}_x/\mathfrak{m}_x^2$. We shall study these concepts in detail in Chapter 6.

4.B.18. Exercise An (affine) scheme which is isomorphic to the spectrum of a local ring is called a l o c a l s c h e m e. Particular examples of local schemes are points and, more generally, fat points. Let $Y := $ Spec R be the local scheme corresponding to a local ring R and let X be an arbitrary scheme.

(1) Show that a morphism $Y \to X$ is given by a point $x \in X$ (which is the image of the only closed point of Y) and a local homomorphism $\mathcal{O}_x \to R$ of local rings.

(2) Let $x \in X$ be a point. By (1) there corresponds a canonical morphism Spec $\mathcal{O}_x \to X$ to the identity of $\mathcal{O}_x$. Show that this morphism is an embedding of topological spaces and that its image is the set of generalizations of the point x. (Cf. Exercise 4.B.8 (1). This set coincides with the intersection of all the neighbourhoods of x in X. The scheme Spec $\mathcal{O}_x$ is called the l o c a l s c h e m e of X in x.)

4.C. Finiteness Conditions on Schemes

The most important schemes satisfy finiteness conditions which we want to discuss in this section.

4.C.1. Definition (1) An affine scheme $(X, \mathcal{O}_X)$ is said to be N o e t h e r i a n, if the ring $R = \Gamma(X, \mathcal{O}_X)$ is Noetherian.

(2) A scheme $(X, \mathcal{O}_X)$ is said to be l o c a l l y N o e t h e r i a n, if X has an open covering of Noetherian affine schemes or, equivalently, if X has a basis of open Noetherian affine subschemes.

(3) A scheme $(X, \mathcal{O}_X)$ is said to be N o e t h e r i a n, if one of the following equivalent conditions holds:

a) $(X, \mathcal{O}_X)$ has a finite open covering of Noetherian affine schemes.

b) $(X, \mathcal{O}_X)$ is locally Noetherian and the topological space X is quasi-compact.

c) $(X, \mathcal{O}_X)$ is locally Noetherian and the topological space X is Noetherian.

For the proof of the equivalence of the three conditions in (3) and also for the proof of the fact that definitions (1) and (3) are consistent we have to show the following:

4.C.2. Lemma *Let R be a ring. If $X := \operatorname{Spec} R$ has a covering by open affine subsets U_i, $i \in I$, with $R_i := \Gamma(U_i, \widetilde{R})$ Noetherian for all $i \in I$, then $R = \Gamma(X, \widetilde{R})$ is Noetherian.*

PROOF. Let $f \in R$ and $D(f) \subseteq U_i$ for some $i \in I$. Then $R_f = (R_i)_f$ is Noetherian. Since X is quasi-compact, we have to prove the following: If the elements $f_1, \ldots, f_n \in R$ generate the unit ideal and if the rings R_{f_i} are Noetherian for $i = 1, \ldots, n$, then R is Noetherian too. But for an ideal $\mathfrak{a}$ in R the ideals $\mathfrak{a} R_{f_i}$ are finitely generated in R_{f_i}. Therefore there exist finitely many elements $h_1, \ldots, h_r \in \mathfrak{a}$ which generate the ideal $\mathfrak{a} R_{f_i}$ for $i = 1, \ldots, n$. Then $\mathfrak{a} R_{\mathfrak{p}} = \mathfrak{b} R_{\mathfrak{p}}$ for all $\mathfrak{p} \in \operatorname{Spec} R$, where $\mathfrak{b} := \sum_{j=1}^{r} Rh_j$, and hence $\mathfrak{a} = \mathfrak{b}$ by the following well-known result. ●

4.C.3. Lemma *Let R be a ring and M an R-module. If U, V are submodules of M with $U_{\mathfrak{p}} \subseteq V_{\mathfrak{p}}$ for all $\mathfrak{p} \in \operatorname{Spec} R$, then $U \subseteq V$.*

PROOF. Because of $((U + V)/V)_{\mathfrak{p}} = (U_{\mathfrak{p}} + V_{\mathfrak{p}})/V_{\mathfrak{p}} = 0$ for all $\mathfrak{p} \in \operatorname{Spec} R$, it is enough to show that $M = 0$ if $M_{\mathfrak{p}} = 0$ for all $\mathfrak{p} \in \operatorname{Spec} R$. Now for $x \in M$ we have $Rx \cong R/\operatorname{Ann}_R x$ and $\operatorname{Ann}_R x = R$, since otherwise $(Rx)_{\mathfrak{p}} = R_{\mathfrak{p}}/(\operatorname{Ann}_R x)R_{\mathfrak{p}} \neq 0$ for all $\mathfrak{p} \in V(\operatorname{Ann}_R x) \neq \emptyset$. ●

4.C.4. Remark Instead of using 4.C.3 to finish the proof of 4.C.2 one can use the following explicit construction: Let $x \in \mathfrak{a}$. From representations $x/1 = \sum_j a_{ij} h_j / f_i^{m_i}$ in R_{f_i} with $a_{ij} \in R$ one obtains $f_i^{k_i} x = \sum_j a_{ij} h_j$ for some $k_i \in \mathbb{N}$. Finally an equation $1 = \sum_{i=1}^{n} f_i^{k_i} g_i$ gives $x = 1 \cdot x = \sum_j (\sum_i g_i g_{ij}) h_j \in \mathfrak{b}$. With the same argument one proves more generally: *Let M be a module over R. If M_{f_i} is a finite R_{f_i}-module for $i = 1, \ldots, n$, then M is a finite R-module.*

4.C.5. Definition (1) A scheme X over a ring A (i. e. an A-scheme) is said to be o f f i n i t e t y p e o v e r A , if there is a finite affine open covering U_i, $i \in I$, of X, such that the A-algebra $\Gamma(U_i, \mathcal{O}_X)$ is of finite type over A for every $i \in I$.

(2) An S-scheme X with the structure morphism $\varphi : X \to S$ is said to be o f f i n i t e t y p e o v e r S, if one of the following equivalent conditions holds:

a) For every open affine $V \subseteq S$ the inverse image $U := \varphi^{-1}(V)$ is a $\Gamma(V, \mathcal{O}_S)$-scheme of finite type.

b) There is an open affine cover V_i, $i \in I$, of S such that for each $i \in I$ the inverse image $U_i := \varphi^{-1}(V_i)$ is a $\Gamma(V_i, \mathcal{O}_S)$-scheme of finite type.

The equivalence of the two conditions in (2) is an immediate consequence of the following lemma.

4.C.6. Lemma *Let $X \to Y$ be a morphism of affine schemes with $A := \Gamma(Y, \mathcal{O}_Y)$ and $B := \Gamma(X, \mathcal{O}_X)$. If X has an open affine cover U_i, $i \in I$, such that $B_i := \Gamma(U_i, \mathcal{O}_X)$ is of finite type over A for each $i \in I$, then B is of finite type over A.*

PROOF. For each $f \in B$ with $D(f) \subseteq U_i$ for some $i \in I$ the algebra $B_f = (B_i)_f$ is of finite type over A. It is thus sufficient to show that if there are elements $f_1, \ldots, f_n \in B$ which generate the unit ideal in B such that B_{f_i} is of finite type over A for all $i = 1, \ldots, n$, then B is also of finite type over A. A set of A-algebra generators of B is obtained in the following way. Let $1 = \sum_{i=1}^{n} g_i f_i$ with $g_i \in B$ and let $b_j \in B$, $j = 1, \ldots, m$, be finitely many elements such that B_{f_i} is generated over A by $1/f_i$ and $b_j/1$, $j = 1, \ldots, m$. Then the elements $f_1, \ldots, f_n, g_1, \ldots, g_n, b_1, \ldots, b_m$ generate the A-algebra B as is easily seen.   •

An affine A-scheme $X = \mathrm{Spec}\, B$ is of finite type over A if and only if B is an A-algebra of finite type. An open subscheme of an S-scheme of finite type is again of finite type over S. By condition a) of Definition 4.C.5 this implies:

4.C.7. Proposition *Let $S = \mathrm{Spec}\, A$ be an affine scheme. For every open affine subset $U \subseteq S$ the algebra $\Gamma(U, \mathcal{O}_S)$ is of finite type over A.*

By Hilbert's basis theorem it is obvious that schemes of finite type over a locally Noetherian scheme are also locally Noetherian and even Noetherian if the base scheme is Noetherian.

4.D. Product of Schemes

Let X and Y be S-schemes. The p r o d u c t of X and Y over S, denoted by

$$X \times_S Y,$$

is defined to be an S-scheme together with S-morphisms $p : X \times_S Y \to X$ and $q : X \times_S Y \to Y$ such that given any S-scheme Z and S-morphisms $f : Z \to X$ and $g : Z \to Y$, there exists a unique S-morphism $F : Z \to X \times_S Y$ such that $f = p \circ F$ and $g = q \circ F$. The morphisms p and q are called the (c a n o n i c a l) p r o j e c t i o n   m o r p h i s m s. We denote this S-morphism F also by (f, g). The product $X \times_S Y$ coincides with the category-theoretical product of X and Y in the category of S-schemes.[4]) If the base scheme $S = \mathrm{Spec}\, A$ is affine, i.e. if we are dealing with A-schemes X and Y, it is customary to use the notation $X \times_A Y$ in place of $X \times_{\mathrm{Spec}\, A} Y$. Products $X \times_{\mathbb{Z}} Y$ over $\mathrm{Spec}\, \mathbb{Z}$ are called a b s o l u t e   p r o d u c t s and are also denoted by $X \times Y$. The same short notation is also used for $X \times_S Y$, if the base scheme S is clear from the context or fixed once for all.

---

[4]) Let us recall that for an arbitrary category $\mathcal{C}$ and for objects $X, Y \in \mathrm{Obj}\, \mathcal{C}$ a product of X and Y in $\mathcal{C}$ is a triple (T, p, q) with $T \in \mathrm{Obj}\, \mathcal{C}$, $p \in \mathrm{Hom}_{\mathcal{C}}(T, X)$ and $q \in \mathrm{Hom}_{\mathcal{C}}(T, Y)$ such that for arbitrary $Z \in \mathrm{Obj}\, \mathcal{C}$ the canonical map $F \mapsto (p \circ F, q \circ F)$ from $\mathrm{Hom}_{\mathcal{C}}(Z, T)$ to $\mathrm{Hom}_{\mathcal{C}}(Z, X) \times \mathrm{Hom}_{\mathcal{C}}(Z, Y)$ is bijective. If a product exists, it is unique up to canonical isomorphism. Products in the opposite category $\mathcal{C}^{\mathrm{op}}$ are called c o p r o d u c t s (or s u m s) in the original category $\mathcal{C}$. For example, for a commutative ring A the tensor product $B \otimes_A C$ is a coproduct of B and C in the category of *commutative A-algebras*. The sum $\biguplus_{i \in I} X_i = \coprod_{i \in I} X_i$ of S-schemes $X_i = (X_i, \mathcal{O}_{X_i})$, $i \in I$, described in Exercise 4.B.4 (1) is a coproduct of the schemes X_i in the category of S-schemes.

From the identification $\mathrm{Hom}_{\mathrm{Spec}\,A}(Z, \mathrm{Spec}\,R) = \mathrm{Hom}_{A\text{-alg}}(R, \Gamma(Z, \mathcal{O}_Z))$ (see Theorem 4.B.9) and the fact that the tensor product is a coproduct in the category of commutative A-algebras we get the following important existence theorem.

4.D.1. Theorem *Let* $X = \mathrm{Spec}\,B$ *and* $Y = \mathrm{Spec}\,C$ *be affine* A-*schemes. Then* $\mathrm{Spec}(B \otimes_A C)$ *is a product* $X \times_A Y$ *(where the canonical projections* $X \times_A Y \to X$ *and* $X \times_A Y \to Y$ *correspond to the canonical* A-*algebra homomorphisms* $B \to B \otimes_A C$, $b \mapsto b \otimes 1$, *and* $C \to B \otimes_A C$, $c \mapsto 1 \otimes c$). *In particular,* $X \times_A Y$ *is also affine.*

The product $X \times_S Y$ *for arbitrary schemes* S, X *and* Y *also exists*, but we will not carry out the construction. It uses a simple but tedious gluing procedure which is based on the affine case 4.D.1 and the following lemma. Its proof is easy and left to the reader.

4.D.2. Lemma *Let* X *and* Y *be* S-*schemes with structure morphisms* $\alpha : X \to S$ *and* $\beta : Y \to S$. *Let* U, V *and* T *be open subsets of* X, Y *and* S *respectively with* $\alpha(U) \subseteq T$ *and* $\beta(V) \subseteq T$. *If* $(X \times_S Y, p, q)$ *is a product of* X *and* Y *over* S, *then* $p^{-1}(U) \cap q^{-1}(V)$ *is a product* $U \times_T V$ *(where the canonical projections are the restrictions of* p *and* q *to* $p^{-1}(U) \cap q^{-1}(V)$). *Furthermore,* $U \times_T V = U \times_S V$ *as* S-*schemes.*

4.D.3. Example (B a s e c h a n g e) One of the most important applications of the product is the passage from S-schemes to T-schemes with respect to a morphism $T \to S$. Let X be an S-scheme. The product $X_{(T)} := T \times_S X$ is a T-scheme with respect to the canonical projection $T \times_S X \to T$. We say that the T-scheme $X_{(T)}$ is obtained from X by c h a n g e o f t h e b a s e from S to T (with respect to the given morphism $T \to S$). $X_{(T)}$ is also called the p u l l b a c k of X (with respect to $T \to S$). $X_{(T)}$ is part of the following commutative so called C a r t e s i a n d i a g r a m :

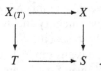

For affine schemes this base change corresponds to the base change for algebras: If B is an A-algebra and if $\varphi : A \to A'$ is a homomorphism of commutative rings, then the algebra $B' := A' \otimes_A B$ is the A'-algebra obtained from B by base change (or base extension) $A \to A'$. If B is commutative, then the A'-scheme $\mathrm{Spec}\,B'$ is the extension of the A-scheme $\mathrm{Spec}\,B$ with respect to the morphism $\mathrm{Spec}\,A' \to \mathrm{Spec}\,A$ corresponding to φ. For an arbitrary A-scheme X the scheme $X_{(\mathrm{Spec}\,A')} = (\mathrm{Spec}\,A') \times_A X$ is also denoted by $X_{(A')}$ (or by $A' \otimes_A X$).

4.D.4. Example (F i b r e s o f a m o r p h i s m) Let s be a point of a scheme S and X be an S-scheme. The $\kappa(s)$-scheme

$$X_s := X_{(\mathrm{Spec}\,\kappa(s))}$$

obtained from X by base change with respect to the canonical inclusion $\mathrm{Spec}\,\kappa(s) \to S$ is called the f i b r e o f X o v e r s.

In the affine case $S = \operatorname{Spec} A$ and $X = \operatorname{Spec} B$ one has $X_s = \operatorname{Spec}(\kappa(s) \otimes_A B) = \operatorname{Spec}(B_{\mathfrak{p}_s}/\mathfrak{p}_s B_{\mathfrak{p}_s})$ and the canonical morphism $X_s \to X$ induces a topological embedding of X_s onto the (set-theoretical) fibre of X over s. For a point $x \in X_s$ corresponding to the prime ideal $\mathfrak{q}_x \subseteq B$ the stalk $\mathcal{O}_{X_s, x}$ is the local ring $B_{\mathfrak{q}_x}/\mathfrak{p}_s B_{\mathfrak{q}_x}$. For a prime ideal $\mathfrak{p} \in \operatorname{Spec} A$ the algebra $B_{\mathfrak{p}}/\mathfrak{p}B_{\mathfrak{p}}$ is often called the f i b r e a l g e b r a of B o v e r $\mathfrak{p}$. The following general result immediately reduces to the affine case:

4.D.5. Proposition *Let X be an S-scheme and let $s \in S$. Then the canonical morphism $X_s \to X$ induces a topological embedding of the fibre X_s onto the (set-theoretic) fibre of X over s. The stalk $\mathcal{O}_{X_s, x}$ of a point $x \in X_s$ ($\subseteq X$) is the local ring $\mathcal{O}_{X,x}/\mathfrak{m}_s \mathcal{O}_{X,x}$.*

4.D.6. Example (P r o d u c t o f m o r p h i s m s) Let X, X', Y and Y' be S-schemes and $(X \times_S X', p, p')$ and $(Y \times_S Y', q, q')$ be the S-products. Let $f : X \to Y$ and $f' : X' \to Y'$ be S-morphisms. Then by the universal property of products there exists a unique morphism $f \times_S f' := (f \circ p, f' \circ p') : X \times_S X' \to Y \times_S Y'$ such that the following diagram is commutative.

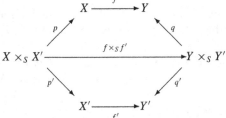

The morphism $f \times_S f'$ is called the p r o d u c t o f t h e m o r p h i s m s f and f'.

4.D.7. Example (G r a p h o f a m o r p h i s m) Let $f : X \to Y$ be an S-morphism of S-schemes. Then $\operatorname{id}_X : X \to X$ and f induce a unique S-morphism $\Gamma_f : X \to X \times_S Y$ called the S-g r a p h of f. For instance, the d i a g o n a l m o r p h i s m $\Delta_X : X \to X \times_S X$ is the graph of the identity id_X of X.

The product of schemes provides a convenient way to discuss the Hausdorff properties of schemes. Let us recall that a topological space X is a Hausdorff space if and only if it has the following property: For an arbitrary topological space Y and continuous maps $f, g : Y \to X$ the set of points where f and g coincide, i. e. the subset $\{ y \in Y \mid f(y) = g(y) \}$ is closed in Y. Since this subset is the preimage of the diagonal $\Delta_X(X) = \{ (x, x) \in X \times X \mid x \in X \}$ with respect to the continuous map $(f, g) : Y \to X \times X$, $y \mapsto (f(y), g(y))$, X is a Hausdorff space if and only if the diagonal $\Delta_X(X)$ is closed in $X \times X$. The diagonal map $\Delta_X : X \to X \times X$ itself is $(\operatorname{id}_X, \operatorname{id}_X)$ and the diagonal $\Delta_X(X) \subseteq X \times X$ is the set of points where the canonical projections $p, q : X \times X \to X$ coincide. There is a relative version of the Hausdorff condition for topological spaces. Let $\varphi : X \to S$ be a continuous map of topological spaces, i. e. let $X = (X, \varphi)$ be a fibre space over S. Then X is called a H a u s d o r f f s p a c e o v e r S if any two distinct points $x, y \in X$ with $\varphi(x) = \varphi(y)$ (i. e. x and y belong to the same fibre) have disjoint neighbourhoods. This is equivalent to the property that the diagonal $\Delta_X(X)$ is closed in the fibre product $X \times_S X := \{ (x, y) \in X \times X \mid \varphi(x) = \varphi(y) \}$ and also to the property that the set of points where two morphisms $f, g : Y \to X$ of fibre spaces over S coincide is always closed in Y.

These remarks are expected to motivate the following definition for S-schemes. To that effect we recall that two morphisms $f, g : Y \to X$ of schemes coincide scheme-theoretically at a point $y \in Y$, denoted as $f(y) \equiv g(y)$, if $f \circ \iota_y = g \circ \iota_y$, where $\iota_y : \operatorname{Spec} \kappa(y) \to Y$ is the canonical inclusion. (See Section 4.B.)

4.D.8. Definition Let X be an S-scheme. X is said to be s e p a r a t e d o v e r S, if for any two S-morphisms $f, g : Y \to X$, the set $\{y \in Y \mid f(y) \equiv g(y)\}$ is closed in Y. A scheme X is said to be a b s o l u t e l y s e p a r a t e d, if it is separated as a $\mathbb{Z}$-scheme, i.e. if for *arbitrary* morphisms $f, g : Y \to X$ of schemes the set $\{y \in Y \mid f(y) \equiv g(y)\}$ is closed in Y. – A morphism $\varphi : X \to S$ of schemes is called s e p a r a t e d, if X is separated as an S-scheme with structure morphism φ.

Similar to topological spaces we have the following criterion for the separatedness of a scheme using the diagonal morphism $\Delta_X = (\operatorname{id}_X, \operatorname{id}_X) : X \to X \times_S X$:

4.D.9. Proposition *An S-scheme X is separated if and only if the image $\Delta_X(X)$ of the diagonal morphism $\Delta_X : X \to X \times_S X$ is closed in $X \times_S X$.*

For the proof see Exercise 4.D.15 (3).

Let $X = \operatorname{Spec} B$ be an affine A-scheme. Then the diagonal morphism Δ_X is the closed embedding corresponding to the multiplication map $\mu : B \otimes_A B \to B$ defined by $b \otimes c \mapsto bc$, and the scheme X can be identified with $\Delta_X(X) = \operatorname{Spec}((B \otimes_A B)/I_B) = \mathrm{V}(I_B) \subseteq \operatorname{Spec}(B \otimes_A B)$, where $I_B := \operatorname{Ker} \mu$. In particular:

4.D.10. Proposition *Every affine scheme is separated.*

The algebra $B \otimes_A B$ is called the e n v e l o p i n g a l g e b r a of B over A. The following lemma describes the ideal $I_B = \operatorname{Ker} \mu$.

4.D.11. Lemma *Let x_i, $i \in I$, be a system of generators of the commutative A-algebra $B = A[x_i \mid i \in I]$. Then the kernel I_B of the multiplication map $B \otimes_A B \to B$, $b \otimes c \mapsto bc$, is the ideal generated by the elements $x_i \otimes 1 - 1 \otimes x_i$, $i \in I$.*

PROOF. Obviously, $b \otimes 1 - 1 \otimes b \in I_B$ for all $b \in B$. Conversely, let $z := \sum_{j \in J} b_j \otimes c_j \in I_B$, i.e. $\sum_{j \in J} b_j c_j = 0$. Then $z = \sum_{j \in J} (1 \otimes c_j)(b_j \otimes 1 - 1 \otimes b_j)$. Therefore, the elements $b \otimes 1 - 1 \otimes b$, $b \in B$, generate the ideal I_B. Now the lemma follows from the fact that for an arbitrary ideal $\mathfrak{A} \subseteq B \otimes_A B$ the set of all elements $b \in B$ with $b \otimes 1 - 1 \otimes b \in \mathfrak{A}$ is an A-subalgebra of B. (Note: $bc \otimes 1 - 1 \otimes bc = (c \otimes 1)(b \otimes 1 - 1 \otimes b) + (1 \otimes b)(c \otimes 1 - 1 \otimes c)$.)   •

Now we define the most important class of schemes used in algebraic geometry:

4.D.12. Definition Let X be an S-scheme. X is said to be an a l g e b r a i c   s c h e m e
o v e r S or an a l g e b r a i c S-s c h e m e, if X is of finite type and separated over
S. A morphism $f : X \to S$ is called a l g e b r a i c if X is an algebraic S-scheme
with respect to f.

An affine A-scheme $\operatorname{Spec} R$ is algebraic (over A or over $\operatorname{Spec} A$) if and only if R is
an A-algebra of finite type. The objects studied in classical algebraic geometry are
the algebraic K-schemes, where K is a field, in particular, an algebraically closed
field, or most importantly the field $\mathbb{C}$ of complex numbers.

We end this section with some exercises intended to gain further familiarity with
products of schemes.

4.D.13. Exercise Let X, Y and Z be S-schemes. Then show that:

(1) $X \times_S S \cong X$ (where S is an S-scheme with respect to id_S).

(2) $X \times_S Y \cong Y \times_S X$ (C o m m u t a t i v i t y of p r o d u c t s).

(3) $(X \times_S Y) \times_S Z \cong X \times_S (Y \times_S Z)$ (A s s o c i a t i v i t y of p r o d u c t s).

(4) If U is an open subset of S, then $\varphi^{-1}(U) \cong X \times_S U$, where $\varphi : X \to S$ is the structure
morphism.

(5) If V and W are open subsets of X, then $V \cap W \cong (V \times_S W) \cap \Delta_X(X)$. (Note that
$V \times_S W$ is open in $X \times_S X$ by Lemma 4.D.2.)

4.D.14. Exercise Let X and Y be S-schemes with structure morphisms $\eta : X \to S$ and
$\xi : Y \to S$. Let Z be a Y-scheme with the structure morphism $\zeta : Z \to Y$. Thus Z is an
S-scheme with respect to $\xi \circ \zeta$. Then $X \times_S Z \cong (X \times_S Y) \times_Y Z$.

4.D.15. Exercise Let X, Y and Z be schemes over S and let $p, q : X \times_S X \to X$ be the
canonical projections.

(1) Let $z = \Delta_X(x)$ be an element of the diagonal $\Delta_X(X)$. Then show that $\Delta_X : X \to X \times_S X$
induces an isomorphism $\kappa(z) \to \kappa(x)$ which is inverse to the homomorphisms $\kappa(x) \to \kappa(z)$
induced by p and q. (**Hint:** If K and L are fields and $\alpha : K \to L$ and $\beta : L \to K$ are
homomorphisms with $\beta \circ \alpha = \mathrm{id}_K$, then α and β are isomorphisms.)

(2) Show that for an S-morphism $f : Z \to X \times_S X$ the following conditions are equivalent:
a) There is an S-morphism $g : Z \to X$ with $\Delta_X \circ g = f$ (i.e. f factorizes through the
diagonal morphism Δ_X). b) $p \circ f = q \circ f$.

(3) For S-morphisms $f, g : Y \to X$ with $(f, g) : Y \to X \times_S X$ and for a point $y \in Y$ show
that $f(y) \equiv g(y)$ if and only if $(f, g)(y) \in \Delta_X(X)$.

(4) For a point $z \in X \times_S X$ show that $z \in \Delta_X(X)$ if and only if $p(z) \equiv q(z)$.

(5) Give an example where $f(y) = g(y)$ for some $y \in Y$ but not $f(y) \equiv g(y)$.

4.D.16. Exercise Let X and Y be S-schemes. If X and Y are separated (resp. of finite type)
over S, then show that the product $X \times_S Y$ is also separated (resp. of finite type) over S. In
particular, *if X and Y are algebraic S-schemes, then the product $X \times_S Y$ is an algebraic
S-scheme too.*

4.D.17. Exercise Let X, S and S' be schemes and let $f : X \to S$ and $h : S \to S'$ be morphisms. Show that if the composition $h \circ f$ is separated, then f is also separated. Conversely, if f and h are separated, then $h \circ f$ is also separated. Particular cases: (1) If an S-scheme X is absolutely separated (i. e. separated over $\mathbb{Z}$), then X is separated over S. (2) By Proposition 4.D.10, an A-scheme scheme X, for which A a commutative ring, is separated if and only if X is absolutely separated.

4.D.18. Exercise Let X be an S-scheme. If X is separated over S, then show that every open subscheme of X is separated over S. Conversely, assume the following: For every pair (x, y) of points in X there exists an open subset $U \subseteq X$ containing x and y and separated over S. Then X is separated over S. (**Hint:** Show that for S-morphisms $f, g : Y \to X$ the set $\{ y \in Y \mid f(y) \not\equiv g(y) \}$ is open in Y.) In particular, X is absolutely separated (and hence separated over S) if for every pair (x, y) of points in X there exists an affine open subset $U \subseteq X$ containing x and y.

4.D.19. Exercise Let X be an S-scheme with structure morphism $\varphi : X \to S$ and let V_i, $i \in I$, be an open cover of S and let $U_i := \varphi^{-1}(V_i)$ be the corresponding open cover of X. Show that X is a separated S-scheme if and only if U_i is a separated V_i-scheme for every $i \in I$. (**Hint:** Note that $U_i \times_{V_i} U_i = U_i \times_S U_i, i \in I$, is an open cover of $X \times_S X$.)

4.D.20. Exercise Let X be an absolutely separated scheme. If $U, V \subseteq X$ are open and affine, then $U \cap V$ is also affine. (**Hint:** $U \times V \subseteq X \times X$. $U \cap V = (U \times V) \cap \Delta_X(X)$. Closed subsets of an affine scheme are affine.)

4.D.21. Exercise Let S be a scheme, X a reduced S-scheme and Y a separated S-scheme. Let f and g be two S-morphisms of X to Y such that there exists an open dense subset U of X with $f(x) = g(x)$ for all $x \in U$. Show that $f = g$.

4.D.22. Exercise Let $k \subseteq K, k \subseteq L$ be field extensions and $X := \operatorname{Spec} K$, $Y := \operatorname{Spec} L$. Show that:

(1) If the field extensions $k \subseteq K$ and $k \subseteq L$ are finitely generated, then $K \otimes_k L$ is a Noetherian ring of (Krull-)dimension $m := \min (\operatorname{trdeg}_k K, \operatorname{trdeg}_k L)$, i. e. the product $X \times_k Y$ of the zero-dimensional k-schemes X and Y is a Noetherian scheme of dimension m. (**Hint:** One can assume that K and L are purely transcendental, i. e. rational function fields over k in finitely many variables.)

(2) If K is not finitely generated over k, then the k-algebra $K \otimes_k K$ is not Noetherian, i. e. the product scheme $X \times_k X$ is not Noetherian. (**Hint:** Discuss the following three cases separately: a) K is purely transcendental over k; b) K is algebraic and separable over k; and c) K is algebraic and purely inseparable over k.)

4.D.23. Exercise Let K be a field, X an algebraic K-scheme and $x \in X$ a point. Show that:

(1) $\dim \overline{\{x\}} = \operatorname{trdeg}_K \kappa(x)$. (**Hint:** Use Theorem 3.B.17.)

(2) The following conditions are equivalent: a) $\{x\}$ is closed in X. b) $\{x\}$ is locally closed in X. c) $\kappa(x)$ is an algebraic field extension of K. d) $\kappa(x)$ is a finite field extension of K.

(3) X is of dimension ≤ 0 if and only if X is affine and $\Gamma(X, \mathcal{O}_X)$ is a finite K-algebra. In this case, $\Gamma(X, \mathcal{O}_X) = \prod_{x \in X} \mathcal{O}_{X,x}$.

4.E. Affine Morphisms

In this section we discuss a type of morphisms of schemes that are the most natural generalizations of morphisms of affine schemes.

4.E.1. Definition A morphism $f : X \to Y$ of schemes is called a f f i n e if one of the following equivalent conditions holds.

(1) For every open affine subset $V \subseteq Y$ the inverse image $f^{-1}(V) \subseteq X$ is also (open and) affine.

(2) There is a cover of Y by open affine subsets V_i, $i \in I$, such that for each $i \in I$ the inverse image $f^{-1}(V_i) \subseteq X$ is also (open and) affine.

The proof that the condition (2) implies condition (1) is postponed to Example 6.E.11. Since every morphism $f : X \to Y$ of affine schemes $X = \operatorname{Spec} B$ and $Y = \operatorname{Spec} A$ is clearly affine by condition (2), condition (1) implies that for *every* open affine subset $V \subseteq Y$ the inverse image $f^{-1}(V) \subseteq X$ is also affine.

Since morphisms of affine schemes are separated, by Exercise 4.D.19 *every affine morphism is separated.*

4.E.2. Example (F r o b e n i u s m o r p h i s m s) Let p be a prime number and A be a (commutative) ring of characteristic p. Then the map $A \to A$, $x \mapsto x^p$, is a ring homomorphism. This homomorphism and the corresponding morphism $\operatorname{Spec} A \to \operatorname{Spec} A$ is called the F r o b e n i u s (m o r p h i s m). The Frobenius is the identity on the topological space of $\operatorname{Spec} A$. Note that A is an integral A-algebra with respect to the Frobenius homomorphism.

More generally, let X be an $\mathbb{F}_p$-scheme, i. e. $\Gamma(X, \mathcal{O}_X)$ and hence all rings $\Gamma(U, \mathcal{O}_X)$, U non-empty and open in X, are of characteristic p. Then the F r o b e n i u s (m o r p h i s m) $F = F_X : X \to X$ is defined as $(\operatorname{id}_X, \Theta_F)$, where $\Theta_F(V) : \Gamma(V, \mathcal{O}_X) \to \Gamma(V, \mathcal{O}_X)$ is the Frobenius homomorphism. It follows directly from the definition that Frobenius morphisms are affine. They commute with every morphism $f : X \to Y$ of $\mathbb{F}_p$-schemes, i. e. $f \circ F_X = F_Y \circ f$.

If $q = p^n$ is a power of p and if X is an $\mathbb{F}_q$-scheme ($\mathbb{F}_q$ the Galois field with q elements), then the n-th iterate $F^n = F \circ \cdots \circ F$ (n times) is an affine $\mathbb{F}_q$-morphism $X \to X$. The fixed points of F_X^n, i. e. the set of points $x \in X$ with $F_X^n(x) \equiv \operatorname{id}_X(x)$, is exactly the set $X(\mathbb{F}_q)$ of $\mathbb{F}_q$-rational points of X. For another $\mathbb{F}_q$-scheme T one has to distinguish carefully between the two endomorphisms $F_{X_{(T)}}^n = (F_{X_{(T)}})^n$ and $(F_X^n)_{(T)}$ of $X_{(T)} = T \times_{\mathbb{F}_q} X$. The first is the n-th iterate of the Frobenius of $X_{(T)}$ and the second is the T-endomorphism of $X_{(T)}$ obtained from F_X^n by base change. Note that both these morphisms are affine (cf. Exercise 4.E.9). For example, let X be the affine space $\mathbb{A}_{\mathbb{F}_q}^m = \operatorname{Spec} \mathbb{F}_q[U_1, \ldots, U_m]$ and let T be the spectrum $\operatorname{Spec} K$ of an arbitrary field extension K of $\mathbb{F}_q$. Then $X_{(T)} = X_{(K)}$ is the affine space $\mathbb{A}_K^m = \operatorname{Spec} K[U_1, \ldots, U_m]$. The $\mathbb{F}_q$-morphism $F_{X_{(K)}}^n$ maps a K-rational point $(u_1, \ldots, u_m) \in K^m \subseteq \mathbb{A}_K^m$ to itself, but the K-morphism $(F_X^n)_{(T)} = (F_X^n)_{(K)}$ maps $(u_1, \ldots, u_m)$ to $(u_1^q, \ldots, u_m^q)$.

Now we define a special class of affine morphisms:

4.E.3. Definition A morphism $f : X \to Y$ of schemes is called f i n i t e , if one of the following equivalent conditions holds.

(1) For every open affine subset $V \subseteq Y$ the inverse image $U := f^{-1}(V) \subseteq X$ is also (open and) affine and $\Gamma(U)$ is finitely generated as a $\Gamma(V)$-module (i. e. the ring homomorphism $\Theta_f(V) : \Gamma(V) \to \Gamma(U)$ is finite).

(2) There is a cover of Y by open affine subsets $V_i, i \in I$, such that for each $i \in I$ the inverse image $U_i := f^{-1}(V_i) \subseteq X$ is also (open and) affine and $\Gamma(U_i)$ is finitely generated as a $\Gamma(V_i)$-module.

In order to show that condition (2) implies condition (1), we note first that f is an affine morphism. Furthermore, condition (2) immediately implies that the open affine subsets $V \subseteq Y$ for which $\Gamma(f^{-1}(V))$ is finite over $\Gamma(V)$ form a basis of the topology on Y. Hence it is enough to show the following: If B is an A-algebra and if $f_1, \ldots, f_n \in A$ with $\sum_{i=1}^{n} Af_i = A$ (i. e. $\bigcup_{i=1}^{n} D(f_i) = \operatorname{Spec} A$) such that B_{f_i} is a finite A_{f_i}-algebra for all $i = 1, \ldots, n$, then B is a finite A-algebra. This simple result has already been mentioned in Remark 4.C.4.

Obviously every finite morphism is separated and of finite type and hence algebraic. Furthermore we claim that:

4.E.4. Proposition *Every finite morphism $f : X \to Y$ of schemes is closed, i. e. f maps closed subsets of X to closed subsets of Y.*

PROOF. Let Z be a closed subset of X and $V_i, i \in I$, be an open affine cover of Y. Because of $f(Z) \cap V_i = f(Z \cap f^{-1}(V_i))$, we may assume that X and Y are affine schemes. In this case the proposition is a special case of the following result.   •

4.E.5. Lemma *Let $\varphi : A \to B$ be an integral extension of rings and let $\varphi^* :$ $\operatorname{Spec} B \to \operatorname{Spec} A$ be the corresponding morphism of schemes. Then $\varphi^*(V(\mathfrak{b})) = V(\varphi^{-1}(\mathfrak{b}))$ for every ideal $\mathfrak{b} \subseteq B$. In particular, φ^* is a closed map.*

PROOF. The induced homomorphism $A/\varphi^{-1}(\mathfrak{b}) \to B/\mathfrak{b}$ is injective and integral. Now apply Lemma 3.B.9 (3).   •

Let $(X, \mathcal{O}_X)$ be a scheme. Recall that open subsets $U \subseteq X$ (with structure sheaves $(U, \mathcal{O}_X | U)$) are the o p e n s u b s c h e m e s of X. A morphism $f : X \to Y$ of schemes is called an o p e n e m b e d d i n g, if f is an isomorphism of X onto an open subscheme V of Y composed with the canonical inclusion $V \hookrightarrow Y$. Closed embeddings are defined in the following way:

4.E.6. Definition A morphism $f : X \to Y$ of schemes is called a c l o s e d e m b e d d i n g if one of the following equivalent conditions holds.

(1) For every open affine subset $V \subseteq Y$ the inverse image $f^{-1}(V)$ is an affine (and open) subset of X and $\Theta_f(V) : \Gamma(V, \mathcal{O}_Y) \to \Gamma(f^{-1}(V), \mathcal{O}_X)$ is surjective.

(2) There is an open affine cover V_i, $i \in I$, of Y such that for each $i \in I$ the inverse image $f^{-1}(V_i)$ is an affine (and open) subset of X and $\Theta_f(V_i) : \Gamma(V_i, \mathcal{O}_Y) \to \Gamma(f^{-1}(V_i), \mathcal{O}_X)$ is surjective.

To prove that condition (2) implies condition (1) we first remark that the morphism f is affine and even finite. Furthermore condition (2) implies that the homomorphism $\Gamma(V, \mathcal{O}_Y) \to \Gamma(f^{-1}(V), \mathcal{O}_X)$ is surjective for a basis of the topology of Y. Therefore, it is enough to show that a ring homomorphism $A \to B$ is surjective if there exist elements $f_1, \ldots, f_n \in A$ with $A = \sum_{i=1}^n A f_i$ such that the induced homomorphisms $A_{f_i} \to B_{f_i}$, $i = 1, \ldots, n$, are surjective. This is simple.

4.E.7. Example (1) Let $X := \operatorname{Spec} R$ be an affine scheme and let $\mathfrak{a}$ be an ideal of R. The canonical surjection $\varphi : R \to R/\mathfrak{a}$ induces a closed embedding $\varphi^* : \operatorname{Spec} R/\mathfrak{a} \hookrightarrow \operatorname{Spec} R$. Usually one identifies $\operatorname{Spec} R/\mathfrak{a}$ with $V(\mathfrak{a})$. The scheme $(V(\mathfrak{a}), \widetilde{R/\mathfrak{a}})$ is called a c l o s e d s u b s c h e m e of X. So a closed subset $Z \subseteq X$ carries as many structure sheaves $\widetilde{R/\mathfrak{a}}$ as a closed subscheme as there are ideals $\mathfrak{a} \subseteq R$ with $Z = V(\mathfrak{a})$. A canonical choice for $\mathfrak{a}$ is the unique radical ideal $\mathfrak{J}(Z)$ with Z as zero-set. $(Z, \widetilde{R/\mathfrak{J}(Z)})$ is called the r e d u c e d c l o s e d s u b s c h e m e of X w i t h s u p p o r t Z.

(2) (R e d u c e d c l o s e d s u b s c h e m e s) The last construction can be generalized to an arbitrary scheme X. Let Z be a closed subset of X. Then Z possesses a canonical structure as a reduced closed subscheme of X in the following way. The subsets $U \cap Z$, U open and affine in X, form a basis for the topology on Z. We define $\Gamma(U \cap Z, \mathcal{O}_Z) = \Gamma(U)/\mathfrak{J}(U \cap Z)$, where $\mathfrak{J}(U \cap Z)$ is the ideal of sections in $\Gamma(U)$, that vanishes on $U \cap Z$. If $U = \operatorname{Spec} R$ and $U \cap Z = V(\mathfrak{a})$ for some ideal $\mathfrak{a} \subseteq R$, then $\mathfrak{J}(U \cap Z) = \mathfrak{J}(V(\mathfrak{a})) = \sqrt{\mathfrak{a}}$ by the formal Hilbert's Nullstellensatz 3.A.5 and $(U \cap Z, \mathcal{O}_Z \mid (U \cap Z)) \cong (\operatorname{Spec}(R/\sqrt{\mathfrak{a}}), (R/\sqrt{\mathfrak{a}})\widetilde{\;})$. This shows that $(Z, \mathcal{O}_Z)$ is a scheme and that the canonical morphism $Z \hookrightarrow X$ is a closed embedding. As in the affine case, the scheme $(Z, \mathcal{O}_Z)$ is called the r e d u c e d c l o s e d s u b s c h e m e of X w i t h s u p p o r t Z. The reduction X_{red} introduced in Exercise 4.B.6 (3) is obtained for $Z := X$.

In general a closed subset Z of a scheme X may carry many natural closed subscheme structures as already seen in the case when X is affine. We discuss about this more elaborately in Example 6.E.9.

4.E.8. Exercise Let $f : X \to Y$ be a morphism of schemes. Then show that f is a closed embedding if and only if f is a homeomorphism of X onto a closed subset of Y and for every open affine subset V of Y the ring homomorphism $\Theta_f(V) : \Gamma(V) \to \Gamma(f^{-1}(V))$ is surjective.

4.E.9. Exercise Show that:

(1) If $f : X \to Y$ and $g : Y \to Z$ are closed embeddings (or affine morphisms or finite morphisms resp.), the composition $g \circ f : X \to Z$ is also a closed embedding (or an affine morphism or a finite morphism resp.).

(2) Closed embeddings (or affine morphisms or finite morphisms resp.) are stable under base change, that is, if $f : X \to S$ is a closed embedding (or an affine morphism or a finite morphism resp.) and if $g : T \to S$ is any morphism, then the canonical projection $T \times_S X \to T$ is also a closed embedding (or an affine morphism or a finite morphism resp.).

(3) A morphism $f : X \to S$ is separated if and only if the diagonal morphism $\Delta_X : X \to X \times_S X$ is a closed embedding.

(4) If $X \to S$ and $Y \to S$ are closed embeddings (or affine morphisms or finite morphisms) then $X \times_S Y \to S$ is a closed embedding (or affine or finite, resp.).

4.E.10. Exercise Let $\mathcal{P}$ be a property of morphisms of schemes such that: a) A closed embedding has $\mathcal{P}$. b) A composition of two morphisms having $\mathcal{P}$ has $\mathcal{P}$. c) $\mathcal{P}$ is stable under base change. Then show that:

(1) The product of two morphisms (of schemes) having $\mathcal{P}$ has $\mathcal{P}$.

(2) If $f : X \to Y$ and $g : Y \to Z$ are two morphisms of schemes, and if $g \circ f$ has $\mathcal{P}$ and g is separated, then f has $\mathcal{P}$.

4.E.11. Exercise A morphism $f : X \to Y$ of schemes is called q u a s i - f i n i t e if for every point $y \in Y$, $f^{-1}(y)$ is a finite set. Show that a finite morphism is quasi-finite. (For a partial converse see Theorem 5.B.10.)

4.E.12. Exercise A morphism $f : X \to Y$ of schemes is called q u a s i - c o m p a c t if for every open quasi-compact subset $V \subseteq Y$ the inverse image $f^{-1}(V) \subseteq X$ is also quasi-compact. Show that f is quasi-compact if f is affine or of finite type.

CHAPTER 5: Projective Schemes

5.A. Projective Schemes

Let K be a field. The n-dimensional p r o j e c t i v e   s p a c e over K is classi-
cally defined as $\mathbb{P}^n(K) := (K^{n+1} \setminus \{0\})/\sim$, where the equivalence relation $\sim$
on $K^{n+1} \setminus \{0\}$ is given by $x \sim y$ if and only if $y = \lambda x$ for some $\lambda \in K^\times$.
The equivalence classes are the punctured lines $Kx \setminus \{0\}$, $x \in K^{n+1} \setminus \{0\}$. So
$\mathbb{P}^n(K)$ can also be viewed as the set of all 1-dimensional subspaces of the K-vector
space K^{n+1}. We denote the equivalence class of $(x_0, \ldots, x_n)$ by $\langle x_0, \ldots, x_n \rangle$. The
space $\mathbb{P}^n(K)$ is canonically covered by the $n+1$ affine spaces $U_0, \ldots, U_n$, where
$U_i := \{\langle x_0, \ldots, x_n \rangle \in \mathbb{P}^n(K) \mid x_i \neq 0\}$ is identified with the affine hyperplane
$x_i = 1$ in K^{n+1} and with the affine space K^n through the maps

$$\langle x_0, \ldots, x_i, \ldots, x_n \rangle \leftrightarrow (x_0/x_i, \ldots, x_i/x_i, \ldots, x_n/x_i) \leftrightarrow (x_l/x_i)_{0 \leq l \leq n, l \neq i} \,.$$

This motivates us to interpret U_i as the K-spectrum of the polynomial subalgebra

$$A_i := K[X_l/X_i \mid 0 \leq l \leq n, \, l \neq i]$$

of the rational function field $K(X_0, \ldots, X_n)$. The intersection $U_i \cap U_j$, $i \neq j$, as
part of U_i is $D_K(X_j/X_i)$ and as part of U_j it is $D_K(X_i/X_j)$. Hence, in both cases
the intersection $U_i \cap U_j$ is identified with the K-spectrum of the K-algebra

$$A_i[(X_j/X_i)^{-1}] = A_j[(X_i/X_j)^{-1}] = K[X_l/X_i, X_{l'}/X_j \mid l \neq i, l' \neq j] \,.$$

Now we consider the full spectra of the K-algebras A_i, $0 \leq i \leq n$, and glue
them together along the open subsets $D(X_j/X_i) \subseteq \operatorname{Spec} A_i$, $j \neq i$, in view of the
equality of the algebras $A_i[(X_j/X_i)^{-1}]$ and $A_j[(X_i/X_j)^{-1}]$. In this way we get the
p r o j e c t i v e   s p a c e

$$\mathbb{P}_K^n$$

as a K-scheme. The classical projective space $\mathbb{P}^n(K)$ is then the space $\mathbb{P}_K^n(K)$
of K-rational points of $\mathbb{P}_K^n$. The construction of this projective space has a far-
reaching generalization. It depends on the observation that the algebras $A_i =
K[X_l/X_i \mid 0 \leq l \leq n, l \neq i]$ used above for the construction of $\mathbb{P}_K^n$ are nothing
but the homogeneous parts of degree 0 of the graded Laurent-polynomial algebras
$K[X_0, \ldots, X_n]_{X_i} = K[X_0, \ldots, X_n, X_i^{-1}] = K[X_0, \ldots, X_i^{\pm 1}, \ldots, X_n]$ (with the
classical grading $\deg X_i = 1$, $i = 0, \ldots, n$).

We start with some simple results on $\mathbb{Z}$-graded (commutative) rings $R = \bigoplus_{m \in \mathbb{Z}} R_m$
and formulate them as exercises. We refer to Section 1.D for the general notations.
The set of homogeneous prime ideals of R is called the h o m o g e n e o u s   s p e c t r u m
of R and is denoted by

$$\mathrm{h\text{-}Spec}\, R \,.$$

It is part of the full spectrum $\operatorname{Spec} R$ and is considered with the topology induced from the Zariski topology of $\operatorname{Spec} R$. For a subset $E \subseteq R$ we denote by $\mathrm{h\text{-}V}(E)$ and $\mathrm{h\text{-}D}(E)$ the intersections of $V(E)$ and $D(E)$ with $\mathrm{h\text{-}Spec}\, R$. R is called p o s i t i v e l y g r a d e d if $R_m = 0$ for $m < 0$. In this case $R_+ := \bigoplus_{m>0} R_m$ is an ideal of R with a canonical isomorphism $R/R_+ \cong R_0$. The ideal R_+ is called the i r r e l e v a n t i d e a l of R. More generally, $R_{(k)} := \bigoplus_{m \geq k} R_m$ is an ideal in R for every $k \in \mathbb{N}$ with the residue class algebra $R/R_{(k)} \cong R_0 \oplus \cdots \oplus R_{k-1}$ and the multiplication modified as $R_i R_j := 0$ if $i + j \geq k$.

A homomorphism $\varphi : R \to S$ of $\mathbb{Z}$-graded rings $R = \bigoplus_m R_m$ and $S = \bigoplus_m S_m$ is called h o m o g e n e o u s (or g r a d e d) if $\varphi(R_m) \subseteq S_m$ for all $m \in \mathbb{Z}$. A homomorphism $h : M \to N$ of graded R-modules $M = \bigoplus_m M_m$ and $N = \bigoplus_m N_m$ is called h o m o g e n e o u s of degree $d \in \mathbb{Z}$ if $h(M_m) \subseteq h(N_{m+d})$ for all $m \in \mathbb{Z}$. Multiplication in M by a homogeneous element $a \in R_d$ of degree d is a homogeneous homomorphism of degree d.

5.A.1. Exercise Let $R = \bigoplus_{m \in \mathbb{Z}} R_m$ be a ($\mathbb{Z}$-)graded ring. Show:

(1) If a homogeneous element $x \in R_d$ is a unit of R, then $x^{-1} \in R_{-d}$.

(2) The set of degrees $m \in \mathbb{Z}$ for which $R_m \cap R^\times \neq \emptyset$ is a subgroup of $\mathbb{Z}$.

(3) Assume that there exists a homogeneous unit x of degree 1 in R (i. e. the subgroup of $\mathbb{Z}$ defined in (2) is $\mathbb{Z}$ itself). Then $R = R_0[x, x^{-1}]$ and x is transcendental over R_0 if $R_0 \neq 0$. In other words: R is then the algebra of Laurent-polynomials in one variable over R_0.

5.A.2. Exercise Let $R = \bigoplus_{m \in \mathbb{Z}} R_m$ be a ($\mathbb{Z}$-)graded ring and let $S \subseteq R$ be a multiplicatively closed subset of homogeneous elements. Show that:

(1) The ring $R_S = S^{-1} R$ of fractions inherits a grading from R and the canonical homomorphism $R \to R_S$ is graded. The homogeneous elements of degree m in R_S are the fractions a/s, where $a \in R$ and $s \in S$ are homogeneous with $\deg a - \deg s = m$.

(2) If $M = \bigoplus_{m \in \mathbb{Z}} M_m$ is a graded R-module, then the module $M_S = S^{-1} M$ inherits a grading from M such that M_S is a graded R_S-module and the canonical homomorphism $M \to M_S$ is homogeneous (of degree 0).

(3) The canonical topological embedding $\operatorname{Spec} R_S \to \operatorname{Spec} R$ corresponding to the canonical homogeneous homomorphism $R \to R_S$ of graded rings induces a topological embedding $\mathrm{h\text{-}Spec}\, R_S \to \mathrm{h\text{-}Spec}\, R$. Its image is the set of all homogeneous prime ideals $\mathfrak{p}$ of R with $\mathfrak{p} \cap S = \emptyset$. In particular, for a homogeneous element $f \in R$ the open set $\mathrm{h\text{-}D}(f) \subseteq \mathrm{h\text{-}Spec}\, R$ can be identified with $\mathrm{h\text{-}Spec}\, R_f$.

(4) If $T \subseteq R$ is another multiplicatively closed set of homogeneous elements, then the canonical homogeneous homomorphism $R_S \to R_T$ is an isomorphism if and only if every element of T divides some element of S, i. e. if and only if T is a subset of the h o m o g e n e o u s s a t u r a t i o n of S (containing all homogeneous elements of R which divide some element of S).

5.A.3. Exercise Let $R = \bigoplus_{m \in \mathbb{Z}} R_m$ be a ($\mathbb{Z}$-)graded ring. Show that:

(1) The nilradical $\mathfrak{n}_R$ of R is a homogeneous ideal of R. (**Hint:** If $\sum_{m \leq m_0} a_m$ is nilpotent, then a_{m_0} is nilpotent.)

(2) If $\mathfrak{a}$ is a homogeneous ideal of R, then the radical ideal $\sqrt{\mathfrak{a}}$ is also homogeneous.

5.A.4. Exercise Let $R = \bigoplus_{m \in \mathbb{Z}} R_m$ be a ($\mathbb{Z}$-)graded ring.

(1) A homogeneous ideal $\mathfrak{p}$ of R is prime if and only if $\mathfrak{p} \neq R$ and for *homogeneous* elements $f, g \in R$, $fg \in \mathfrak{p}$ implies $f \in \mathfrak{p}$ or $g \in \mathfrak{p}$.

(2) Let $\mathfrak{p}$ be a prime ideal in R (not necessarily homogeneous). Then the homogeneous elements of $\mathfrak{p}$ generate a (homogeneous) prime ideal in R. In particular, all the minimal prime ideals of R are homogeneous. (Since the nilradical $\mathfrak{n}_R$ is the intersection of the minimal prime ideals of R, this solves Exercise 5.A.3 once again. This also shows that the intersection of all homogeneous prime ideals of R is the same as the intersection $\mathfrak{n}_R$ of all prime ideals of R. From this follows the graded version of the formal Hilbert's Nullstellensatz 3.A.5: *If* $\mathfrak{a} \subseteq R$ *is a homogeneous ideal, then* $\sqrt{\mathfrak{a}} = \bigcap_{\mathfrak{p} \in h\text{-}V(\mathfrak{a})} \mathfrak{p}$.)

5.A.5. Exercise Let $R = \bigoplus_{m \in \mathbb{Z}} R_m$ be a ($\mathbb{Z}$-)graded ring. Assume that there is a homogeneous unit x of degree $d \neq 0$ in R (i. e. the subgroup of $\mathbb{Z}$ introduced in Exercise 5.A.1 (2) is not the zero group). Show that:

(1) If R_0 is an integral domain, then the nilradical $\mathfrak{n}_R$ is a prime ideal. If $\mathfrak{q}$ is a prime ideal in R_0, then $\sqrt{\mathfrak{q}R}$ is a prime ideal of R.

(2) The inclusion $\iota : R_0 \to R$ induces a homeomorphism $\iota^* : h\text{-}\mathrm{Spec}\, R \to \mathrm{Spec}\, R_0$ (with $\iota^*(\mathfrak{p}) = \mathfrak{p} \cap R_0 = \mathfrak{p}_0$). (**Hint:** The inverse homeomorphism is given by $\mathfrak{q} \mapsto \sqrt{\mathfrak{q}R}$.)

5.A.6. Exercise Let f be a homogeneous element of degree 1 in the ($\mathbb{Z}$-)graded ring $R = \bigoplus_{m \in \mathbb{Z}} R_m$. Show that:

(1) The projection $R \to R/R(f-1)$ induces a ring homomorphism $R_f \to R/R(f-1)$ and by restriction a canonical homomorphism $\varphi : (R_f)_0 \to R/R(f-1)$.

(2) The additive homomorphism $R \to (R_f)_0$ with $a_m \mapsto a_m/f^m$ for $m \in \mathbb{Z}$ and $a_m \in R_m$ is a ring homomorphism and induces a ring homomorphism $\psi : R/R(f-1) \to (R_f)_0$.

(3) φ and ψ are isomorphisms inverse to each other. In particular, $(R_f)_0 \cong R/R(f-1)$.

(4) If $g \in R_d$, $d > 0$, then $(R_g)_0 \cong R^{[d]}/R^{[d]}(g-1)$, where $R^{[d]} = \bigoplus_{m \in \mathbb{Z}} R_{dm}$ is the so-called d-th Veronese transform of R.

(**Remark:** The ring $R/R(f-1)$ in (1) is said to be obtained from R by setting $f = 1$. Note that $R_f = (R_f)_0[f/1, 1/f]$ and $f/1$ is transcendental over $(R_f)_0$ (if f is not nilpotent, cf. Exercise 5.A.1 (3)).)

Now let $R = \bigoplus_{m \in \mathbb{N}} R_m$ be a positively graded ring. Then by Exercises 5.A.2 (3) and 5.A.5 (2) we have canonical identifications

$$h\text{-}\mathrm{D}(f) = h\text{-}\mathrm{Spec}\, R_f = \mathrm{Spec}((R_f)_0).$$

Therefore, the open set

$$\mathrm{Proj}\, R := \bigcup_f h\text{-}\mathrm{D}(f) = h\text{-}\mathrm{D}(R_+) \subseteq h\text{-}\mathrm{Spec}\, R,$$

where f runs through the set of homogeneous elements of R of *positive* degrees, is covered by the open affine spectra $h\text{-}\mathrm{D}(f) = \mathrm{Spec}((R_f)_0)$.

5.A.7. Definition Let $R = \bigoplus_{m \in \mathbb{N}} R_m$ be a positively graded ring. Then

$$\mathrm{Proj}\, R = \{\mathfrak{p} \in h\text{-}\mathrm{Spec}\, R \mid R_+ \not\subseteq \mathfrak{p}\} = h\text{-}\mathrm{D}(R_+) \subseteq h\text{-}\mathrm{Spec}\, R$$

is called the projective spectrum of R.

For a family $\mathcal{F} = (f_j)_{j \in J}$ of elements $f_j \in R$ we set

$$V_+(\mathcal{F}) := V(\mathcal{F}) \cap \mathrm{Proj}\, R = \bigcap_{j \in J} V_+(f_j)$$

and

$$D_+(\mathcal{F}) := D(\mathcal{F}) \cap \mathrm{Proj}\, R = \bigcup_{j \in J} D_+(f_j).$$

Note that $V_+(\mathcal{F}) = V_+(\mathcal{F}')$ and $D_+(\mathcal{F}) = D_+(\mathcal{F}')$, where $\mathcal{F}'$ is the family of homogeneous components of all the elements of $\mathcal{F}$. The irreducible closed subsets of $\mathrm{Proj}\, R$ are the sets $V_+(\mathfrak{p})$, $\mathfrak{p} \in \mathrm{Proj}\, R$, where $\mathfrak{p}$ runs through the set of homogeneous prime ideals of R. The standard open sets $D_+(f) = \mathrm{h\text{-}D}(f)$, f homogeneous of positive degree, are a basis of the topology of $X := \mathrm{Proj}\, R$. In view of $D_+(f) = \mathrm{Spec}((R_f)_0)$, we want to define a structure sheaf $\mathcal{O}_X$ on X with the property $\mathcal{O}_X|D_+(f) = \widetilde{(R_f)_0}$. This is done by setting

$$\Gamma(D_+(f), \mathcal{O}_X) := (R_f)_0$$

for the standard open sets $D_+(f)$ and the canonical restrictions $(R_f)_0 \to (R_g)_0$ in case $D_+(g) \subseteq D_+(f)$. Note that for homogeneous elements $f, g \in R$ of positive degrees the condition $D_+(g) (= \mathrm{h\text{-}D}(g)) \subseteq D_+(f) (= \mathrm{h\text{-}D}(f))$ is equivalent to both the conditions $V_+(f) \subseteq V_+(g)$ and $\mathrm{h\text{-}V}(f) \subseteq \mathrm{h\text{-}V}(g)$ and hence to $\sqrt{Rg} \subseteq \sqrt{Rf}$ by Hilbert's Nullstellensatz (cf. Exercise 5.A.4 (2)), i. e. to the condition that the homogeneous saturation of the multiplicatively closed set S_f is contained in the homogeneous saturation of S_g. This gives the canonical homogeneous homomorphism $R_f \to R_g$ and hence the restriction map $(R_f)_0 \to (R_g)_0$.

Obviously $\mathcal{O}_X$ is a presheaf of rings. But it is actually a sheaf. More precisely:

5.A.8. Theorem *Let R be a positively graded ring and $X := \mathrm{Proj}\, R$. Then the presheaf $\mathcal{O}_X$ with $\mathcal{O}_X(D_+(f)) = (R_f)_0$, f homogeneous of positive degree, is a sheaf. Moreover,*

$$(D_+(f), \mathcal{O}_X|D_+(f)) = (\mathrm{Spec}((R_f)_0), \widetilde{(R_f)_0}).$$

In particular, $(X, \mathcal{O}_X)$ is a scheme.

PROOF. Since $\widetilde{(R_f)_0}$ is a sheaf (and not only a presheaf), it is enough to prove the equality $\mathcal{O}_X|D_+(f) = \widetilde{(R_f)_0}$. But this is a simple consequence of the definitions.●

The scheme $(X, \mathcal{O}_X)$ in Theorem 5.A.8 is called the p r o j e c t i v e   s c h e m e belonging to the graded ring R. Since all the restriction maps $\rho_{D_+(g)}^{D_+(f)} (R_f)_0 \to (R_g)_0$ are R_0-algebra homomorphisms, the projective scheme $(X, \mathcal{O}_X)$ is an R_0-scheme. The s t r u c t u r e   m o r p h i s m

$$X = \mathrm{Proj}\, R \to \mathrm{Spec}\, R_0$$

maps a point $\mathfrak{p} \in \mathrm{Proj}\, R$ to $\mathfrak{p} \cap R_0 = \mathfrak{p}_0 \in \mathrm{Spec}\, R_0$. By Exercise 5.A.6 (4) the open affine subset $D_+(f) \subseteq \mathrm{Proj}\, R$, f homogeneous of degree $d > 0$, is isomorphic to $\mathrm{Spec}\, R^{[d]}/(f - 1)$. The stalk of the structure sheaf $\mathcal{O}_{\mathrm{Proj}\, R}$ at a point $\mathfrak{p} \in \mathrm{Proj}\, R$

is the local ring $(R_{[\mathfrak{p}]})_0$, where $R_{[\mathfrak{p}]}$ is the ring of fractions with respect to the multiplicatively closed set of all *homogeneous* elements of R not in $\mathfrak{p}$.

The projective A-scheme belonging to the polynomial algebra $A[X_0, \ldots, X_n]$ with the standard grading is called the p r o j e c t i v e   s p a c e over A of (relative) dimension n and is denoted by
$$\mathbb{P}_A^n.$$
Since $\mathbb{P}_A^n = \operatorname{Proj} A[X_0, \ldots, X_n] = \bigcup_{i=0}^n \operatorname{D}_+(X_i)$, the projective scheme $\mathbb{P}_A^n$ is covered by the $n+1$ open affine schemes $\operatorname{D}_+(X_i)$. Their algebras of global sections are the 0-th components $A[X_0, \ldots, X_i^{\pm 1}, \ldots, X_n]_0 = A[X_j/X_i \mid j \neq i] \cong A[X_0, \ldots, X_n]/(X_i - 1)$. Therefore, each of these affine schemes is isomorphic to the affine space $\mathbb{A}_A^n$. The open cover $\bigcup_{i=0}^n \operatorname{D}_+(X_i)$ is called the s t a n d a r d a f f i n e   c o v e r of $\mathbb{P}_A^n$.

5.A.9. Exercise Let K be a field, $n \in \mathbb{N}^*$ and $D_n := \det(X_{ij})_{1 \le i, j \le n}$. Let $R = \bigoplus_{m \in \mathbb{Z}} R_m$ be the graded K-algebra $K\big[X_{ij}, 1 \le i, j \le n; 1/D_n\big] = K\big[X_{ij}, 1 \le i, j \le n\big]_{D_n}$, where all the indeterminates have degree 1 and hence D_n the degree n. The general linear group $\mathrm{GL}_n(K)$ can be identified with the set $\mathbf{GL}_{n,K}(K) = \operatorname{Hom}_K(\operatorname{Spec} K, \mathbf{GL}_{n,K})$ of K-rational points of the affine K-scheme
$$\mathbf{GL}_{n,K} := \operatorname{Spec} R = \operatorname{D}(D_n) \subseteq \mathbb{A}_K^{n^2}.$$

(1) Show that in a similar way the projective linear group $\mathrm{PGL}_n(K)$ can be identified with the set $\mathbf{PGL}_{n,K}(K) := \operatorname{Hom}_K(\operatorname{Spec} K, \mathbf{PGL}_{n,K})$, where $\mathbf{PGL}_{n,K}$ is the affine K-scheme
$$\mathbf{PGL}_{n,K} := \operatorname{Spec} R_0 = \operatorname{D}_+(D_n) \subseteq \mathbb{P}_K^{n^2 - 1}.$$

By Exercise 5.A.6 (4) one has $R_0 \cong K\big[X_{ij}, 1 \le i, j \le n\big]^{[n]}/(D_n - 1)$.

(2) Try to describe $K\big[X_{ij}, 1 \le i, j \le 2\big]^{[2]}/(D_2 - 1)$ explicitly, for example, for $K = \mathbb{C}$.

For a non-standard grading of the polynomial algebra $A[X_0, \ldots, X_n]$ with $\deg X_i = \gamma_i > 0$, $i = 0, \ldots, n$, the projective scheme $\operatorname{Proj} A[X_0, \ldots, X_n]$ is called the w e i g h t e d (or anisotropic) p r o j e c t i v e   s p a c e over A defined by the t u p l e   o f   w e i g h t s $\gamma = (\gamma_0, \ldots, \gamma_n)$ and is denoted by
$$\mathbb{P}_{\gamma,A}^n.$$
Again the scheme $\mathbb{P}_{\gamma,A}^n$ is covered by the open affine schemes $\operatorname{D}_+(X_i)$, $i = 0, \ldots, n$, but now the algebras of sections $\Gamma(\operatorname{D}_+(X_i), \mathcal{O}_{\mathbb{P}_{\gamma,A}^n}) = A[X_0, \ldots, X_i^{\pm 1}, \ldots, X_n]_0$ are, in general, not isomorphic to polynomial algebras over A. However, $\mathbb{P}_{\gamma,A}^1 \cong \mathbb{P}_A^1$ for arbitrary $\gamma = (\gamma_0, \gamma_1) \in (\mathbb{N}^*)^2$, since $A[X_0^{\pm 1}, X_1]_0 = A[T]$, $A[X_0, X_1^{\pm 1}]_0 = A[T^{-1}]$, $T := X_1^{\gamma_0/d}/X_0^{\gamma_1/d}$, $d := \operatorname{GCD}(\gamma_0, \gamma_1)$.

5.A.10. Proposition *For an arbitrary positively graded ring R the projective scheme $\operatorname{Proj} R$ is absolutely separated.*

PROOF. We use Exercise 4.D.18. Since the standard open sets $\operatorname{D}_+(f)$, $f \in R$ homogeneous of positive degree, are affine by Theorem 5.A.8, the proposition is a consequence of the following more general lemma.　　　　　　　　　　　　　　●

5.A.11. Lemma *Let R be a positively graded ring and let $x_1, \ldots, x_m \in \operatorname{Proj} R$.*
Then there exists a homogeneous element $f \in R$ of positive degree such that
$x_1, \ldots, x_m \in D_+(f)$.

PROOF. We have to show: If $\mathfrak{p}_1, \ldots, \mathfrak{p}_m$ are homogeneous prime ideals of R
with $R_+ \not\subseteq \mathfrak{p}_j$, then there exists a homogeneous element $f \in R_+$ with $f \notin$
$\mathfrak{p}_1 \cup \cdots \cup \mathfrak{p}_m$. But the proof of this statement is the same as that of Lemma 3.B.31.
In the present situation the elements a, b used in the proof of Lemma 3.B.31 can
be chosen homogeneous of the same positive degree.   •

Let $\varphi : R \to S$ be a homogeneous homomorphism of positively graded rings.
The corresponding map $\varphi^* : \operatorname{Spec} S \to \operatorname{Spec} R$ induces by restriction a continuous
map $\operatorname{h-}\varphi^* : \operatorname{h-Spec} S \to \operatorname{h-Spec} R$. The preimage of the open set $\operatorname{Proj} R =$
$\operatorname{h-D}(R_+) \subseteq \operatorname{h-Spec} R$ is the open set $\operatorname{h-D}(R_+ S)$.[1]) Hence $\operatorname{h-}\varphi^*$ induces a
continuous map $\operatorname{Proj} \varphi$ of the open subset $U := \operatorname{h-D}(R_+ S) \cap \operatorname{Proj} S = D_+(R_+ S)$
of $\operatorname{Proj} S$ in $\operatorname{Proj} R$:

$$\operatorname{Proj} \varphi : U = D_+(R_+ S) \to \operatorname{Proj} R .$$

$\operatorname{Proj} \varphi$ *extends in a natural way to a morphism of schemes*: For an arbitrary homogeneous element $f \in R_+$ the preimage of the standard open set $D_+(f) \subseteq \operatorname{Proj} R$
is the standard open set $D_+(\varphi(f)) \subseteq U$ and the ring homomorphism φ extends to
a homomorphism $R_f \to S_f = S_{\varphi(f)}$. By restriction this gives the desired homomorphism

$$\Theta_{\operatorname{Proj} \varphi}(D_+(f)) : \Gamma(D_+(f), \mathcal{O}_{\operatorname{Proj} R}) = (R_f)_0 \to (S_f)_0 = \Gamma(D_+(\varphi(f)), \mathcal{O}_U)$$

of the rings of sections. Thus we have shown:

5.A.12. Theorem *Let $\varphi : R \to S$ be a homogeneous homomorphism of positively
graded rings. Then φ induces in a canonical way a morphism*

$$\operatorname{Proj} \varphi = (\operatorname{Proj} \varphi, \Theta_{\operatorname{Proj} \varphi}) : D_+(R_+ S) \to \operatorname{Proj} R$$

of schemes which is affine.

The closed complement $V_+(R_+ S) = \operatorname{Proj} S \setminus D_+(R_+ S)$ in $\operatorname{Proj} S$ *where the morphism* $\operatorname{Proj} \varphi$ *is not defined* is called the c e n t r e of the homomorphism $\varphi : R \to S$
or of the map $\operatorname{Proj} \varphi$. Therefore, $\operatorname{Proj} \varphi$ is defined everywhere on $\operatorname{Proj} S$ if and only
if its centre is empty. For a charaterization of this situation see Exercise 5.A.16.

5.A.13. Example Let K be a field and let $L_0, \ldots, L_r$ be K-linearly independent linear
forms in the standardly graded polynomial algebra $K[X_0, \ldots, X_n]$. Then the homogeneous
inclusion homomorphism $h : K[L_0, \ldots, L_r] \hookrightarrow K[X_0, \ldots, X_n]$ induces a morphism of
schemes $\mathbb{P}_K^n \setminus V_+(L_0, \ldots, L_r) \to \operatorname{Proj}(K[L_0, \ldots, L_r]) \cong \mathbb{P}_K^r$. Its centre $V_+(L_0, \ldots, L_r)$
is the linear closed subscheme $\mathbb{P}_K^{n-r-1} \cong \operatorname{Proj}(K[X_0, \ldots, X_n]/(L_0, \ldots, L_r)) \subseteq \mathbb{P}_K^n$. This
is the classical situation discussed in undergraduate courses. If $r = n - 1$, then the center
is just the K-rational singleton $L_0 = \cdots = L_{n-1} = 0$. Up to linear transformation one can
always assume that the linear forms $L_0, \ldots, L_r$ are the indeterminates $X_0, \ldots, X_r$.

---

[1]) Recall that we denote by $R_+ S$ the extended ideal $\varphi(R_+)S \subseteq S_+$.

5.A.14. Example (P r o j e c t i v e   a n d   a f f i n e   c o n e s) Let $R = \bigoplus_{m \geq 0} R_m$ be a positively graded ring and let $R[X]$ be the polynomial algebra over R whose grading is given by $\deg a_m X^i = m + i$ for $a_m \in R_m$. The canonical inclusion $h : R \hookrightarrow R[X]$ induces the morphism $\operatorname{Proj} h : D_+(R_+[X]) = \operatorname{Proj} R[X] \setminus V_+(R_+[X]) \to \operatorname{Proj} R$. Its center $V_+(R_+[X]) = \operatorname{Proj}((R/R_+)[X]) = \operatorname{Proj}(R_0[X]) = \mathbb{P}^0(R_0) = \operatorname{Spec} R_0$ is called the v e r t e x (s c h e m e). Note that for an arbitrary commutative ring A the structure morphism $\mathbb{P}^0_A = \operatorname{Proj} A[X_0] \to \operatorname{Spec} A$ is an isomorphism of schemes. The projective scheme $\operatorname{Proj} R[X]$ associated in this way with $\operatorname{Proj} R$ is called the p r o j e c t i v e   c o n e $\operatorname{Proj} R[X]$ over $\operatorname{Proj} R$ with respect to the coordinate ring R.

The projective cone $\operatorname{Proj} R[X]$ contains $D_+(X) = \operatorname{Spec} R[X^{\pm 1}]_0 = \operatorname{Spec}(R[X]/(X - 1)) = \operatorname{Spec} R$ as an affine open subscheme which is called the a f f i n e   c o n e of $\operatorname{Proj} R$ with respect to the given coordinate ring R. Since $V_+(R_+[X]) \cap V_+(X) = V_+(R_+[X] + (X)) = V_+(R[X]_+) = \emptyset$, the affine cone $D_+(X)$ contains the vertex of the projective cone. If $D_+(X)$ is identified with $\operatorname{Spec} R$, then the vertex is identified with $V(R_+) \subseteq \operatorname{Spec} R$. The projection $\operatorname{Proj} h$ induces a morphism $D_+(X) \setminus V_+(R_+[X]) = D_+(R_+[X]X) \to \operatorname{Proj} R$ which is also affine. Over a standard open affine set $D_+(f) \subseteq \operatorname{Proj} R$, $f \in R_+$ homogeneous, it identifies as the morphism of affine schemes corresponding to the canonical inclusion $(R_f)_0 \hookrightarrow R_f$.

For $f \in R_1$ the ring R_f can be identified as the ring $(R_f)_0[f/1, 1/f]$ of Laurent polynomials and the ring $(R_f[X])_0$ as the ring of polynomials $(R_f)_0[1/f] \subseteq R_f$. (In general, for $f \in R_d$, $d > 0$, $(R_f[X])_0$ can be identified as $\bigoplus_{m \leq 0}(R_f)_m \subseteq R_f$.) It follows that in case R is standardly graded, the fibre of $\operatorname{Proj} h$ over a point $x \in \operatorname{Proj} R$ is the affine line $\operatorname{Spec}(\kappa(x)[X^{-1}]) \cong \mathbb{A}^1_{\kappa(x)}$ and the fibre over x of $\operatorname{Proj} h$ restricted to the affine cone is the punctured affine line $\operatorname{Spec}(\kappa(x)[X, X^{-1}]) \cong \mathbb{A}^1_{\kappa(x)} \setminus \{0\}$. The reader should also describe the fibres for non-standard gradings.

We mention that in general *the cones depend on the chosen graded coordinate ring R for the representation of the scheme* $\operatorname{Proj} R$. For instance take a field K and the two representations $\operatorname{Proj} K[X_0, X_1] \cong \operatorname{Proj}(K[Y_0, Y_1, Y_2]/(Y_1^2 - Y_0 Y_2)) \cong \mathbb{P}^1_K$ of $\mathbb{P}^1_K$ (cf. the end of Exercise 5.A.23). The two projective cones $\operatorname{Proj} K[X_0, X_1, X_2] \cong \mathbb{P}^2_K$ and $\operatorname{Proj}(K[Y_0, Y_1, Y_2, Y_3]/(Y_1^2 - Y_0 Y_2)) = V_+(Y_1^2 - Y_0 Y_2) \subseteq \mathbb{P}^3_K$ are non-isomorphic K-schemes. The vertex $v \in D_+(Y_3) \cap V_+(Y_1^2 - Y_0 Y_2) \cong \operatorname{Spec}(K[Y_0, Y_1, Y_2]/(Y_1^2 - Y_0 Y_2))$ of the second cone is the 'singular' point $(0, 0, 0) \in V(Y_1^2 - Y_0 Y_2) \subseteq \mathbb{A}^3_K$. Its stalk is the non-regular local ring $K[Y_0, Y_1, Y_2]_\mathfrak{m}/(Y_1^2 - Y_0 Y_2)$, where $\mathfrak{m} = (Y_0, Y_1, Y_2)$ is the maximal ideal corresponding to the zero point. (Cf. Example 2.C.5 for $K = \mathbb{R}$ or $K = \mathbb{C}$.) The affine cones are the affine plane $\operatorname{Spec} K[X_0, X_1] \cong \mathbb{A}^2_K$ and the 'singular' affine scheme $\operatorname{Spec}(K[Y_0, Y_1, Y_2]/(Y_1^2 - Y_0 Y_2))$ respectively.

5.A.15. Exercise Let $R = \bigoplus_{m \in \mathbb{N}} R_m$ be a positively graded ring.

(1) Show that the following conditions are equivalent: a) $\operatorname{Proj} R = \emptyset$. b) $\operatorname{h-V}(R_+) = \operatorname{h-Spec} R$. c) $R_+ \subseteq \mathfrak{n}_R$. – If the ideal R_+ is finitely generated (in particular, if R is Noetherian), these conditions are equivalent to the following conditions: d) There is a $k \in \mathbb{N}$ with $R_{(k)} = 0$. e) R is a finite R_0-module. (**Hint**: If R_+ is finitely generated and if $R_+ \subseteq \mathfrak{n}_R$, then $R_+^n = 0$ for some $n \in \mathbb{N}$ and the modules R_+^k/R_+^{k+1} are also finitely generated and annihilated by R_+ and hence finite R_0-modules. – **Remark**: The condition that the ideal R_+ is finitely generated is discussed in Exercise 5.A.19.)

(2) Let $\mathfrak{a} = \bigoplus_{m \in \mathbb{N}} \mathfrak{a}_m$ be a homogeneous ideal in R. The closed set $V_+(\mathfrak{a}) \subseteq \operatorname{Proj} R$ can be identified with the space $\operatorname{Proj}(R/\mathfrak{a})$. Show that the following conditions are equivalent: a) $\operatorname{Proj}(R/\mathfrak{a}) = \emptyset$. b) $V_+(\mathfrak{a}) = \emptyset$. c) $R_+ \subseteq \sqrt{\mathfrak{a}}$. – If the ideal R_+ is finitely generated,

these conditions are equivalent to: d) There is an $m_0 \in \mathbb{N}$ with $R_m = \mathfrak{a}_m$ for $m \geq m_0$. e) $R/\mathfrak{a}$ is a finite R_0-module.

5.A.16. Exercise Let $\varphi : R \to S$ be a homogeneous homomorphism of positively graded rings. Show that the following conditions are equivalent: a) $\operatorname{Proj}\varphi$ is defined on the whole of $\operatorname{Proj} S$, i.e. the centre of φ is empty. b) $V_+(R_+S) = \emptyset$. c) $S_+ \subseteq \sqrt{R_+S}$. – If the ideal S_+ is finitely generated, these conditions are equivalent to: d) There is an $m_0 \in \mathbb{N}$ with $S_m = (R_+S)_m$ for $m \geq m_0$. e) S/R_+S is a finite S_0-algebra. – If, in addition, S_0 is a finite R_0-algebra, these conditions are equivalent to: f) S/R_+S is a finite R_0-algebra. g) S is a finite R-algebra. (**Hint:** For the implication f) $\Rightarrow$ g) use Nakayama's lemma described in the next exercise.)

5.A.17. Exercise (Nakayama's lemma for graded modules) Let R be a positively graded ring and let $M = \bigoplus_{m \in \mathbb{Z}} M_m$ be a $\mathbb{Z}$-graded R-module such that, for some $m_0 \in \mathbb{Z}$, $M_m = 0$ for all $m < m_0$. Let x_i, $i \in I$, be homogeneous elements in M. Show that x_i, $i \in I$, generate M as an R-module if and only if the residue classes of x_i, $i \in I$, generate M/R_+M as an $R/R_+ (= R_0)$-module. In particular, M is a finitely generated R-module if and only if M/R_+M is a finitely generated R_0-module. (**Hint:** If the $\overline{x}_i \in M/R_+M, i \in I$, generate M/R_+M, show $M_m \subseteq \sum_i Rx_i$ by induction on m.)

5.A.18. Exercise Let $\varphi : R \to S$ be a surjective homogeneous homomorphism of positively graded rings. Show that $\operatorname{Proj}\varphi$ is a closed embedding (see Definition 4.E.6) with image $V_+(\operatorname{Ker}\varphi)$. In particular, for an arbitrary homogeneous ideal $\mathfrak{a} \subseteq R$ the projective scheme $\operatorname{Proj}(R/\mathfrak{a})$ can be identified as a closed subscheme of $\operatorname{Proj} R$ with support $V_+(\mathfrak{a})$. In particular, if $R = A[x_0, \ldots, x_n]$ is finitely generated over $A := R_0$ and if $\gamma_i := \deg x_i > 0$, $i = 0, \ldots, n$, then the canonical homogeneous substitution homomorphism $\varepsilon : A[X_0, \ldots, X_n] \to R$ with $X_i \mapsto x_i$ identifies R as the algebra $A[X_0, \ldots, X_n]/\mathfrak{a}$, $\mathfrak{a} := \operatorname{Ker}\varepsilon$, and the projective scheme $\operatorname{Proj} R$ as the closed subscheme $\operatorname{Proj}(A[X_0, \ldots, X_n]/\mathfrak{a}) = V_+(\mathfrak{a})$ of the weighted projective space $\mathbb{P}^n_{\gamma, A} = \operatorname{Proj} A[X_0, \ldots, X_n]$, $\gamma = (\gamma_0, \ldots, \gamma_n)$. (**Remark:** The closed embedding $\operatorname{Proj}\varphi$ can be an isomorphism without φ itself being an isomorphism. See Exercise 5.A.26 (4) below. Note that this phenomenon cannot happen in the affine case.)

5.A.19. Exercise Let $R = \bigoplus_{m \in \mathbb{N}} R_m$ be a positively graded ring. In Exercise 5.A.15 the condition that the irrelevant ideal R_+ is finitely generated plays an important role. In this exercise, among other things, we want to characterize this property.

(1) Prove that for a given family x_i, $i \in I$, of homogeneous elements of positive degrees in R the following conditions are equivalent: a) The elements x_i, $i \in I$, generate the ideal R_+. b) $R = R_0[x_i \mid i \in I]$, i.e. the x_i, $i \in I$, generate the algebra R over R_0. (**Hint:** For the implication a) $\Rightarrow$ b) use induction on $\deg f$ to prove that every homogeneous element $f \in R$ belongs to $R_0[x_i \mid i \in I]$.)

If these conditions hold, then for $m \in \mathbb{N}$ the R_0-module R_m is generated by the monomials $x^\nu = \prod_{i \in I} x_i^{\nu_i}$, $\nu \in \mathbb{N}^{(I)}$, with $\deg x^\nu = \langle \gamma, \nu \rangle = \sum_{i \in I} \gamma_i \nu_i = m$, $\gamma_i := \deg x_i (> 0)$.

(2) Show that the following conditions are equivalent: a) The ideal R_+ is finitely generated. b) R is an R_0-algebra of finite type.

If these conditions hold and if the homogeneous elements $x_1, \ldots, x_n$ of positive degrees $\gamma_1, \ldots, \gamma_n$ generate the ideal R_+, then for any $m \in \mathbb{N}$ the R_0-module R_m is generated by the monomials x^ν, $\langle \gamma, \nu \rangle = m$, which are finite in number. In particular, R_m is a finite

R_0-module. Furthermore, for every $k \in \mathbb{N}$, one has the inclusions $R_{(k\delta)} \subseteq R_+^k \subseteq R_{(k)}$ where $\delta := \max\{\gamma_1, \ldots, \gamma_n\}$, and the ideal $R_{(k)} = \sum_{m \geq k} R_m$ is generated by $R_k + R_{k+1} + \cdots + R_{k+\delta-1}$, i.e. by the monomials x^ν, $k \leq \langle \gamma, \nu \rangle < k + \delta$.

(3) Show that the following conditions are equivalent: a) R is Noetherian. b) R_0 ($= R/R_+$) is Noetherian and the ideal R_+ is finitely generated. c) R_0 ($= R/R_+$) is Noetherian and R is an R_0-algebra of finite type.

(4) If R is Noetherian, then the d-th Veronese transform $R^{[d]} = \bigoplus_{m \in \mathbb{N}} R_{dm}$ is also Noetherian. (**Hint:** Since $R^{[d]}$ is a direct summand of R as an $R^{[d]}$-module, this follows already from Exercise 1.C.6 (7).)

(5) If R is an R_0-algebra of finite type, then $R^{[d]}$ is also an R_0-algebra of finite type and R is finite over $R^{[d]}$. (This is a consequence of (4) even in the case that R, i.e. R_0, is not Noetherian.)

The algebraic results of the last exercise have the following immediate consequences for the projective scheme $\operatorname{Proj} R$.

5.A.20. Theorem *Let $R = \bigoplus_{m \in \mathbb{N}} R_m$ be a positively graded ring. If R is an R_0-algebra of finite type, then $\operatorname{Proj} R$ is an algebraic R_0-scheme. In particular, if R is Noetherian, then $\operatorname{Proj} R$ is an algebraic scheme over the Noetherian ring R_0.*

PROOF. The second part follows from the first part and Exercise 5.A.19 (3). To prove the first part let $x_1, \ldots, x_n$ be a system of homogeneous generators of the ideal R_+ (cf. Exercise 5.A.19 (2)). Then $\operatorname{Proj} R$ is covered by the finitely many open affine sets $D_+(x_i) = \operatorname{Spec}((R_{x_i})_0)$, $i = 1, \ldots, n$. Since $\operatorname{Proj} R$ is separated by Proposition 5.A.10, it remains to show that the R_0-algebras $(R_{x_i})_0$ are of finite type. Let x be an arbitrary homogeneous element of R of positive degree d. By Exercise 5.A.6 (4) $(R_x)_0 = R^{[d]}/R^{[d]}(x-1)$, and $R^{[d]}$ is of finite type over R_0 by Exercise 5.A.19 (5). ●

5.A.21. Definition Let A be a ring. An A-scheme X is called a p r o j e c t i v e a l g e b r a i c A-s c h e m e if $X \cong \operatorname{Proj} R$, where R is a positively graded algebra of finite type over $R_0 = A$.

By Theorem 5.A.20 a projective algebraic A-scheme is an algebraic A-scheme. By Exercise 5.A.18 projective algebraic A-schemes can be identified as closed subschemes of the weighted projective spaces $\mathbb{P}^n_{\gamma, A}$.

As already mentioned, the graded algebra R in Definition 5.A.18 is not uniquely determined by X. For instance, by the next proposition one can always assume that the algebra R has a s t a n d a r d g r a d i n g, i.e. R is generated by R_1 as an R_0-algebra. A standardly graded algebra is obviously positively graded. By Exercise 5.A.19 (1) a positively graded ring R has a standard grading if and only if its irrelevant ideal R_+ is generated by the homogeneous elements of degree 1. If x_i, $i \in I$, are generators of degree 1, then $\operatorname{Proj} R$ is covered by the affine subsets $D_+(x_i) = \operatorname{Spec}(R_{x_i})_0 = \operatorname{Spec} R/R(x_i - 1)$, $i \in I$. If I is finite, for instance $I = \{0, \ldots, n\}$, then $\operatorname{Proj} R$ is a projective algebraic R_0-scheme and can be identified as a closed subscheme of the standard projective space $\mathbb{P}^n_{R_0}$. For a standardly graded ring R the ideals $R_{(k)}$ and R^k_+ coincide for every $k \in \mathbb{N}$.

The following proposition is a consequence of Exercises 5.A.23 and 5.A.25.

5.A.22. Proposition *A projective algebraic A-scheme is always of the form* $\text{Proj } R$, *where R is a standardly graded algebra of finite type over $A = R_0$.*

5.A.23. Exercise Let $R = \bigoplus_{m \in \mathbb{N}} R_m$ be a positively graded algebra and $d \in \mathbb{N}^*$. Then the canonical inclusion $h : R^{[d]} \hookrightarrow R$ induces an isomorphism $\text{Proj } h : \text{Proj } R \to \text{Proj } R^{[d]}$. (**Hint:** For $f \in R_{dm}$, $m \in \mathbb{N}^*$, one has $(R_f)_0 = (R^{[d]}_f)_0 \;(\cong R^{[dm]}/R^{[dm]}(f-1))$. – It is convenient to change the grading of the d-th Veronese transform $R^{[d]} = \bigoplus_{m \in \mathbb{N}} R_{dm}$ by setting $R^{[d]}_m := R_{dm}$. Then R_d is the R_0-module of homogeneous elements of degree 1 in $R^{[d]}$. Of course, this change of grading has no influence on $\text{Proj } R^{[d]}$. (But the inclusion $R^{[d]} \hookrightarrow R$ no longer remains a homogeneous homomorphism.))

As explicit examples of the Veronese isomorphism consider a ring A and the polynomial algebra $R = A[X_0, \ldots, X_n]$ with the standard grading. Then for $d \in \mathbb{N}^*$ the Veronese transform $R^{[d]}$ is generated by the $\binom{n+d}{n}$ monomials X^ν, $|\nu| = d$. By Exercise 5.A.18 the surjective substitution homomorphism $S := A[Y_\nu \,|\, |\nu| = d] \to R^{[d]}$, $Y_\nu \mapsto X^\nu$, defines a closed embedding $\text{Proj } R^{[d]} = \text{Proj } R = \mathbb{P}^n_A \to \text{Proj } S = \mathbb{P}^{\binom{n+d}{n}-1}_A$. This embedding is called the d-th **Veronese embedding**. The simplest non-trivial example is the case $n = 1$ and $d = 2$, where the kernel of the substitution homomorphism $A[U, V, W] \to R^{[2]} = A[X_0^2, X_0 X_1, X_1^2]$, $U \mapsto X_0^2$, $V \mapsto X_0 X_1$, $W \mapsto X_1^2$, is the principal ideal $(UW - V^2)$. Therefore, the Veronese embedding embeds $\mathbb{P}^1_A \cong \text{Proj}(A[U, V, W]/(UW - V^2))$ into $\mathbb{P}^2_A$ with image $V_+(UW - V^2)$.

5.A.24. Exercise Let $\varphi : R \to S$ be a homogeneous homomorphism of positively graded rings. If $\varphi \,|\, R_{(k)} : R_{(k)} \to S_{(k)}$ is a bijection for some $k \in \mathbb{N}$, then φ induces a scheme isomorphism $\text{Proj } \varphi : \text{Proj } S \to \text{Proj } R$.

5.A.25. Exercise Let $R = \bigoplus_{m \in \mathbb{N}} R_m$ be a positively graded algebra of finite type over R_0. Then there is a $d \in \mathbb{N}^*$ such that the d-th Veronese transform $R^{[d]}$ is generated by R_d over R_0 and hence is standardly graded (if the grading of $R^{[d]}$ is changed as indicated in the remark of Exercise 5.A.23). (**Hint:** Let $R = R_0[x_0, \ldots, x_n]$ with $\gamma_i := \deg x_i > 0$, $i = 0, \ldots, n$, and let $m := \text{LCM}(\gamma_0, \ldots, \gamma_n)$. Then one can take for d the maximum of m and the degrees $\langle \gamma, \nu \rangle$ with $0 \le \gamma_i \nu_i < m$ for all i and $m|\langle \gamma, \nu \rangle$. For the proof first note that $d = sm$ is a positive multiple of m. Let $y_i := x_i^{m/\gamma_i}$ and $S := R_0[y_0, \ldots, y_n]$. Then $S_{dt} = R_0[S_d]_{dt} \subseteq R_0[R_d]_{dt}$ for all $t \in \mathbb{N}$. Now let x^ν be a monomial of degree $\langle \gamma, \nu \rangle = rd = rsm$, $r > 1$, and let $\nu_i := \alpha_i m/\gamma_i + \mu_i$, where $\alpha_i \in \mathbb{N}$ and $0 \le \mu_i < m/\gamma_i$. Then $rd = rsm = \langle \gamma, \nu \rangle = \langle \gamma, \mu \rangle + m|\alpha|$, $|\alpha| = \sum_{i=0}^n \alpha_i$. Hence $m|\langle \gamma, \mu \rangle$ and, therefore, $\langle \gamma, \mu \rangle \le d$. Furthermore, $d - \langle \gamma, \mu \rangle = m(|\alpha| - s(r-1)) = mu$ with $u < |\alpha|$. Consequently, there exists $\alpha' \le \alpha$ with $|\alpha'| = u$ and $\deg(y^{\alpha'} x^\mu) = m|\alpha'| + \langle \gamma, \mu \rangle = d$. It follows $x^\nu = y^{\alpha - \alpha'}(y^{\alpha'} x^\mu) \in S_{d(r-1)} R_d \subseteq R_0[R_d]$.)

5.A.26. Exercise (Ideal of inertia forms) Let R be a positively graded ring. We set

$$\mathfrak{I}(R) := \Gamma_{R_+}(R) := \{ g \in R \,|\, R_+ \subseteq \sqrt{\text{Ann}_R \, g} \}$$

$$= \{ g \in R \,|\, g/1 = 0/1 \text{ in } R_f \text{ for all } f \in R_+ \}.$$

$\mathfrak{I}(R)$ is a homogeneous ideal of R and is called the i d e a l   o f   i n e r t i a   f o r m s or the H u r w i t z   i d e a l of R. If x_i, $i \in I$, is a system of generators of the ideal R_+, then $\mathfrak{I}(R) = \{ g \in R \mid g/1 = 0/1 \text{ in } R_{x_i} \text{ for all } i \in I \}$. For an arbitrary homogeneous ideal $\mathfrak{a}$ of R the ideal

$$\mathfrak{a}_{\text{sat}} := \pi^{-1}(\mathfrak{I}(R/\mathfrak{a})),$$

where $\pi : R \to R/\mathfrak{a}$ is the canonical surjection, is called the (H u r w i t z)   s a t u r a t i o n of $\mathfrak{a}$. By definition, $0_{\text{sat}} = \mathfrak{I}(R)$. Show that:

(1) If the ideal R_+ is generated by the homogeneous elements $x_1, \dots, x_n$ (see Exercise 5.A.19 (1)), then

$$\mathfrak{I}(R) = \{ g \in R \mid R_+^k g = 0 \text{ for some } k \in \mathbb{N} \} = \bigcup_{k \in \mathbb{N}} \operatorname{Ann}_R R_+^k$$

$$= \{ g \in R \mid R_{(k)} g = 0 \text{ for some } k \in \mathbb{N} \} = \bigcup_{k \in \mathbb{N}} \operatorname{Ann}_R R_{(k)}$$

$$= \{ g \in R \mid x_i^k g = 0 \text{ for } i = 1, \dots, n \text{ and some } k \in \mathbb{N} \} = \bigcup_{k \in \mathbb{N}} \operatorname{Ann}_R \left(\sum_{i=1}^n R x_i \right).$$

(2) For a homogeneous ideal $\mathfrak{a} \subseteq R$ the following conditions are equivalent: a) $\mathfrak{a} \subseteq \mathfrak{I}(R)$. a') $\mathfrak{a}_{\text{sat}} = \mathfrak{I}(R)$. b) For all homogeneous elements $f \in R_+$ the canonical homomorphism $R_f \to R_f/\mathfrak{a}_f$ is an isomorphism, i.e. $\mathfrak{a}_f = 0$ in R_f. – If R is standardly graded, then these conditions are equivalent to: c) For all homogeneous elements $f \in R_+$ the canonical homomorphism $(R_f)_0 \to (R_f/\mathfrak{a}_f)_0$ is an isomorphism. (**Hint:** Note that for a homogeneous element f of degree 1 the homomorphism $R_f \to R_f/\mathfrak{a}_f$ is an isomorphism if and only if $(R_f)_0 \to (R_f/\mathfrak{a}_f)_0$ is an isomorphism. – In general, condition c) does not imply condition a). To give an example, let A be an arbitrary nonzero commutative ring and $R := A[X, Y]/(X^2) = A[x, y]$, where $\deg X := 1$ and $\deg Y := 2$. Then the principal ideal $\mathfrak{a} := Rx$ fulfills condition c) but not condition a). Indeed $\mathfrak{I}(R) = 0$ in this case, because y is a non-zero divisor in R.)

If the ideals R_+ and $\mathfrak{a}$ are finitely generated (for instance, if R is Noetherian), then conditions a) and b) are equivalent to: d) $\mathfrak{a} \cap R_{(k)} = 0$ for some $k \in \mathbb{N}$, i.e. $\mathfrak{a}$ is a finitely generated R_0-module.

(3) For homogeneous ideals $\mathfrak{a}$, $\mathfrak{b}$ of R the following conditions are equivalent: a) $\mathfrak{a}_{\text{sat}} = \mathfrak{b}_{\text{sat}}$. b) For all homogeneous elements $f \in R_+$ the kernels of the canonical homomorphisms $R_f \to R_f/\mathfrak{a}_f$ and $R_f \to R_f/\mathfrak{b}_f$ coincide. – If R is standardly graded, then these conditions are equivalent to: c) For all homogeneous elements $f \in R_+$ the kernels of the canonical homomorphisms $(R_f)_0 \to (R_f/\mathfrak{a}_f)_0$ and $(R_f)_0 \to (R_f/\mathfrak{b}_f)_0$ coincide.

If the ideals R_+, $\mathfrak{a}$ and $\mathfrak{b}$ are finitely generated (for instance, if R is Noetherian), then these conditions are equivalent to: d) $\mathfrak{a} \cap R_{(k)} = \mathfrak{b} \cap R_{(k)}$ for some $k \in \mathbb{N}$. (**Hint:** Note that $\mathfrak{a}_{\text{sat}} = \mathfrak{b}_{\text{sat}}$ if and only if $\mathfrak{a} \subseteq \mathfrak{b}_{\text{sat}}$ and $\mathfrak{b} \subseteq \mathfrak{a}_{\text{sat}}$.)

(4) Let $\mathfrak{a}$ be a homogeneous ideal of R and $\iota : \operatorname{Proj}(R/\mathfrak{a}) \to \operatorname{Proj} R$ be the canonical closed embedding. Show that if $\mathfrak{a} \subseteq \mathfrak{I}(R)$, then ι is an isomorphism. Conversely, if ι is an isomorphism and *if R is standardly graded*, then $\mathfrak{a} \subseteq \mathfrak{I}(R)$. *Thus, for standardly graded algebras R the Hurwitz ideal of inertia forms $\mathfrak{I}(R)$ is the largest homogeneous ideal $\mathfrak{a}$ such that the closed embedding* $\operatorname{Proj} R/\mathfrak{a} \to \operatorname{Proj} R$ *is an isomorphism.* (**Hint:** ι is an isomorphism if and only if for all homogeneous elements $f \in R_+$ the canonical homomorphism $(R_f)_0 \to (R_f/\mathfrak{a}_f)_0$ is an isomorphism.)

(5) Let $\mathfrak{a}$ and $\mathfrak{b}$ be homogeneous ideals of R. Show that if $\mathfrak{a}_{\text{sat}} = \mathfrak{b}_{\text{sat}}$, then the canonical inclusions $\operatorname{Proj}(R/\mathfrak{a}) \to \operatorname{Proj} R$ and $\operatorname{Proj}(R/\mathfrak{b}) \to \operatorname{Proj} R$ define the same subscheme of

Proj R, i. e. for every homogeneous element $f \in R_+$ the kernels of $(R_f)_0 \to (R_f/\mathfrak{a}_f)_0$ and $(R_f)_0 \to (R_f/\mathfrak{b}_f)_0$ coincide. The converse is true, if R is standardly graded.

5.A.27. Exercise Let $\varphi : R \to S$ be a finite homogeneous homomorphism of positively graded rings and let R be of finite type over R_0. Show that φ induces a finite morphism Proj $S \to$ Proj R.

5.A.28. Exercise Let $R = \bigoplus_{m \geq 0}$ be a positively graded ring and let $R_0 \to R_0'$ be a ring homomorphism. Then show that Spec $R_0' \times_{\mathrm{Spec}\, R_0}$ Proj $R = \mathrm{Proj}(R_0' \otimes_{R_0} R)$. (**Hint:** For a homogeneous element $f \in R_+$ one has $R_0' \otimes_{R_0} (R_f)_0 = (R_0' \otimes_{R_0} R_f)_0$.) In particular, the fibre of the structure morphism Proj $R \to$ Spec R_0 over a point $s \in$ Spec R_0 is the projective scheme $\mathrm{Proj}(\kappa(s) \otimes_{R_0} R) = \mathrm{Proj}(R_{\mathfrak{p}_s}/\mathfrak{p}_s R_{\mathfrak{p}_s})$.

5.A.29. Exercise (Segre products) Let $R = \bigoplus_{m \geq 0} R_m$ and $S = \bigoplus_{m \geq 0} S_m$ be positively graded rings with $R_0 = S_0 =: A$. Define the graded ring $R \#_A S := \bigoplus_{m \geq 0} (R \#_A S)_m$ with $(R \#_A S)_m := R_m \otimes_A S_m$ for all $m \in \mathbb{N}$ and with the obvious multiplication. Show that $\mathrm{Proj}(R \#_A S) \cong \mathrm{Proj}\, R \times_{\mathrm{Spec}\, A} \mathrm{Proj}\, S$. (**Hint:** Show that for $f \in R_m$, $g \in S_m$ and $h := f \otimes g \in (R \#_A S)_m$ one has $\left(R_f\right)_0 \otimes_A \left(S_g\right)_0 \cong \left((R \#_A S)_{f \otimes g}\right)_0$.) – The ring $R \#_A S$ is called the Segre product of R and S. If R and S are standardly graded, then so is $R \#_A S$. If R and S are A-algebras of finite type, then so is $R \#_A S$. In particular, the product of two projective algebraic A-schemes is again a projective algebraic A-scheme. (**Hint:** If R is generated (over A) by the homogeneous elements $x_1, \ldots, x_m$ of positive degrees $\gamma_1, \ldots, \gamma_m$ and if S is generated by the homogeneous elements $y_1, \ldots, y_n$ of positive degrees $\delta_1, \ldots, \delta_n$, then $R \#_A S$ is generated by the homogeneous elements $x^\mu \otimes y^\nu$, $\langle \gamma, \mu \rangle = \langle \delta, \nu \rangle < \max\{m, n\} \max\{\gamma_1, \ldots, \gamma_m\} \max\{\delta_1, \ldots, \delta_n\}$.)

For an explicit example consider the polynomial algebras $R := A[X_0, \ldots, X_m]$ and $S := A[Y_0, \ldots, Y_n]$ with standard grading. Then the Segre product $R \#_A S$ can be identified as the subalgebra $A[X_i Y_j \mid 0 \leq i \leq m, 0 \leq j \leq n]$ of the polynomial algebra in the variables X_i, Y_j (where deg $X_i Y_j := 1$). The surjective substitution homomorphism $A[Z_{ij}]_{i,j} \to R \#_A S$ with $Z_{ij} \mapsto X_i Y_j$ defines a closed embedding

$$\mathrm{Proj}(R \#_A S) = \mathbb{P}_A^m \times_A \mathbb{P}_A^n \to \mathrm{Proj}\, A[Z_{ij}]_{i,j} = \mathbb{P}_A^{mn+m+n}$$

called the Segre embedding.

5.A.30. Exercise Let $R = \bigoplus_{m \in \mathbb{N}} R_m$ be a positively graded algebra of finite type over $C := R_0$. If C is a finite algebra over A, then Proj R is also a projective algebraic A-scheme. (**Hint:** Proj $R \cong \mathrm{Proj}(A \oplus R_+)$ by Exercise 5.A.24 and $A \oplus R_+$ is of finite type over A.) In particular, if C is a finite A-algebra, then Spec C is a projective algebraic A-scheme with respect to the finite morphism Spec $C \to$ Spec A. (Cf. also Exercise 5.B.9 (2) and Theorem 5.B.10.)

5.A.31. Exercise Let R be a positively graded Noetherian ring $\neq 0$. Show that dim Proj $R + 1$ is the dimension of the open subset $D(R_+) \subseteq$ Spec R. (**Hint:** Use dim $A[X, X^{-1}] =$ dim $A + 1$ for an arbitrary Noetherian ring $A \neq 0$. See Exercise 3.B.41.) Moreover, if R_0 is a field, then dim Proj $R + 1 =$ dim R. (**Hint:** Use Theorem 3.B.22 and the fact that the minimal prime ideals of R are homogeneous.)

5.A.32. Example (Projective closures of an affine algebraic scheme) Let $X = \mathrm{Spec}\, A[y_1, \ldots, y_n] = \mathrm{Spec}\, A[Y_1, \ldots, Y_n]/\mathfrak{a} = V(\mathfrak{a}) \subseteq \mathbb{A}_A^n$ be an affine algebraic A-scheme. We want to find a projective algebraic A-scheme $Y = \mathrm{Proj}\, S \subseteq \mathbb{P}_A^n$ such that X can be identified as a basic open affine subscheme $D_+(f) \subseteq Y$. Let $\widehat{S} := A[X_0, \ldots, X_n]$ with the standard grading. We identify $(\widehat{S}_{X_0})_0 = A[X_1/X_0, \ldots, X_n/X_0]$ with the polynomial algebra $A[Y_1, \ldots, Y_n]$ by setting $Y_i = X_i/X_0$ for $1 \leq i \leq n$ and hence the affine space $\mathbb{A}_A^n$ with the basic open affine subscheme $D_+(X_0) \subseteq \mathbb{P}_A^n$.

For the ideal $\mathfrak{a} \subseteq A[Y_1, \ldots, Y_n] = A[X_1/X_0, \ldots, X_n/X_0]$ given above we call the homogeneous ideal $\mathfrak{a}^h := \bigoplus_{m \geq 0} \mathfrak{a}_m^h$, where

$$\mathfrak{a}_m^h := \left\{ G \in \widehat{S}_m \mid G/X_0^m = G(1, X_1/X_0, \ldots, X_n/X_0) \in \mathfrak{a} \right\}$$
$$= \left\{ G \in \widehat{S}_m \mid G(1, y_1, \ldots, y_n) = 0 \right\},$$

the h o m o g e n i z a t i o n of $\mathfrak{a}$. For instance, if A is an integral domain and if $\mathfrak{a} = (F)$ is a non-zero principal ideal, then the homogenization of $\mathfrak{a}$ is the principal ideal $\mathfrak{a}^h = (F^h)$, where

$$F^h := X_0^{\deg F} F(X_1/X_0, \ldots, X_n/X_0)$$

is the so-called h o m o g e n i z a t i o n of F. *Proof.* If $F = \sum_{j=0}^d F_j$ is the decomposition of F into homogeneous parts with $F_d \neq 0$, then $F^h = \sum_{j=0}^d X_0^{d-j} F_j(X_1, \ldots, X_n)$. In particular, F^h is homogeneous of degree $d = \deg F$ and hence $F^h \in \mathfrak{a}^h$. Conversely, if $G \in \mathfrak{a}^h$ is homogeneous of degree m, then by definition

$$G = X_0^m H(X_1/X_0, \ldots, X_n/X_0) F(X_1/X_0, \ldots, X_n/X_0) = X_0^{m - \deg H - \deg F} H^h F^h \in (F^h)$$

for some $H \in A[Y_1, \ldots, Y_n]$, since $\deg H + \deg F = \deg HF \leq m$.

Note that the formula

$$X_0^{\deg H + \deg F}(HF)^h = X_0^{\deg HF} H^h F^h$$

holds for arbitrary base rings A. But for an ideal $\mathfrak{a}$ generated by polynomials $F_1, \ldots, F_r$ the homogenization $\mathfrak{a}^h$ only contains but not necessarily equals the ideal $(F_1^h, \ldots, F_r^h)$ generated by the homogenizations of the generators $F_1, \ldots, F_r$ of $\mathfrak{a}$ (even if A is an integral domain). By definition, the ideal $\mathfrak{a}^h$ is the largest of all homogeneous ideals $\mathfrak{b} \subseteq \widehat{S}$ for which $(\mathfrak{b}\widehat{S}_{X_0})_0 = \sum_{m \geq 0} \mathfrak{b}_m/X_0^m \subseteq \mathfrak{a}$. Since $(\mathfrak{a}^h \widehat{S}_{X_0})_0 = \mathfrak{a}$, we get

$$A[y_1, \ldots, y_n] = A[Y_1, \ldots, Y_n]/\mathfrak{a} = (\widehat{S}_{X_0})_0/(\mathfrak{a}^h \widehat{S}_{X_0})_0 = ((\widehat{S}/\mathfrak{a}^h)_{X_0})_0.$$

This means $X = \operatorname{Spec} A[y_1, \ldots, y_n]$ can be identified in a canonical way with the basic open affine subscheme $D_+(x_0)$ of $\operatorname{Proj} S$, where S is the standardly graded ring

$$S := \widehat{S}/\mathfrak{a}^h = A[X_0, \ldots, X_n]/\mathfrak{a}^h = A[x_0, \ldots, x_n].$$

The projective algebraic A-scheme

$$\operatorname{Proj} S = \operatorname{Proj}(A[X_0, \ldots, X_n]/\mathfrak{a}^h) \subseteq \mathbb{P}_A^n$$

is called the p r o j e c t i v e   c l o s u r e $\overline{X}$ of $X = \operatorname{Spec}(A[Y_1, \ldots, Y_n]/\mathfrak{a}) \subseteq \mathbb{A}_A^n$. X is dense in $\overline{X}$. But, *the projective closure $\overline{X}$ depends essentially on the choice of the generators $y_1, \ldots, y_n$ of the A-algebra $\Gamma(X, \mathcal{O}_X) = A[Y_1, \ldots, Y_n]/\mathfrak{a} = A[y_1, \ldots, y_n]$, i.e. on the embedding $X \hookrightarrow \mathbb{A}_A^n$.* The complement $\overline{X} \setminus X = V_+(x_0) = \operatorname{Proj}(S/(x_0))$ of X in $\overline{X}$ is called the s c h e m e   a t   i n f i n i t y of X (with respect to the A-generators $y_1, \ldots, y_n$).

Let K be a field. If X is the hypersurface $V(F) = \operatorname{Spec}(K[Y_1, \ldots, Y_n]/(F)) \subseteq \mathbb{A}_K^n$ of degree $d = \deg F > 0$, then the projective closure $\overline{X}$ is the hypersurface $V_+(F^h) = \operatorname{Proj}(K[X_0, \ldots, X_n]/(F^h)) \subseteq \mathbb{P}_K^n$. The scheme at infinity is $\operatorname{Proj} K[Y_1, \ldots, Y_n]/(F_d)$, where F_d is the leading homogeneous form of $F = \sum_{i=0}^d F_i$, $F_d \neq 0$.

Let $X = \operatorname{Spec} K[Y] = \mathbb{A}_K^1$. Then the projective closure $\overline{X}$ of X is determined as $\operatorname{Proj} S = \mathbb{P}_K^1$, $S = K[X_0, X_1]$, with $V_+(X_0) = \operatorname{Proj} K[X_1] = \operatorname{Spec} K$ as the scheme at infinity. If we choose the generators $y_1 := Y$ and $y_2 := Y^2$, we get the representation $K[Y] = K[y_1, y_2] \cong K[Y_1, Y_2]/(Y_2 - Y_1^2)$ and the projective closure $\overline{X} = \operatorname{Proj} S'$ with $S' = K[X_0, X_1, X_2]/(X_0 X_2 - X_1^2)$. The graded K-algebras S and S' are not isomorphic, but the projective closures $\operatorname{Proj} S$ and $\operatorname{Proj} S'$ are. The scheme at infinity in the second case is $\operatorname{Proj}(K[X_1, X_2]/(X_1^2)) \cong \operatorname{Spec}(K[X_1]/(X_1^2))$ which is not reduced. Now we take the generators $y_1 := Y$ and $y_2 := Y^3$. Then the projective closure $\overline{X}$ is $\operatorname{Proj} S''$ with $S'' = K[X_0, X_1, X_2]/(X_0^2 X_2 - X_1^3)$. Now even $\operatorname{Proj} S''$ is not isomorphic to $\mathbb{P}_K^1 = \operatorname{Proj} S$. (Look at $D_+(x_2) \cong \operatorname{Spec} K[Z_0, Z_1]/(Z_0^2 - Z_1^3) \subseteq \operatorname{Proj} S''$.) The scheme at infinity is $\operatorname{Proj}(K[X_1, X_2]/(X_1^3)) \cong \operatorname{Spec}(K[X_1]/(X_1^3))$.

The graded A-algebra S which defines the projective closure $\overline{X} = \operatorname{Proj} S$ of the affine algebraic A-scheme $X = \operatorname{Spec} B$, $B = A[y_1, \ldots, y_n]$, can be described directly without using the ideal $\mathfrak{a}$ of relations of the generators $y_1, \ldots, y_n$ as the graded A-subalgebra

$$S := A[X_0, y_1 X_0, \ldots, y_n X_0] = \bigoplus_{m \geq 0} B_m X_0^m \subseteq B[X_0]$$

of the polynomial algebra $B[X_0]$. Obviously,

$$B_m = \{F(y_1, \ldots, y_n) \mid \deg F \leq m\}, \quad m \in \mathbb{N}.$$

Therefore, the sequence B_m, $m \in \mathbb{N}$, is an ascending filtration $\mathcal{F}$ of B by A-submodules with the properties

$$B_0 \subseteq B_1 \subseteq \cdots \subseteq B, \quad \bigcup_{m \in \mathbb{N}} B_m = B, \quad 1 \in B_0 = A \cdot 1, \quad B_i B_j \subseteq B_{i+j} \text{ for all } i, j \in \mathbb{N},$$

and S is the graded ring defined by this filtration. It is obvious that the kernel of the homogeneous substitution homomorphism $\widehat{S} \to S$, $X_0 \mapsto X_0$, $X_i \mapsto y_i X_0$, $i = 1, \ldots, n$, is the homogenization $\mathfrak{a}^h$ of the ideal $\mathfrak{a}$ of relations in $y_1, \ldots, y_n$. So indeed the algebra S is the same as above. Moreover, the equality $S_{X_0} = B[X_0]_{X_0} = B[X_0, X_0^{-1}]$ gives immediately the desired identification $D_+(X_0) = \operatorname{Spec}(S_{X_0})_0 = \operatorname{Spec} B = X$. The scheme at infinity has the nice interpretation as the projective scheme

$$\operatorname{Proj}(S/SX_0) = \operatorname{Proj}(\bigoplus_{m \geq 0}(B_m/B_{m-1}))$$

$(B_{-1} := 0)$ given by the graded algebra $\bigoplus_{m \geq 0}(B_m/B_{m-1})$ associated to the filtration $\mathcal{F}$.

One generalizes this construction by taking to given positive weights $\gamma_1, \ldots, \gamma_n$ the graded A-algebra

$$S := A[X_0, y_1 X_0^{\gamma_1}, \ldots, y_n X_0^{\gamma_n}] \subseteq B[X_0].$$

This defines a projective closure $\overline{X} = \operatorname{Proj} S = V_+(\mathfrak{a}_\gamma^h)$ which is embedded in the weighted projective space $\mathbb{P}_{\gamma', A}^n$, $\gamma' := (1, \gamma) = (1, \gamma_1, \ldots, \gamma_n)$, where $\mathfrak{a}_\gamma^h = \bigoplus_{m \in \mathbb{N}} \mathfrak{a}_{\gamma, m}^h \subseteq \widehat{S} := A[X_0, X_1, \ldots, X_n]$ is the γ-h o m o g e n i z a t i o n of $\mathfrak{a}$ with

$$\mathfrak{a}_{\gamma, m}^h := \{G \in \widehat{S}_m \mid G(1, y_1, \ldots, y_n) = 0\}.$$

Here $\widehat{S}$ carries the γ'-grading.

5.A.33. Example (B l o w i n g u p) Another useful construction in projective geometry is the b l o w i n g u p. We start with an affine scheme $X = \operatorname{Spec} A$ and a closed subscheme $Z = \operatorname{Spec} A/\mathfrak{a}$ with defining ideal $\mathfrak{a} \subseteq A$. The standardly graded A-subalgebra $R_\mathfrak{a}(A) := A[\mathfrak{a}T] = \bigoplus_{m \in \mathbb{N}} \mathfrak{a}^m T^m$ of the polynomial algebra $A[T]$ in one variable is called the R e e s a l g e b r a o f A with respect to $\mathfrak{a}$. The projective scheme $\operatorname{Bl}_Z(X) := \operatorname{Proj} R_\mathfrak{a}(A)$ with structure morphism

$$\sigma : \mathrm{Bl}_Z(X) \to X = \mathrm{Spec}\, A$$

is called the **b l o w i n g   u p   o f** X along the **c e n t r e** Z. For $a \in \mathfrak{a}$, we have $A[\mathfrak{a}T]_a = A_a[T]$, hence outside the center Z the blowing up is an isomorphism, i. e. $\sigma^{-1}(X \setminus Z) = \mathrm{Bl}_Z(X) \setminus \sigma^{-1}(Z) \to X \setminus Z$ is an isomorphism. The restriction $Z \times_X \mathrm{Bl}_Z(X)$ of $\mathrm{Bl}_Z(X)$ to Z is the projective scheme $\mathrm{Proj}\, \mathrm{R}_\mathfrak{a}(A)/\mathfrak{a}\mathrm{R}_\mathfrak{a}(A)$ belonging to the so called **a s s o c i a t e d   g r a d e d   r i n g**

$$G_\mathfrak{a}(A) := \mathrm{R}_\mathfrak{a}(A)/\mathfrak{a}\,\mathrm{R}_\mathfrak{a}(A) = \bigoplus_{m \in \mathbb{N}} (\mathfrak{a}^m/\mathfrak{a}^{m+1})T^m$$

of A with respect to $\mathfrak{a}$. For $a \in \mathfrak{a}$, the algebra $(A[\mathfrak{a}T]_{aT})_0$ of global sections of the affine part $\mathrm{Spec}(A[\mathfrak{a}T]_{aT})_0$ of $\mathrm{Bl}_Z(X) = \mathrm{Proj}\, \mathrm{R}_\mathfrak{a}(A)$ can be identified with the subalgebra $A[\mathfrak{a}/a]$ of A_a. This is immediate from the canonical graded inclusion $A[\mathfrak{a}T]_{aT} \subseteq A[T]_{aT} = A_a[T^{\pm 1}]$. From this it follows directly that the closed subscheme $Z \times_X \mathrm{Bl}_Z(X)$ of $\mathrm{Bl}_Z(X)$ is defined by an invertible ideal sheaf in $\mathcal{O}_{\mathrm{Bl}_Z(X)}$, i. e. locally this ideal sheaf is generated by a non-zero divisor, namely, for $a \in \mathfrak{a}$, we have $\mathfrak{a}A[\mathfrak{a}/a] = aA[\mathfrak{a}/a]$ and a is a non-zero divisor in $A[\mathfrak{a}/a]$, it is even a unit in the extension A_a. For this reason the closed subscheme $Z \times_X \mathrm{Bl}_Z(X) \subseteq \mathrm{Bl}_Z(X)$ is called the **e x c e p t i o n a l   d i v i s o r** of the blowing up $\mathrm{Bl}_Z(X)$. (With the notations of Example 6.E.14, the ideal sheaf of the exceptional divisor is isomorphic to $\mathcal{O}_{\mathrm{Bl}_Z(X)}(1)$.) It follows that the blowing up $\mathrm{Bl}_Z(X) \to X$ is an isomorphism if and only if $\mathfrak{a}$ is already an invertible ideal in A. Furthermore, if X is locally Noetherian, then by Krull's Principal Ideal Theorem 3.B.28 the exceptional divisor is purely 1-codimensional in $\mathrm{Bl}_Z(X)$ (if it is non-empty).

For an arbitrary scheme X and a closed subscheme Z defined by an ideal sheaf $\mathcal{J}(Z)$ (cf. Example 4.E.7 (1)), the blowing up $\mathrm{Bl}_Z(X)$ is obtained by choosing an affine cover $U_i, i \in I$, of X and glueing together the blowing ups $\mathrm{Bl}_{Z \cap U_i}(U_i), i \in I$, which are constructed above.

The blowing up construction simplifies the structure of the ideal sheaf $\mathcal{J}(Z)$, but instead of X one has to deal with the more complicated geometric structure of $\mathrm{Bl}_Z(X)$. The most classical example is the blowing up of the origin $\{0\} \cong \mathrm{Spec}\, K$ in the affine space $\mathbb{A}_K^n, n \geq 2$, over a field K. In this case $A = K[X_1, \ldots, X_n]$, $\mathfrak{a} = \mathfrak{m}_0 = (X_1, \ldots, X_n)$ and the origin is replaced by the projective space $\mathrm{Proj}\, G_{\mathfrak{m}_0}(A) \cong \mathbb{P}_K^{n-1}$, since the associated graded ring $G_{\mathfrak{m}_0}(A) = \bigoplus_{m \in \mathbb{N}} (\mathfrak{m}_0^m/\mathfrak{m}_0^{m+1})T^m$ is (canonically) isomorphic to $A = K[X_1, \ldots, X_n]$ itself. The affine part $U_i := \mathrm{Spec}\, A[\mathfrak{m}_0 T]_{X_i T}$ of the blowing up has the algebra of global sections $\Gamma(U_i) = K[X_1, \ldots, X_n, X_1/X_i, \ldots, X_n/X_i] = K[X_1/X_i, \ldots, X_{i-1}/X_i, X_i, X_{i+1}/X_i, \ldots, X_n/X_i]$ and hence is isomorphic to the affine space $\mathbb{A}_K^n$. The reader should try to understand the glueing of these affine spaces, in particular, for $n = 2$. For $K = \mathbb{R}$, one should try to sketch the set of real points of the blowing up $\mathrm{Bl}_{\{0\}}(\mathbb{A}_\mathbb{R}^2)$. In general, if one blows up the closed point $\{\mathfrak{m}_A\}$ (with its reduced scheme structure) in $\mathrm{Spec}\, A$ of a Noetherian local ring A of (Krull-)dimension d then one replaces this point by the projective $(A/\mathfrak{m}_A)$-scheme $\mathrm{Proj}(G_{\mathfrak{m}_A}(A))$ of dimension $d - 1$ which is called **Z a r i s k i ' s   t a n g e n t   s p a c e** of A. Its d-dimensional cone $\mathrm{Spec}\, G_{\mathfrak{m}_A}(A)$ is called **Z a r i s k i ' s   t a n g e n t   c o n e** of A.

We emphasize that the blowing up $\mathrm{Bl}_Z(X)$ depends strongly on the ideal sheaf $\mathcal{J}(Z)$ which defines $Z = \mathrm{V}(\mathcal{J}(Z))$; for instance, take $Z = \{0\} \subseteq \mathbb{A}_K^2$ with defining ideal (X_1^r, X_2^r), $r \geq 2$. Then the two canonical affine pieces of the blowing up have structure algebras $K[X_1, X_2, X_1^r/X_2^r]$ and $K[X_1, X_2, X_2^r/X_1^r]$ both isomorphic to $K[U, V, W]/(WU^r - V^r)$ and hence have singularities along the line $u = v = 0$. (For $r = 2$ the spectrum of this algebra is Whitney's umbrella, at least for $K = \mathbb{R}$.) If we replace the ideal sheaf $\mathcal{J}(Z)$ by a power $\mathcal{J}(Z)^d$, $d \in \mathbb{N}^*$, the blowing up $\mathrm{Bl}_Z(X)$ is not changed, since in the affine case the Rees algebra $R_{\mathfrak{a}^d}(A)$ is (up to a scaling of the grading) the d-th Veronese transform of $R_\mathfrak{a}(A)$, cf. Exercise 5.A.23. Generally, the blowing up $\mathrm{Bl}_Z(X)$ is reduced (resp. integral) if the base space X is reduced (resp. integral).

Let X be a locally Noetherian scheme. Then $\mathrm{Bl}_Z(X)$ is also locally Noetherian. In the next section (see Theorem 5.B.1) we show that in this situation the projection $\mathrm{Bl}_Z(X) \to X$ is closed. (For this property one needs only that the ideal sheaf $\mathcal{J}(Z)$ is locally finitely generated.) In particular, it is surjective if $X \setminus Z$ is dense in X.

5.B. Main Theorem of Elimination

The most important result of this section is the following theorem 5.B.1. It uses for a positively graded ring R the Hurwitz ideal $\mathcal{J}(R)$ of inertia forms and especially its component $\mathcal{J}_0(R) := \mathcal{J}(R)_0$ of degree 0 which is called the e l i m i n a t i o n i d e a l of R. See Exercise 5.A.26 for the definition and general properties of $\mathcal{J}(R)$. Recall, in particular, that the canonical closed embedding $\mathrm{Proj}(R/\mathcal{J}(R)) \to \mathrm{Proj}\, R$ is an isomorphism (Exercise 5.A.26 (4)).

5.B.1. Main Theorem of Elimination *The structure morphism of a projective algebraic scheme is closed, i. e. if $R = \bigoplus_{m \geq 0} R_m$ is a positively graded algebra of finite type over $A := R_0$, then the structure morphism $h : \mathrm{Proj}\, R \to \mathrm{Spec}\, A$ is a closed map. More precisely, $\mathrm{Im}\, h = \mathrm{V}(\mathcal{J}_0(R))$, where $\mathcal{J}_0(R)$ is the elimination ideal of R.*

PROOF. It is enough to prove the last statement, because for an arbitrary homogeneous ideal $\mathfrak{a} \subseteq R$ we have the following commutative diagram:

$$V_+(\mathfrak{a}) = \mathrm{Proj}(R/\mathfrak{a}) \xrightarrow{\;h\,|\,V_+(\mathfrak{a})\;} \mathrm{Spec}\, A$$
$$\bar{h} \searrow \qquad \nearrow$$
$$\mathrm{Spec}\, A/\mathfrak{a}_0$$

First note that a point $s \in \mathrm{Spec}\, A$ does not belong to $\mathrm{Im}\, h$ if and only if the fibre $(\mathrm{Proj}\, R)_s = \mathrm{Proj}(R_{\mathfrak{p}_s}/\mathfrak{p}_s R_{\mathfrak{p}_s})$ (cf. Exercise 5.A.28) of h over s is empty. By Exercise 5.A.15 (1) this is equivalent to $(R_{\mathfrak{p}_s}/\mathfrak{p}_s R_{\mathfrak{p}_s})_{(k)} = \bigoplus_{m \geq k}(R_m)_{\mathfrak{p}_s}/\mathfrak{p}_s(R_m)_{\mathfrak{p}_s} = 0$ for some $k \in \mathbb{N}$. But all the $A_{\mathfrak{p}_s}$-modules $(R_m)_{\mathfrak{p}_s}$ are finitely generated. So by Nakayama's lemma (for local rings) this condition is equivalent to $(R_{\mathfrak{p}_s})_{(k)} = 0$ for some $k \in \mathbb{N}$.

By definition (see Exercise 5.A.26 (1))

$$\mathcal{J}_0(R) = \bigcup_{k \geq 0} \mathrm{Ann}_A\, R_{(k)} = \bigcup_{k \geq 0}\left(\bigcap_{m \geq k} \mathrm{Ann}_A\, R_m\right).$$

Let $R = A[x_0, \ldots, x_n]$ with homogeneous elements x_i of positive degrees $\gamma_i := \deg x_i$, $i = 0, \ldots, n$, and $\mu := \max\{\gamma_0, \ldots, \gamma_n\} > 0$. Then, since every monomial x^ν of degree $\geq k + \mu$ has a monomial of degree between k and $k + \mu - 1$ as a factor, we have

$$\operatorname{Ann}_A R_{(k)} = \bigcap_{m=k}^{k+\mu-1} \operatorname{Ann}_A R_m = \operatorname{Ann}_A (R_k \oplus \cdots \oplus R_{k+\mu-1}).$$

Because of Supp $M = \mathrm{V}(\operatorname{Ann}_A M)$ for any finite A-module M (cf. Exercise 3.A.24) the following conditions are equivalent:

(1) $\mathfrak{p} \notin \mathrm{V}(\operatorname{Ann}_A(R_{(k)}))$. (2) $\bigoplus_{m=k}^{k+\mu-1}(R_m)_{\mathfrak{p}} = 0$. (3) $(R_{(k)})_{\mathfrak{p}} = (R_{\mathfrak{p}})_{(k)} = 0$.

Now we can complete the proof in the following way: By the remark at the beginning of the proof, for $\mathfrak{p} \in \operatorname{Spec} A$ the condition $\mathfrak{p} \notin \operatorname{Im} h$ is equivalent to $(R_{\mathfrak{p}})_{(k)} = 0$, i.e. to

$$\mathfrak{p} \notin \operatorname{Supp}_A R_{(k)} = \operatorname{Supp}_A \Big(\bigoplus_{m=k}^{k+\mu-1} R_k\Big) = \mathrm{V}\Big(\operatorname{Ann}_A \Big(\bigoplus_{m=k}^{k+\mu-1} R_k\Big)\Big) = \mathrm{V}\Big(\operatorname{Ann}_A R_{(k)}\Big)$$

for some $k \in \mathbb{N}$ (using the result of Exercise 3.A.24) or to

$$\mathfrak{p} \notin \bigcap_k \mathrm{V}(\operatorname{Ann}_A R_{(k)}) = \mathrm{V}\Big(\bigcup_k \operatorname{Ann}_A R_{(k)}\Big) = \mathrm{V}(\mathfrak{I}_0(R)). \qquad \bullet$$

We give an important application of the Main Theorem of Elimination 5.B.1. Let $F_1, \ldots, F_r$ be homogeneous polynomials of positive degrees $\delta_1, \ldots, \delta_r$ in the graded polynomial algebra $A[X_0, \ldots, X_n]$ with $\gamma_i := \deg X_i > 0$ and let R be the residue class algebra $R := A[X_0, \ldots, X_n]/(F_1, \ldots, F_r)$. Then by Theorem 5.B.1 the image of $\mathrm{V}_+(F_1, \ldots, F_r) = \operatorname{Proj} R \subseteq \operatorname{Proj} A[X_0, \ldots, X_n] = \mathbb{P}^n_{\gamma, A}$ in $\operatorname{Spec} A$ is the closed set $\mathrm{V}(\mathfrak{I}_0(R))$, i.e. the polynomials $F_1, \ldots, F_r$ have a common zero in $\mathbb{P}^n_{\gamma, A}$ which lies over a given point $s \in \operatorname{Spec} A$ if and only if s is a zero of the ideal $\mathfrak{I}_0(R)$. The polynomials

$$F_\rho = \sum_{\langle \gamma, \nu \rangle = \delta_\rho} a_{\rho\nu} X^\nu, \qquad \rho = 1, \ldots, r,$$

can be viewed as specializations of the generic polynomials

$$F_\rho = \sum_{\langle \gamma, \nu \rangle = \delta_\rho} U_{\rho\nu} X^\nu, \qquad \rho = 1, \ldots, r,$$

with indeterminate coefficients $U_{\rho\nu}$ under the substitution $\mathsf{A} := \mathbb{Z}[U_{\rho\nu}]_{\rho,\nu} \to A$, $U_{\rho\nu} \mapsto a_{\rho\nu}$. It follows $R = A \otimes_{\mathsf{A}} \mathsf{R}$ and $\operatorname{Proj} R = \operatorname{Spec} A \times_{\mathsf{A}} \operatorname{Proj} \mathsf{R}$ (cf. Exercise 5.A.28), where R is the generic algebra

$$\mathsf{R} := \mathsf{A}[X_0, \ldots, X_n]/(\mathsf{F}_1, \ldots, \mathsf{F}_r).$$

From the cartesian diagram

$$
\begin{array}{ccc}
\operatorname{Proj} R = \operatorname{Spec} A \times_{\mathsf{A}} \operatorname{Proj} \mathsf{R} & \longrightarrow & \operatorname{Proj} \mathsf{R} \\
{\scriptstyle h}\downarrow & & \downarrow{\scriptstyle \mathsf{h}} \\
\operatorname{Spec} A & \longrightarrow & \operatorname{Spec} \mathsf{A}
\end{array}
$$

it follows that Im h is the preimage of Im h with respect to the substitution morphism Spec $A \to$ Spec A , i. e.

$$\text{Im } h = V(\mathfrak{J}_0(R)) = V(\mathfrak{J}_0(R)A) .$$

So the elimination ideal $\mathfrak{J}_0(R)$ and the extension $\mathfrak{J}_0(R)A$ of the generic elimination ideal $\mathfrak{J}_0(R) \subseteq A$ have the same zero set in Spec A. (But, in general, these ideals do not coincide! Of course, one has the inclusion $\mathfrak{J}_0(R)A \subseteq \mathfrak{J}_0(R)$.) Since A is Noetherian, the generic elimination ideal $\mathfrak{J}_0(R)$ has a finite set of generators

$$R_1, \ldots, R_t \in A = \mathbb{Z}[U_{\rho\nu}]_{\rho,\nu} .$$

Every such system of polynomials is called a r e s u l t a n t   s y s t e m for the weights $\gamma_0, \ldots, \gamma_n$ and the degrees $\delta_1, \ldots, \delta_r$. We summarize this discussion in the following theorem:

5.B.2. Theorem *Let* $R_1, \ldots, R_t$ *be a resultant system for the positive weights* $\gamma_0, \ldots, \gamma_n$ *and the positive degrees* $\delta_1, \ldots, \delta_r$. *Then the homogeneous polynomials* $F_\rho = \sum_{\langle \gamma, \nu \rangle = \delta_\rho} a_{\rho\nu} X^\nu \in A[X_0, \ldots, X_n]$, $\rho = 1, \ldots, r$, *have a common zero in the projective space* $\mathbb{P}^n_{\gamma,A}$ *which lies over a given point* $s \in$ Spec A *if and only if* s *is a common zero of the specialized resultant system* $R_\tau := R_\tau(a_{\rho\nu}) \in A$, $\tau = 1, \ldots, t$. *In particular,* Proj $R = \emptyset$, *i. e.* $R = A[X_0, \ldots, X_n]/(F_1, \ldots, F_r)$ *is a finite* A-*algebra, if and only if the elements* R_τ, $\tau = 1, \ldots, t$, *generate the unit ideal in* A.

Note that the last statement in Theorem 5.B.2 gives a numerical criterion for the positively graded A-algebra R to be finite. Usually it is not so easy to decide whether an arbitrary A-algebra of finite type is finite or not.

To find a resultant system for given weights and degrees is, in general, rather difficult. The first interesting case is the so called m a i n   c a s e   o f   e l i m i n a t i o n, i. e. the case $r = n + 1$, where the number of polynomials F_ρ coincides with the number of indeterminates X_i (cf. Exercise 5.B.3 below). In this case the generic elimination ideal $\mathfrak{J}_0(R)$ is a principal ideal a generator R^δ_γ $(\delta = (\delta_1, \ldots, \delta_{n+1}))$ of which is called the r e s u l t a n t. It is defined up to sign (and can be the zero polynomial in the anisotropic case). A detailed discussion in a more general set-up can be found in [9] .

5.B.3. Exercise In the generic situation assume that $F_1, \ldots, F_r \in A[X_0, \ldots, X_n]$ and $r \leq n$. Then the generic elimination ideal $\mathfrak{J}_0(R)$ is the zero ideal. In particular, for an arbitrary commutative ring A and homogeneous polynomials $F_1, \ldots, F_r \in A[X_0, \ldots, X_n]$ of positive degrees with $r \leq n$ and for every point $s \in$ Spec A the polynomials $F_1, \ldots, F_r$ have a common zero in $\mathbb{P}^n_{\gamma,A}$ which lies over s (where $\gamma := (\deg X_i)_{0 \leq i \leq n}$). If A is an algebraically closed field, this means that the polynomials $F_1, \ldots, F_r$ have a common non-trivial zero in A^{n+1}. (**Hint:** For an arbitrary field K the ideal $(F_1, \ldots, F_r) \subseteq K[X_0, \ldots, X_n]$ has codimension $\leq r < n + 1$ by Theorem 3.B.29.)

For a positively graded Noetherian ring R the Hurwitz ideal of inertia forms $\mathfrak{J}(R)$ and the elimination ideal $\mathfrak{J}_0(R) = \mathfrak{J}(R)_0$ can be described in the following way: Let $(0) = \mathfrak{q}_1 \cap \cdots \cap \mathfrak{q}_{r_0} \cap \mathfrak{q}_{r_0+1} \cap \cdots \cap \mathfrak{q}_r$, be a primary decomposition of the zero ideal of R with associated prime ideals $\mathfrak{p}_i = \sqrt{\mathfrak{q}_i}, i = 1, \ldots, r$. Suppose $R_+ \subseteq \mathfrak{p}_j$ for and only for $r_0 + 1 \leq j \leq r$, i. e. $V_+(\mathfrak{q}_j) = V_+(\mathfrak{p}_j) = \emptyset$ if and only if $r_0 + 1 \leq j \leq r$. Then $\text{Proj } R = V_+(0) = V_+(\mathfrak{q}_1) \cup \cdots \cup V_+(\mathfrak{q}_{r_0}) = V_+(\mathfrak{p}_1) \cup \cdots \cup V_+(\mathfrak{p}_{r_0})$ and $V_+(\mathfrak{p}_j), 1 \leq j \leq r_0$, are the irreducible components of $\text{Proj } R$. The prime ideals $\mathfrak{p}_{r_0+1}, \ldots, \mathfrak{p}_r$ are called the i r r e l e v a n t p r i m e i d e a l s of R and the intersection $\mathfrak{Q}(R) := \bigcap_{j=r_0+1}^r \mathfrak{q}_j$ is called the i r r e l e v a n t c o m p o n e n t of R. Because of $R_+ \subseteq \sqrt{\mathfrak{Q}(R)}$, the ideal $\mathfrak{Q}(R)$ contains a power of R_+ or, equivalently, an ideal $R_{(k)}$ for some $k \in \mathbb{N}$.

5.B.4. Proposition *With the notations just introduced we have*

$$\mathfrak{J}(R) = \mathfrak{q}_1 \cap \cdots \cap \mathfrak{q}_{r_0}.$$

The simple p r o o f of this proposition is left to the reader as an exercise. Note that $\mathfrak{Q}(R) \cap \mathfrak{J}(R) = (0)$. It follows that $\mathfrak{J}(R)_{(k)} = 0$ for some $k \in \mathbb{N}$ or, equivalently, that $\mathfrak{J}(R)$ is a finite R_0-module.

Let X be a projective algebraic A-scheme, i. e. $X = \text{Proj } R$, where R is a graded algebra of finite type over $A = R_0$. Then for an arbitrary A-algebra A' the A'-scheme $X_{(A')} = \text{Spec } A' \times_A X = \text{Proj}(A' \otimes_A R)$ is also projective algebraic and hence the structure morphism $\text{Proj } X_{(A')} \to \text{Spec } A'$ is a closed mapping by Theorem 5.B.1. From this it follows that for an arbitrary base change $T \to \text{Spec } A$ the structure morphism is closed. So the structure morphism $X \to \text{Spec } A$ is universally closed in the following sense.

5.B.5. Definition A morphism $f : X \to Y$ of schemes is called u n i v e r s a l l y c l o s e d if one of the following (obviously) equivalent conditions holds.

(1) For any scheme Y' and morphism $g : Y' \to Y$, the canonical projection $Y' \times_Y X \to Y'$ is a closed map.

(2) For any *affine* scheme Y' and morphism $g : Y' \to Y$, the canonical projection $Y' \times_Y X \to Y'$ is a closed map.

A morphism $f : X \to Y$ is universally closed if there is an open cover V_i, $i \in I$, of Y such that its restrictions $f^{-1}(V_i) \to V_i$, $i \in I$, are universally closed.

In topology, a universally closed continuous map is called a proper map. I.e., a continuous map $f : X \to Y$ of topological spaces X, Y is called p r o p e r if for an arbitrary continuous map $g : Y' \to Y$ the projection $Y' \times_Y X \to Y'$ is closed, where $Y' \times_Y X$ is the fibre product $\{(y', x) \in Y' \times X \mid g(y') = f(x)\} \subseteq Y' \times X$ of topological spaces. For f to be proper it is sufficient to assume that for an arbitrary topological space Z the map $\text{id}_Z \times f : Z \times X (= (Z \times Y) \times_Y X) \to Z \times Y$ is closed, since the general fibre product $Y' \times_Y X$ can be identified with the preimage of the graph $\Gamma(g) \subseteq Y' \times Y$ in $Y' \times X$ with respect to the map $\text{id}_{Y'} \times f : Y' \times X \to Y' \times Y$

and since for a closed map $h : S \to T$ of topological spaces and for an arbitrary subspace $T' \subseteq T$ the restriction $h^{-1}(T') \to T'$ is also closed. A proper map of topological spaces can be characterized in the following way: *A continuous map* $f : X \to Y$ *is proper if and only if* f *is closed and all the fibres of* f *are quasi-compact.* (See, for instance, Bourbaki, N.: General Topology, Chapter 1.) In particular, *a Hausdorff space* X *is compact if and only if the map* $X \to \{P\}$ *of* X *into a singleton* $\{P\}$ *is proper.*

In algebraic geometry properness is defined in a more restrictive sense.

5.B.6. Definition A morphism $f : X \to Y$ of schemes is called p r o p e r, if f is algebraic (i. e. separated and of finite type) and universally closed. An S-scheme X is called p r o p e r, if the structure morphism $X \to S$ is proper.

The main theorem of elimination 5.B.1 (together with the discussions before Definition 5.B.5) can now be reformulated in the following way:

5.B.7. Theorem *Let A be a commutative ring. Then every projective algebraic A-scheme is proper.*

With respect to the characterization of compact spaces X as Hausdorff spaces for which the map $X \to \{P\}$ is proper, Theorem 5.B.7 (or the Main Theorem of Elimination 5.B.1) can be considered as the algebraic counterpart to the fact that the projective spaces $\mathbb{P}^n(\mathbb{K})$ and their closed subspaces are compact (endowed with their strong topology).

5.B.8. Exercise A morphism $f : X \to S$ of schemes is called a l o c a l l y p r o j e c t i v e m o r p h i s m, if there exists an affine open cover V_i, $i \in I$, of S, such that for each $i \in I$, $U_i := f^{-1}(V_i)$ is a projective algebraic $\Gamma(V_i)$-scheme.

(1) Show that any locally projective morphism is proper.

(2) If the morphism $X \to S$ is locally projective, then for an arbitrary base change $T \to S$ the morphism $X_{(T)} \to T$ is also locally projective.

(3) If $X \to S$ and $Y \to S$ are locally projective morphisms, then $X \times_S Y \to S$ is also locally projective. (**Hint:** Use Exercise 5.A.29.)

5.B.9. Exercise Let X be an affine A-scheme.

(1) The structure morphism $X \to \operatorname{Spec} A$ is universally closed if and only if $\Gamma(X, \mathcal{O}_X)$ is integral over A. (**Hint:** Use Exercise 3.B.42 (3).)

(2) X is a projective algebraic A-scheme if and only if $\Gamma(X, \mathcal{O}_X)$ is finite over A.

5.B.10. Theorem *A projective algebraic A-scheme* $\pi : X \to \operatorname{Spec} A$ *is affine if and only if all the fibres* $\pi^{-1}(y)$, $y \in \operatorname{Spec} A$, *are finite sets. In this case the structure morphism* π *is even finite* (i. e. X *is affine and* $\Gamma(X)$ *is finite over* $A = \Gamma(\operatorname{Spec} A)$).

PROOF. By assumption $X = \operatorname{Proj} R$ with $R = \bigoplus_{m \in \mathbb{N}} R_m$ of finite type over $A = R_0$. (1) Let π be affine and $\mathfrak{p} = \mathfrak{p}_y \subseteq A$ for a point $y \in \operatorname{Spec} A$. Then $\pi^{-1}(y) = \operatorname{Proj}(R_\mathfrak{p}/\mathfrak{p}R_\mathfrak{p})$ is projective and affine over $\kappa(y) = A_\mathfrak{p}/\mathfrak{p}A_\mathfrak{p}$, hence finite by Exercise 5.B.9(2). (2) Conversely, let all the fibres of π be finite. By Exercise 5.B.9(2) it suffices to show that π is affine, i.e. for any $y \in \operatorname{Spec} A$ there exists an open affine neighbourhood U of y such that $\pi^{-1}(U)$ is affine. By 5.A.11, there exists a homogeneous element $f \in R$ of positive degree such that the set $D_+(f) \subseteq X$ contains $\pi^1(y)$. By the Main Theorem of Elimination 5.B.1 the image $\pi(V_+(f)) \subseteq \operatorname{Spec} A$ is closed, hence $\pi^{-1}(D(a)) = D_+(f) \cap \pi^{-1}(D(a)) = D_+(af)$ is affine for any $a \in A$, $a \notin \mathfrak{p}_y$, with $D(a) \cap \pi(V_+(f)) = \emptyset$. •

5.B.11.Exercise Let $R = \bigoplus_{m \in \mathbb{N}} R_m$ be a positively graded R_0-algebra of finite type, $\mathfrak{p} = \bigoplus_{m \in \mathbb{N}} \mathfrak{p}_m$ a homogeneous prime ideal representing a closed point $x \in X := \operatorname{Proj} R$ with $\kappa(x) =: K$ and let $S = \bigoplus_{m \in \mathbb{N}} S_m := R/\mathfrak{p} = \bigoplus_{m \in \mathbb{N}} R_m/\mathfrak{p}_m$. Then $\operatorname{Proj} S = \operatorname{Spec} K$.

(1) Show that $k := S_0 = R_0/\mathfrak{p}_0$ is a field, i.e. $\mathfrak{p}_0$ is a maximal ideal in R_0, and that K is a finite extension of k. (**Hint:** By 5.B.1 the morphism $\operatorname{Proj} S \to \operatorname{Spec} S_0$ is closed.)

(2) The (graded) normalization of S (cf. Exercise 1.E.10(3)b)) has the form $K[t]$ with a homogeneous element $t \neq 0$ of positive degree $d := \operatorname{GCD}(m \in \mathbb{N}^* \mid S_m \neq 0)$ and $S_{rd} = K_r t^r$ with k-subvector spaces K_r of K such that $K_r K_s \subseteq K_{r+s}$ for all $r, s \in \mathbb{N}$ and $K_r = K$ for $r \gg 0$. If $S_d \neq 0$ one may assume that $k = K_0 \subseteq K_1$ and hence $k = K_0 \subseteq K_1 \subseteq K_2 \subseteq \cdots$. If $K = k$ (for example, if k is algebraically closed), then $S \cong k[N] \subseteq k[\mathbb{N}] = k[T]$ for some non-zero submonoid N of $(\mathbb{N},+)$.

(3) Let k be a field and let $\gamma = (\gamma_0, \ldots, \gamma_n) \in (\mathbb{N}^*)^{n+1}$ with $\operatorname{GCD}(\gamma_0, \ldots, \gamma_n) = 1$. The k-rational points of $\mathbb{P}^n_{\gamma,k} = \operatorname{Proj} k[X_0, \ldots, X_n]$ (with $\deg X_i = \gamma_i$ for $i = 0, \ldots, n$) correspond to the non-zero $(n+1)$-tuples $\alpha = (\alpha_0, \ldots, \alpha_n) \in k^{n+1}$ where two such tuples α, β have to be identified if $\operatorname{supp}\alpha \ (= \{i \mid \alpha_i \neq 0\}) = \operatorname{supp}\beta$ and if there is a $\lambda \in k^\times$ with $\alpha_i = \lambda^{\gamma_i/\delta}\beta_i$ for all $i \in \operatorname{supp}\alpha$, $\delta := \operatorname{GCD}(\gamma_i \mid i \in \operatorname{supp}\alpha)$. – How many $\mathbb{F}_q$-rational points has $\mathbb{P}^n_{\gamma,\mathbb{F}_q}$ for a finite field $\mathbb{F}_q$ with q elements?

5.C. Mapping Theorem of Chevalley

The image of a projective morphism is closed. For an arbitrary continuous map of topological spaces the image can be quite weird, even for differentiable or analytic maps. As an example consider the image of the map $\mathbb{R} \to \mathbb{R}^2/\mathbb{Z}^2$, $t \mapsto [t, \alpha t]$, where α is a fixed irrational number. On the contrary the image of an algebraic morphism (of Noetherian schemes) is rather well-behaved. This is described by the so called mapping theorem of Chevalley.

Let X be an arbitrary topological space. A subset $Y \subseteq X$ is called c o n s t r u c t i b l e, if it belongs to the smallest set algebra containing all the open sets of X. Here, a subset $\mathcal{A} \subseteq \mathfrak{P}(X)$ is called an a l g e b r a (o f s e t s) if $X \in \mathcal{A}$ and if $\mathcal{A}$ is closed under *finite* intersection and under complement. Then $\mathcal{A}$ is also closed under finite union. If we identify $\mathcal{A}$ with a subset of the Boolean algebra $\mathfrak{P}(X) = \mathbb{F}_2^X$, then $\mathcal{A}$ *is a set algebra if and only if $\mathcal{A}$ is a subring of $\mathfrak{P}(X)$.* We denote the algebra of constructible subsets of X by $\mathcal{C}_X$. It contains all the open and closed subsets of X, and hence all the l o c a l l y   c l o s e d   s u b s e t s $U \cap F$, U open in X and F closed in X, and the finite unions $Y_1 \cup \cdots \cup Y_r$ of locally closed subsets $Y_1, \ldots, Y_r$. Since these unions already form a set algebra, we get:

5.C.1. Proposition *Let X be a topological space. Then every constructible subset of X is a finite union of locally closed subsets of X.*

5.C.2. Corollary *Let X be a topological space. If $Y \subseteq X$ is a constructible subset of X and if $Z \subseteq Y$ is a constructible subset of Y, then Z is a constructible subset of X. In particular, if $Y_1, \ldots, Y_r$ form a finite cover of X by constructible subsets and if for $Z \subseteq X$ the intersections $Z \cap Y_i$ are constructible in Y_i, $i = 1, \ldots, r$, then Z is constructible in X.*

With these notations the theorem of Chevalley says:

5.C.3. Mapping Theorem of Chevalley *Let $h : X \to Y$ be a morphism of finite type of Noetherian schemes. If $Z \subseteq X$ is constructible in X, then $h(Z) \subseteq Y$ is constructible in Y.*

PROOF. We can assume that $Z = U \cap F$ is locally closed. Then Z is a closed subscheme of the open subscheme U of X and hence also Noetherian and of finite type over Y. Hence we can assume that $Z = X$. Furthermore, by Corollary 5.C.2 we can assume that X and Y are affine and irreducible and that the image of X is dense in Y. So we have to prove: If A is a Noetherian integral domain and if $B \supseteq A$ is of finite type over A, then the image of $h :$ Spec $B \to$ Spec A is constructible. Using Noetherian induction we can also assume that the image of every proper closed subset of Spec B is constructible in Spec A.

By 1.F.12 there exist an element $f \in A$, $f \neq 0$, and elements $z_1, \ldots, z_m \in B$ such that $z_1, \ldots, z_m$ are algebraically independent over A and B_f is finite over $A_f[z_1, \ldots, z_m]$. Then Spec $B_f \to$ Spec $A_f[z_1, \ldots, z_m]$ is surjective by the lying-over theorem 3.B.9(3) and Spec $A_f[z_1, \ldots, z_m] \to$ Spec A_f is obviously surjective. It follows that $h(\mathrm{D}(f)) = h(\mathrm{Spec}\, B \setminus \mathrm{V}(f)) = \mathrm{D}(f) \subseteq \mathrm{Spec}\, A$. Since $h(\mathrm{V}(f))$ is constructible by induction hypothesis, Im $h = \mathrm{D}(f) \cup h(\mathrm{V}(f)) \subseteq \mathrm{Spec}\, A$ is constructible too.   •

5.C.4. Corollary *In the situation of* Theorem 5.C.3 *if* Im h *is dense in Y, then* Im h *contains a dense open subset.*

The corollary is an immediate consequence of Theorem 5.C.3 and the following simple exercise.

5.C.5. Exercise Let Y be a topological space with only finitely many irreducible components. If Z is a dense constructible subset of Y, then Z contains a dense open subset of Y. (**Hint:** Reduce to the case that Y is irreducible.)

5.C.6. Exercise (1) Let Y be a Noetherian scheme and let $Z \subseteq Y$ be a constructible subset which contains with every element y all its generalizations (cf. Exercise 4.B.8 (1)). Show that Z is open in Y. (**Hint:** One may assume that Y is irreducible and $Z \neq \emptyset$. Then Z contains the generic point of Y and hence, by Exercise 5.C.5, a non-empty open subset U of Y. By Noetherian induction $Z \cap (Y \setminus U)$ is open in $Y \setminus U$ and hence, altogether, Z is open in Y.)

(2) With part (1) and the Going-down Theorem for flat ring homomorphisms mentioned at the end of Example 3.B.10 prove the following O p e n M a p p i n g T h e o r e m: Let $h : X \to Y$ be a morphism of finite type of (locally) Noetherian schemes such that for every $x \in X$ the homomorphism $\mathcal{O}_{Y,h(x)} \to \mathcal{O}_{X,x}$ is flat (such morphisms are called f l a t). Then h is an open mapping. – Derive an open mapping theorem from the Going-down Theorem 3.B.12.

5.C.7. Exercise Let Y be a Noetherian scheme and let $Z \subseteq Y$ be a constructible subset. Construct a morphism $h : X \to Y$ of finite type such that $Z = \mathrm{Im}\, h$.

CHAPTER 6: Regular, Normal and Smooth Points

In this chapter, we will discuss the normalization of a scheme; we will also introduce the concepts of smooth and regular points. Before we plunge into these let us first develop some necessary Commutative Algebra which we will use quite heavily. We start with a discussion on regular local rings and normal domains.

6.A. Regular Local Rings

Let R be a Noetherian local ring with maximal ideal $\mathfrak{m} = \mathfrak{m}_R$ and let $k := R/\mathfrak{m}$ be its residue field. The dimension of the k-vector space $\mathfrak{m}/\mathfrak{m}^2$ is called the e m - b e d d i n g d i m e n s i o n of R and is denoted by $\operatorname{emdim} R$. Then by Nakayama's lemma

$$\operatorname{emdim} R = \mu(\mathfrak{m}),$$

where $\mu(\mathfrak{m})$ is the minimal number of generators of $\mathfrak{m}$. From the generalization 3.B.29 of Krull's Principal Ideal Theorem, we get

$$\dim R = \operatorname{ht} \mathfrak{m} \leq \operatorname{emdim} R.$$

By Proposition 3.B.30 there exist elements $f_1, \ldots, f_d$, $d := \dim R$, such that the radical of the ideal $Rf_1 + \cdots + Rf_d$ is $\mathfrak{m}$. Every such system $f_1, \ldots, f_d$ is called a s y s t e m o f p a r a m e t e r s of R. For arbitrary elements $a_1, \ldots, a_r \in \mathfrak{m}$ and $\mathfrak{a} := (a_1, \ldots, a_r)$ we obviously have

$$\operatorname{emdim} R/\mathfrak{a} = \operatorname{Dim}_k \mathfrak{m}/(\mathfrak{m}^2 + \mathfrak{a}) = \operatorname{emdim} R - \operatorname{Dim}_k (\mathfrak{m}^2 + \mathfrak{a})/\mathfrak{m}^2 \geq \operatorname{emdim} R - r$$

and $\dim R/\mathfrak{a} \geq \dim R - r$. To prove the last inequality let $f_1, \ldots, f_d \in \mathfrak{m}$ represent a system of parameters of $R/\mathfrak{a}$. Then the radical of the ideal generated by $a_1, \ldots, a_r, f_1, \ldots, f_d$ is $\mathfrak{m}$. Hence $r + d \geq \dim R$ by Theorem 3.B.29.

6.A.1. Definition A Noetherian local ring R is called a r e g u l a r l o c a l r i n g if $\operatorname{emdim} R = \dim R$.

By definition, regular local rings are those local rings for which the maximal ideal is generated by a system of parameters. Any system of parameters which generates the maximal ideal of a regular local ring is called a r e g u l a r s y s t e m o f p a r a - m e t e r s.

6.A.2. Example (1) *Fields are the regular local rings of dimension 0.*

(2) *The regular local rings of dimension 1 are the local principal ideal domains which are not fields,* i.e. the d i s c r e t e v a l u a t i o n r i n g s. For the proof one has to show that a local ring R is a principal ideal domain if $\dim R = 1$ and if the maximal ideal $\mathfrak{m}$ is principal. Let $\mathfrak{m} = Rx$ and let $\mathfrak{p} \subset \mathfrak{m}$ be a minimal prime ideal. Then $\mathfrak{p} = \mathfrak{a}x$ for some ideal $\mathfrak{a} \subseteq R$.

Since $\mathfrak{p}$ is prime and $x \notin \mathfrak{p}$, by Nakayama's lemma, we get $\mathfrak{a} = \mathfrak{p}$ and $\mathfrak{p} = 0$. [1]) Thus R is indeed an integral domain. Now let $\mathfrak{a} \subseteq R$ be an arbitrary non-zero ideal of R. Since $R/\mathfrak{a}$ is zero-dimensional, there exists an $n \in \mathbb{N}$ such that $\mathfrak{m}^n = Rx^n \subseteq \mathfrak{a}$ and hence $Rx^{n+1} \subset \mathfrak{a}$. Therefore, there exists an $m \in \mathbb{N}$, $m \le n$, with $\mathfrak{a} \subseteq Rx^m$, but $\mathfrak{a} \not\subseteq Rx^{m+1}$. It follows that $\mathfrak{a} = \mathfrak{b}x^m$ for some ideal $\mathfrak{b} \subseteq R$ and $\mathfrak{b} = R$, since otherwise $\mathfrak{b} \subseteq \mathfrak{m} = Rx$ and $\mathfrak{a} \subseteq Rx^{m+1}$.

In a discrete valuation ring R each prime element π constitutes a regular system of parameters and is also called a u n i f o r m i z i n g   p a r a m e t e r. Every element $x \in Q(R)$, $x \ne 0$, has a unique representation $x = \varepsilon \pi^m$ with a unit $\varepsilon \in R^\times$ and an integer $m \in \mathbb{Z}$. The mapping $x \mapsto m =: \mathrm{v}(x)$ is the normalized discrete valuation of $Q(R)$.

(3) *The formal power series ring* $k[[X_1, \ldots, X_n]]$ *over a field* k *is a regular local ring.* That the power series algebra is Noetherian follows from Exercise 1.C.6 (9). The indeterminates form a regular system of parameters.

6.A.3. Proposition *Let* R *be a regular local ring of dimension* d. *For elements* $a_1, \ldots, a_r \in \mathfrak{m} = \mathfrak{m}_R$ *the following conditions are equivalent*:

(1) $a_1, \ldots, a_r$ *is part of a regular system of parameters of* R.

(2) *The residue classes* $\bar{a}_1, \ldots, \bar{a}_r \in \mathfrak{m}/\mathfrak{m}^2$ *are linearly independent over* $k = R/\mathfrak{m}$.

(3) $R/(a_1, \ldots, a_r)$ *is a regular local ring of dimension* $d - r$.

PROOF. The equivalence "(1) $\Leftrightarrow$ (2)" follows directly from Nakayama's Lemma.

(1) $\Rightarrow$ (3): Let $a_1, \ldots, a_r, a_{r+1}, \ldots, a_d$ be a regular system of parameters for R and let $\mathfrak{a} := (a_1, \ldots, a_r)$. Then the maximal ideal $\mathfrak{m}/\mathfrak{a}$ of the local ring $R/\mathfrak{a}$ is generated by the images of $a_{r+1}, \ldots, a_d$ in $R/\mathfrak{a}$. Therefore, $d - r = \mathrm{emdim}\, R/\mathfrak{a} \ge \dim R/\mathfrak{a} \ge \dim R - r = d - r$ and so $R/\mathfrak{a}$ is regular of dimension $d - r$.

(3) $\Rightarrow$ (1): Let $b_1, \ldots, b_{d-r} \in \mathfrak{m}$ be such that the images of $b_1, \ldots, b_{d-r}$ in $R/\mathfrak{a}$ generate the maximal ideal $\mathfrak{m}/\mathfrak{a}$ of the local ring $R/\mathfrak{a}$. Then $a_1, \ldots, a_r$, $b_1, \ldots, b_{d-r}$ generate $\mathfrak{m}$ and form a regular system of parameters of R.   •

6.A.4. Corollary *Let* R *be a regular local ring with maximal ideal* $\mathfrak{m}$. *For an ideal* $\mathfrak{a} \subseteq \mathfrak{m}$ *the following conditions are equivalent*:

(1) $R/\mathfrak{a}$ *is regular.*

(2) $\mathfrak{a}$ *is generated by part of a regular system of parameters of* R.

PROOF. The implication "(2) $\Rightarrow$ (1)" follows from Proposition 6.A.3.

For the converse we use Lemma 6.A.5 below which says that any regular local ring is an integral domain. Let $d := \dim R = \mathrm{Dim}_k\, \mathfrak{m}/\mathfrak{m}^2$ and let the residue classes of $a_1, \ldots, a_r \in \mathfrak{a}$ form a basis of the vector space $(\mathfrak{a} + \mathfrak{m}^2)/\mathfrak{m}^2$. By Proposition 6.A.3 the ring $R/\mathfrak{a}'$, $\mathfrak{a}' := (a_1, \ldots, a_r) \subseteq \mathfrak{a}$, is a regular local ring of

---

[1]) This argument gives, quite generally, a simple proof of the following special case of Krulls's Principal Ideal Theorem: *If* Ax *is a principal prime ideal in a Noetherian ring* A, *then* $\mathrm{ht}\, Ax \le 1$. – The reader should also prove that any Noetherian local ring with principal maximal ideal is a principal ideal ring (not necessarily a domain), see Remark 3.B.44.

dimension $d - r = \text{Dim}_k \, \mathfrak{m}/(\mathfrak{a} + \mathfrak{m}^2) = \text{emdim} \, R/\mathfrak{a} = \dim R/\mathfrak{a}$. Since $R/\mathfrak{a}'$ is an integral domain, $\mathfrak{a} = \mathfrak{a}'$. ●

6.A.5. Lemma *A regular local ring R is an integral domain.*

PROOF. We use induction on $d := \dim R$. Let $d > 0$ and let $\mathfrak{p}_1, \ldots, \mathfrak{p}_r$ be the minimal prime ideals of R. By Lemma 3.B.31 there exists an element $a \in \mathfrak{m}$, $a \notin \mathfrak{m}^2 \cup \mathfrak{p}_1 \cup \cdots \cup \mathfrak{p}_r$. R/Ra is a regular local ring of dimension $d - 1$ by Proposition 6.A.3 and hence an integral domain by induction hypothesis. Since Ra is a prime ideal, there exists a minimal prime ideal $\mathfrak{p} \subset Ra$. Then $\mathfrak{p} = \mathfrak{a}a$ for some ideal $\mathfrak{a} \subseteq R$. Because of $a \notin \mathfrak{p}$ we have $\mathfrak{a} = \mathfrak{p}$ and $\mathfrak{p} = 0$ by Nakayama's Lemma. ●

One of the most important theorems on regular local rings is the syzygy theorem. To describe this theorem let R be an arbitrary commutative ring and let M be an R-module. A p r o j e c t i v e   r e s o l u t i o n of M is an exact sequence

$$\cdots \to F_2 \to F_1 \to F_0 \to M \to 0$$

with projective R-modules $F_0, F_1, F_2, \ldots$. If the R-modules F_i, $i = 0, 1, 2, \ldots$, are even free, then the resolution is called a f r e e   r e s o l u t i o n of M. If $F_i = 0$ for $i > i_0$ with $F_{i_0} \neq 0$, then i_0 is called the l e n g t h of the resolution. If M has a projective resolution of finite length, then M is called a module of f i n i t e p r o j e c t i v e   d i m e n s i o n. In this case the p r o j e c t i v e   d i m e n s i o n of M is the infimum of the lengths of its projective resolutions. It is denoted by

$$\text{pd}_R M \,.$$

By convention $\text{pd}_R 0 = -\infty$.

Projective modules over a local ring R are free. We use this result only for finite R-modules M. In this case, its *proof* is very simple: Let $x_1, \ldots, x_n$ be a minimal system of generators of M, $f : R^n \to M$ the surjective homomorphism of R-modules sending the elements e_i of the standard basis of R^n to x_i, $i = 1, \ldots, n$, and let $K := \text{Ker} f$. Because $x_1, \ldots, x_n$ is minimal, one has $K \subseteq \mathfrak{m}_R R^n$ and since M is projective, K is a direct summand of R^n, i.e. $R^n \cong K \oplus M$. This implies $K = K \cap \mathfrak{m}_R R^n = \mathfrak{m}_R K$ and thus $K = 0$ by Nakayama's Lemma and $R^n \cong M$.

Now let M be an arbitrary finite module over a Noetherian local ring R with residue class field $k = R/\mathfrak{m}$. Then M has an almost canonical projective, i.e. free resolution. Take a minimal system $x_1, \ldots, x_{n_0}$ of generators of M and let $f_0 : R^{n_0} \to M$ be as above the surjective homomorphism of R-modules with $e_i \mapsto x_i$, $i = 1, \ldots, n_0$. The kernel M_1 of f_0 is the (first) syzygy module of M (with respect to $x_1, \ldots, x_{n_0}$). It is finitely generated (since R is Noetherian) and a submodule of $\mathfrak{m}R^{n_0}$ (since the generating system is minimal). Now take a minimal system of generators $y_1, \ldots, y_{n_1}$ of M_1. This defines a surjective homomorphism $f_1 : R^{n_1} \to M_1 \subseteq R^{n_0}$. Repeating this way we get the desired free resolution

$$\cdots \longrightarrow R^{n_1} \overset{f_1}{\longrightarrow} R^{n_0} \overset{f_0}{\longrightarrow} M \longrightarrow 0,$$

which is called a m i n i m a l   f r e e   r e s o l u t i o n of M. The integers $n_i \in \mathbb{N}$ in a minimal free resolution are uniquely determined by M. *Proof.* In the induced complex $\cdots \to R^{n_1}/\mathfrak{m}R^{n_1} \to R^{n_0}/\mathfrak{m}R^{n_0} \to 0$ all the homomorphisms are zero and its i-th homology is, therefore,

$$\operatorname{Tor}_i^R(k, M) \cong k^{n_i}, \quad i \in \mathbb{N}.$$

The integers $n_i = \operatorname{Dim}_k \operatorname{Tor}_i^R(k, M)$ are called the B e t t i   n u m b e r s of M. In particular, M has finite projective dimension if and only if there is an integer $i_0 \in \mathbb{N}$ such that $\operatorname{Tor}_i^R(k, M) = 0$ for $i > i_0$. In this case $\operatorname{pd}_R M$ is the supremum of $i \in \mathbb{N}$ with $\operatorname{Tor}_i^R(k, M) \neq 0$. The Betti numbers of a finite direct sum are the sum of the corresponding Betti numbers of the summands. In particular, a finite direct sum $M = M_1 \oplus \cdots \oplus M_m$ of finite R-modules $M_1, \ldots, M_m$ has finite projective dimension if and only if each M_i, $i = 1, \ldots, r$, has finite projective dimension. In this case $\operatorname{pd}_R M = \max \{\operatorname{pd}_R M_i \mid i = 1, \ldots, m\}$. Since the Tor-vector spaces $\operatorname{Tor}_i^R(k, M)$ can be computed from a free resolution $\cdots \to F_1 \to F_0 \to k \to 0$ of k as the homology of the derived complex $\cdots \to F_1 \otimes_R M \to F_0 \otimes_R M \to 0$, one gets the following:

If the residue field k of a Noetherian local ring R has finite projective dimension, then for any finite R-module M one has $\operatorname{pd}_R M \leq \operatorname{pd}_R k$.

Now we can state the syzygy theorem:

6.A.6. Syzygy Theorem of Hilbert–Serre *For a Noetherian local ring R the following conditions are equivalent*:

(1) *R is a regular local ring.*

(2) *Any finite R-module has finite projective dimension.*

(3) *The residue field $k = R/\mathfrak{m}$ of R has finite projective dimension (as R-module).*

In this case, for every finite R-module M

$$\operatorname{pd}_R M \leq \operatorname{pd}_R k = \dim R.$$

PROOF. $(1) \Rightarrow (2)$: (Induction on $d := \dim R$) If $d = 0$, then R is a field. Let $d > 0$ and take $a \in \mathfrak{m} \setminus \mathfrak{m}^2$. Then by Proposition 6.A.3 $\overline{R} := R/Ra$ is a regular local ring of dimension $d - 1$ and by Lemma 6.A.5 a is a non-zero divisor of R. Let M be a non-zero finite R-module, $\cdots \to F_1 \to F_0 \to M \to 0$ a minimal free resolution of M and let M' be the kernel of $F_0 \to M$. If M is not free, i.e. if $M' \neq 0$, then the sequence $\cdots \to F_2/aF_2 \to F_1/aF_1 \to M'/aM' \to 0$ is also exact (see Exercise 6.A.10) and hence a minimal free resolution of the $\overline{R}$-module $\overline{M}' := M'/aM'$. By induction hypothesis $\operatorname{pd}_{\overline{R}} \overline{M}' \leq d - 1$ and hence $\operatorname{pd}_R M = \operatorname{pd}_R M' + 1 = \operatorname{pd}_{\overline{R}} \overline{M}' + 1 \leq d$.

The implication $(2) \Rightarrow (3)$ is clear. To prove the implication $(3) \Rightarrow (1)$ by induction on $\dim R$ we use some elementary facts about the zero-divisors of a Noetherian ring which will be discussed in Exercises 6.A.10 and 6.A.11 below and prove simultaneously the equation $\operatorname{pd}_R k = \dim R$. Let $0 \to F_r \to \cdots \to F_0 = R \to k \to 0$

be a minimal free resolution of k. If $r = 0$, then $R = k$ is a field. Let $r > 0$. Then F_r is a non-zero free submodule of $\mathfrak{m}F_{r-1}$ and the maximal ideal $\mathfrak{m}$ is not an associated prime ideal of R, since otherwise by Exercise 6.A.11 there would exist an element $x \in R \setminus \{0\}$ with $x\mathfrak{m} = 0$ and hence $xF_r = 0$. In particular, $\dim R > 0$. By Lemma 3.B.31 there exists a (non-zero divisor) $a \in \mathfrak{m}$, $a \notin \mathfrak{m}^2 \cup (\bigcup_{\mathfrak{p} \in \mathrm{Ass}\, R} \mathfrak{p})$. Then $0 \to F_r/aF_r \to \cdots \to F_1/aF_1 \to \mathfrak{m}/a\mathfrak{m} \to 0$ is a minimal free resolution of the $\overline{R}$-module $\mathfrak{m}/a\mathfrak{m}$ (by Exercise 6.A.10), where $\overline{R} := R/Ra$. But $k = R/\mathfrak{m} \cong Ra/a\mathfrak{m}$ is a direct summand of $\mathfrak{m}/a\mathfrak{m}$ (with complement $(Rb_1 + \cdots + Rb_s + a\mathfrak{m})/a\mathfrak{m}$, where $a, b_1, \ldots, b_s$ is a minimal generating system of $\mathfrak{m}$). Hence k is also an $\overline{R}$-module of finite projective dimension $\leq r - 1$. By induction hypothesis, $\overline{R}$ is a regular local ring of dimension $\leq r - 1$. Hence, because $\dim R = \dim \overline{R} + 1$ and $\mathrm{emdim}\, R = \mathrm{emdim}\, \overline{R} + 1$, R is a regular local ring of dimension r.

The proof of the implication $(1) \Rightarrow (2)$ showed that $\mathrm{pd}_R M \leq \dim R$ for all finite R-modules M and in the proof of the implication "$(3) \Rightarrow (1)$" we also showed that $\mathrm{pd}_R k = \dim R$.   ●

Two of the most spectacular corollaries of the homological characterization of regular local rings in Theorem 6.A.6 are the following theorems for which (in this generality) no non-homological proofs are known till now.

6.A.7. Theorem (Auslander–Buchsbaum–Serre) *Let R be a regular local ring. Then $R_\mathfrak{p}$ is regular for every $\mathfrak{p} \in \mathrm{Spec}\, R$.*

PROOF. For a given prime ideal $\mathfrak{p} \in \mathrm{Spec}\, R$ the R-module $R/\mathfrak{p}$ has a finite free resolution $0 \to F_r \to \cdots \to F_0 \to R/\mathfrak{p} \to 0$ by Theorem 6.A.6. Localizing at $\mathfrak{p}$ yields the free resolution $0 \to (F_r)_\mathfrak{p} \to \cdots \to (F_0)_\mathfrak{p} \to R_\mathfrak{p}/\mathfrak{p}R_\mathfrak{p} \to 0$ of the residue field $\kappa(\mathfrak{p}) = R_\mathfrak{p}/\mathfrak{p}R_\mathfrak{p}$ of $R_\mathfrak{p}$. Again by Theorem 6.A.6, $R_\mathfrak{p}$ is regular.   ●

6.A.8. Theorem (Auslander–Buchsbaum) *Every regular local ring R is factorial.*

PROOF. The following proof by induction on $\dim R$ is due to I. Kaplansky. Let $\dim R > 1$ and $a \in \mathfrak{m}_R \setminus \mathfrak{m}_R^2$. Then a is a prime element in R since R/Ra is also regular. Therefore, it is enough to show that the ring of fractions R_a is factorial, i.e. that for every prime ideal $\mathfrak{p} \in D(a) \subseteq \mathrm{Spec}\, R$ with $\mathrm{ht}\, \mathfrak{p} = 1$ the extended ideal $\mathfrak{p}R_a$ is principal. Since for every $\mathfrak{q} \in D(a)$ the localization $R_\mathfrak{q}$ is regular by Theorem 6.A.7 and moreover of dimension $< \dim R$, $R_\mathfrak{q}$ is factorial by induction hypothesis and hence the ideal $\mathfrak{p}R_\mathfrak{q}$ principal. It follows that the ideal $\mathfrak{p}R_a$ is an invertible ideal of R_a (i.e. a projective R_a-module of rank 1). Now from a free resolution $0 \to F_r \to \cdots \to F_0 \to \mathfrak{p} \to 0$ of $\mathfrak{p}$ which exists by Theorem 6.A.6, we derive a free resolution $0 \to (F_r)_a \to \cdots \to (F_0)_a \to \mathfrak{p}R_a \to 0$ of the projective R_a-module $\mathfrak{p}R_a$. Then $\mathfrak{p}R_a \oplus (F_1)_a \oplus (F_3)_a \oplus \cdots \cong (F_0)_a \oplus (F_2)_a \oplus \cdots$, i.e. $\mathfrak{p}R_a \oplus R_a^n \cong R_a^{n+1}$ for some $n \in \mathbb{N}$. Taking the $(n+1)$-th exterior power one obtains $(\mathfrak{p}R_a) \otimes \Lambda^n(R_a^n) \cong \Lambda^{n+1}(R_a^{n+1})$ and finally $\mathfrak{p}R_a \cong R_a$.   ●

6.A.9. Exercise Let $R \subseteq S$ be a faithfully flat extension of local Noetherian rings (i. e. S is flat over R and the inclusion $R \subseteq S$ is a local homomorphism). Show that, if S is a regular local ring, then R is also a regular local ring. (**Hint:** Use the criterion of Theorem 6.A.6 and the following general result: Let $R \subseteq S$ be a faithfully flat extension of local Noetherian rings and M a finitely generated R-module. If $\cdots \to F_2 \to F_1 \to F_0 \to M \to 0$ is a minimal free resolution of M over R, then $\cdots \to S \otimes_R F_2 \to S \otimes_R F_1 \to S \otimes_R F_0 \to S \otimes_R M \to 0$ is a minimal free resolution of $S \otimes_R M$ over S. In particular, $\mathrm{pd}_R M = \mathrm{pd}_S(S \otimes_R M)$.)

6.A.10. Exercise Let R be a commutative ring, $0 \to M' \to M \to M'' \to 0$ an exact sequence of R-modules and let $a \in R$ be a non-zero divisor for M''. Then the sequence $0 \to M'/aM' \to M/aM \to M''/aM'' \to 0$ is also exact. (**Hint:** Apply the Snake Lemma to the following commutative diagram with exact rows.)

$$
\begin{array}{ccccccccc}
0 & \longrightarrow & M' & \longrightarrow & M & \longrightarrow & M'' & \longrightarrow & 0 \\
 & & a\downarrow & & a\downarrow & & a\downarrow & & \\
0 & \longrightarrow & M' & \longrightarrow & M & \longrightarrow & M'' & \longrightarrow & 0
\end{array}
$$

6.A.11. Exercise (A s s o c i a t e d   p r i m e s) Let R be a commutative ring. A prime ideal $\mathfrak{p} \in \operatorname{Spec} R$ is called an a s s o c i a t e d   p r i m e   i d e a l of an R-module M if there is an element $x \in M$ such that $\operatorname{Ann}_R x = \mathfrak{p}$, i. e. if the R-module $R/\mathfrak{p}$ can be embedded into M. The set of associated prime ideals of M is denoted by

$$\operatorname{Ass} M = \operatorname{Ass}_R M \,.$$

We remind that an element $a \in R$ is called a n o n - z e r o - d i v i s o r of M, if the multiplication by a in M is injective. Show that:

(1) $\operatorname{Ass} M' \subseteq \operatorname{Ass} M$ for every submodule M' of M.

(2) $\operatorname{Ass} M' = \{\mathfrak{p}\}$ for every $\mathfrak{p} \in \operatorname{Spec} R$ and every submodule $M' \neq 0$ of $R/\mathfrak{p}$.

(3) For an exact sequence $0 \to M' \to M \to M''$ of R-modules $\operatorname{Ass} M \subseteq \operatorname{Ass} M' \cup \operatorname{Ass} M''$. (**Hint:** If $\mathfrak{p} \in \operatorname{Ass} M$ and $\mathfrak{p} \notin \operatorname{Ass} M'$, then an inclusion $R/\mathfrak{p} \to M$ composed with the homomorphism $M \to M''$ gives an inclusion $R/\mathfrak{p} \to M''$.)

(4) If $\mathfrak{p}$ is maximal in the set of ideals $\operatorname{Ann} x$, $x \in M \setminus \{0\}$, then $\mathfrak{p}$ is a prime ideal and hence $\mathfrak{p} \in \operatorname{Ass} M$. (**Hint:** If $ab \in \mathfrak{p} = \operatorname{Ann} x_0$ and $ax_0 \neq 0$, then $b \in \operatorname{Ann} ax_0 \supseteq \operatorname{Ann} x_0$. Hence $b \in \mathfrak{p} = \operatorname{Ann} x_0$, since $\operatorname{Ann} ax_0 = \operatorname{Ann} x_0$ by the choice of $\mathfrak{p}$.)

(5) For a Noetherian ring R and an R-module M the following conditions are equivalent: a) $M = 0$. b) $\operatorname{Ass} M = \emptyset$.

(6) For a Noetherian ring R and an R-module M the set of zero-divisors of M is $\bigcup_{\mathfrak{p} \in \operatorname{Ass} M} \mathfrak{p}$.

(7) For a Noetherian ring R and a *finite* R-module M the set $\operatorname{Ass} M$ is finite. In particular, the set of zero-divisors of M is a finite union of prime ideals of R. Moreover, if $\mathfrak{a} \subseteq R$ is an ideal consisting only of zero-divisors of M, then there is an $x \in M$, $x \neq 0$, such that $\mathfrak{a}x = 0$. (**Hint:** By Noetherian induction one proves that there is a sequence $0 = M_0 \subset M_1 \subset \cdots \subset M_r = M$ of submodules such that $M_i/M_{i-1} \cong R/\mathfrak{p}_i$ with $\mathfrak{p}_i \in \operatorname{Spec} R$, $i = 1, \ldots, r$. Then $\operatorname{Ass} M \subseteq \{\mathfrak{p}_1, \ldots, \mathfrak{p}_r\}$ by (3). Furthermore, $\mathfrak{a} \subseteq \bigcup_{\mathfrak{p} \in \operatorname{Ass} M} \mathfrak{p}$ and by Lemma 3.B.31 $\mathfrak{a} \subseteq \mathfrak{p}$ for some $\mathfrak{p} \in \operatorname{Ass} M$.)

(8) For a Noetherian ring R, an R-module M and a multiplicatively closed subset S in R

$$\operatorname{Ass}_{R_S} M_S = \operatorname{Spec} R_S \cap \operatorname{Ass}_R M$$

where Spec R_S is identified in the canonical way with a subspace of Spec R. In particular, $\mathfrak{p} \in \mathrm{Ass}\,M$ if and only if $\mathfrak{p}R_\mathfrak{p} \in \mathrm{Ass}_{R_\mathfrak{p}}M_\mathfrak{p}$. (**Hint:** Use the canonical identification $\mathrm{Hom}_{R_S}(R_S/\mathfrak{p}R_S, M_S) = \mathrm{Hom}_R(R/\mathfrak{p}, M)_S$.)

(9) For a Noetherian ring R and a finite R-module M, if $\mathfrak{p} \in \mathrm{Spec}\,R$ is a minimal prime ideal in $\mathrm{V}(\mathrm{Ann}_R M) = \mathrm{Spec}(R/\mathrm{Ann}_R M) \subseteq \mathrm{Spec}\,R$, then $\mathfrak{p} \in \mathrm{Ass}\,M$. (**Hint:** $\mathfrak{p}R_\mathfrak{p} \in \mathrm{Ass}_{R_\mathfrak{p}}M_\mathfrak{p}$.)

(10) For a graded module M over a $\mathbb{Z}$-graded Noetherian ring A, the elements of $\mathrm{Ass}_A M$ are homogeneous prime ideals of A. (**Hint:** If M is finite over A, prove that there is a sequence $0 = M_0 \subset M_1 \subset \cdots \subset M_r = M$ of *homogeneous* submodules of M such that $M_i/M_{i-1} \cong (A/\mathfrak{p}_i)(-m_i)$, $i = 1, \ldots, r$, with homogeneous prime ideals $\mathfrak{p}_i \subseteq A$ and integers $m_i \in \mathbb{Z}$ (in particular, $M_1 = Ax_1$ with a homogeneous element $x_1 \in M$ of degree m_1 and $\mathrm{Ann}_A x_1 = \mathfrak{p}_1$). For a graded module N and an integer n, the module $N(n)$ is the so called s h i f t o f N with $N(n)_m := N_{n+m}$. – An analogous statement is true for $\mathbb{Z}^n$-graded Noetherian rings or, more generally, for D-graded Noetherian rings, where the grading group D has the following property: A D-graded ring $\neq 0$ is an integral domain if its *homogeneous* elements $\neq 0$ are non-zero divisors.)

(11) Let R be an A-algebra and let $\varphi^* : \mathrm{Spec}\,R \to \mathrm{Spec}\,A$ be the corresponding map to the structure homomorphism $\varphi : A \to R$. Assume that R is Noetherian (but A not necessarily). Show for an arbitrary R-module M the equality $\mathrm{Ass}_A M = \varphi^*(\mathrm{Ass}_R M)$. In particular (by (7)), $\mathrm{Ass}_A M$ is finite if M is a finite R-module. (**Hint:** For the inclusion $\mathrm{Ass}_A M \subseteq \varphi^*(\mathrm{Ass}_R M)$ one may assume that M is a finite (even cyclic) R-module. Then consider $M_\mathfrak{p}$ for a given $\mathfrak{p} \in \mathrm{Ass}_A M$ and use (8).) As an example conclude that, for a finite graded module $M = \oplus_{m \in \mathbb{Z}} M_m$ over a positively graded noetherian ring $A = \oplus_{m \in \mathbb{N}} A_m$, the set $\mathrm{Ass}_{A_0} M = \bigcup_{m \in \mathbb{Z}} \mathrm{Ass}_{A_0} M_m$ is finite. Even more is true: The sequence $\mathrm{Ass}_{A_0} M_m, m \in \mathbb{Z}$, is periodic with period $\gamma := \mathrm{LCM}\,(\gamma_i \mid i \in I)$ if A is generated as an A_0-algebra by (finitely many) homogeneous elements x_i of positive degrees $\gamma_i := \deg x_i, i \in I$, i.e. there is an $m_0 \in \mathbb{Z}$ with $\mathrm{Ass}_{A_0} M_{m+\gamma} = \mathrm{Ass}_{A_0} M_m$ for all $m \geq m_0$. (For this use part (10) above.)

(12) Let M be a finite module over the Noetherian ring R. If the ideal $\mathfrak{a} \subseteq R$ contains a non-zero divisor of M, then $\mathfrak{a}$ is generated by non-zero-divisors of M. (**Hint:** Use Lemma 3.B.31. – Show by example that the finiteness of M is necessary in general.)

6.A.12. Definition Let X be a locally Noetherian scheme. A point $x \in X$ is called a r e g u l a r p o i n t of X, if the stalk $\mathcal{O}_{X,x}$ is a regular local ring. X is called r e g u l a r, if all its points are regular. – A Noetherian ring R is called r e g u l a r, if the affine scheme Spec R is regular, i.e. if all the localizations $R_\mathfrak{p}, \mathfrak{p} \in \mathrm{Spec}\,R$, are regular local rings.

A point x of a locally Noetherian scheme X with non-regular stalk $\mathcal{O}_{X,x}$ is sometimes called a s i n g u l a r p o i n t or a s i n g u l a r i t y of X. However, this terminology is not universally accepted. For instance, some authors prefer to call non-smooth points of an algebraic K-scheme, K a field, singular. For them a singular point may be regular. See Section 6.D, in particular, Proposition 6.D.23.

By Theorem 6.A.7 a Noetherian ring R is regular if and only if the localizations $R_\mathfrak{m}$ are regular for all $\mathfrak{m} \in \mathrm{Spm}\,R$. More generally, all the generalizations of a regular point of a locally Noetherian scheme are regular.

6.A.13. Example The polynomial algebra $R[T_1, \ldots, T_n]$ over a regular Noetherian ring R is also regular. More geometrically, *if X is a regular locally Noetherian scheme, then the affine* *space* $\mathbb{A}_X^n := X \times_{\mathbb{Z}} \mathbb{A}_{\mathbb{Z}}^n$ *over* X *is regular.* For the *proof* we may assume that $n = 1$. Let $\mathfrak{P}$ be a prime ideal in $R[T]$ and $\mathfrak{p} := R \cap \mathfrak{P}$. Then $R[T]_{\mathfrak{P}} = R_{\mathfrak{p}}[T]_{\mathfrak{P}R_{\mathfrak{p}}[T]}$. Therefore, we can assume that R is a regular local ring of dimension d and $\mathfrak{p} = \mathfrak{m}$ is the maximal ideal of R. In this case, $\mathfrak{P}$ is the extension $\mathfrak{m}R[T]$ or an ideal $\mathfrak{m}R[T] + FR[T]$, where $F \in R[T]$ is a polynomial which is irreducible modulo $\mathfrak{m}$. In the first case, ht $\mathfrak{P} = d$ and $\mathfrak{P}$ like $\mathfrak{m}$ is generated by d elements. In particular, $R[T]_{\mathfrak{P}}$ is a regular local ring. In the second case, ht $\mathfrak{P} = d + 1$ and $\mathfrak{P}$ is generated by $d + 1$ elements and hence again $R[T]_{\mathfrak{P}}$ is regular.•

Also the projective space $\mathbb{P}_X^n := X \times_{\mathbb{Z}} \mathbb{P}_{\mathbb{Z}}^n$ *over a regular locally Noetherian scheme* X *is again regular.* This follows from the fact that $\mathbb{P}_X^n$ has a cover by $n + 1$ open subschemes which are isomorphic to $\mathbb{A}_X^n$.

As particular cases the affine space $\mathbb{A}_K^n$ and the projective space $\mathbb{P}_K^n$ over a field K are regular schemes. For the affine case we know already from Exercise 2.B.14 (1b) that a maximal ideal $\mathfrak{m} \subseteq K[T_1, \dots, T_n]$ has height n and is generated by n elements. The stalk

$$K[T_1, \dots, T_n]_{\mathfrak{m}} = \{F/G \mid F, G \in K[T_1, \dots, T_n], G(0) \neq 0\},$$

at the origin $0 \in K^n = \mathbb{A}_K^n(K)$ is a typical example of a regular local ring of dimension n. Here the maximal ideal is $\mathfrak{m} := (T_1, \dots, T_n)$. The indeterminates $T_1, \dots, T_n$ form a regular system of parameters. More generally, polynomials $F_1, \dots, F_r \in \mathfrak{m} := (T_1, \dots, T_n)$ are part of a regular system of parameters of $K[T_1, \dots, T_n]_{\mathfrak{m}}$ if and only if the linear parts of $F_1, \dots, F_r$ are linearly independent, i.e. if and only if the J a c o b i a n $\partial(F_1, \dots, F_r)/\partial(T_1, \dots, T_n)$ has in 0 the maximal rank r. Together with Corollary 6.A.4 this indicates the strong connection between the concept of regularity and the concept of smoothness described in the case $K = \mathbb{K}$ in Example 2.C.4. See also Section 6.D.

6.A.14. Exercise Let R be a positively graded Noetherian ring. If R is standardly graded, then Proj R is regular if and only if $D(R_+) \subseteq$ Spec R is a regular open subscheme of the cone Spec R. If R is not standardly graded, then Spec R can be regular without Proj R being regular. For examples consider the weighted projective spaces $\mathbb{P}_{\gamma, K}^n = \text{Proj } K[X_0, \dots, X_n]$ with K a field and deg $X_i = \gamma_i > 0$, $i = 0, \dots, n$. (**Remark:** $\mathbb{P}_{\gamma, K}^n$ is always a normal scheme, cf. Exercise 6.B.26.)

6.B. Normal Domains

Let us recall that an integral domain R is by definition a n o r m a l domain, if R is integrally closed in its field of fractions $Q(R)$. Indeed this is a local property: *An integral domain* R *is normal if and only if all the localizations* $R_{\mathfrak{p}}$, $\mathfrak{p} \in$ Spec R, *are normal.* For R to be normal it is even sufficient to have $R_{\mathfrak{m}}$ to be normal for all $\mathfrak{m} \in$ Spm R. For, if all the $R_{\mathfrak{m}}$, $\mathfrak{m} \in$ Spm R, are normal and if $x \in Q(R)$ is integral over R, then x is integral over $R_{\mathfrak{m}}$ for all $\mathfrak{m} \in$ Spm R and hence $x \in \bigcap_{\mathfrak{m} \in \text{Spm } R} R_{\mathfrak{m}} = R$.

Regular local rings and hence regular Noetherian integral domains are normal. Of course, this follows from the fact that regular local rings are even factorial (cf. Theorem 6.A.8 and Example 1.E.8 (3)). But the normality of a regular local ring is much more elementary than its factoriality. See Exercise 6.B.11 below.

In dimension ≤ 1 regularity and normality for Noetherian local rings coincide.

6.B.1. Theorem *Let R be a Noetherian local ring of dimension 1. Then, R is regular if and only if R is a normal domain.*

PROOF. We have already proved in Example 6.A.2 (2) that any regular local ring of dimension 1 is a discrete valuation ring and hence factorial and normal. Conversely, let R be a normal domain and $0 \neq a \in \mathfrak{m}_R$. Let n be the positive integer such that $\mathfrak{m}_R^n \subseteq Ra$, $\mathfrak{m}_R^{n-1} \not\subseteq Ra$. Let $b \in \mathfrak{m}_R^{n-1}$ and $b \notin Ra$, and let $x := a/b \in Q(R)$. Then $x^{-1} = b/a \notin R$ and hence x^{-1} is not integral over R. From this it follows that $x^{-1}\mathfrak{m}_R \not\subseteq \mathfrak{m}_R$, because otherwise $\mathfrak{m}_R$ would be a faithful $R[x^{-1}]$-module and a finite R-module which contradicts Proposition 1.E.2. But $b\mathfrak{m}_R \subseteq \mathfrak{m}_R^{n-1}\mathfrak{m}_R = \mathfrak{m}_R^n \subseteq Ra$ so that $x^{-1}\mathfrak{m}_R \subseteq R$ and thus $x^{-1}\mathfrak{m}_R = R$. Hence, $\mathfrak{m}_R = Rx$.   •

6.B.2. Proposition *For a Noetherian local ring R the following conditions are equivalent:*

(1) *R is a normal domain.*

(2) *R is reduced and integrally closed in its total quotient ring $Q(R)$.*

PROOF. (1) implies (2) by definition. For the converse, we only have to show that R is an integral domain, if condition (2) holds. Since R is Noetherian and reduced, $(0) = \mathfrak{p}_1 \cap \cdots \cap \mathfrak{p}_r$, where $\mathfrak{p}_1, \ldots, \mathfrak{p}_r$ are the minimal prime ideals of R. Therefore, we have the canonical inclusion $R \hookrightarrow R_1 \times \cdots \times R_r$, where $R_i := R/\mathfrak{p}_i$, $1 \leq i \leq r$. Obviously, $Q(R) = Q(R_1 \times \cdots \times R_r) = Q(R_1) \times \cdots \times Q(R_r)$. Now $R_1 \times \cdots \times R_r$ is a finite and hence integral extension of R. Therefore, by assumption, $R = R_1 \times \cdots \times R_r$. But, R is local, so its only idempotents are 0 and 1. This implies that $r = 1$ and so R is a domain.   •

6.B.3. Exercise The proof of the last proposition gives the following more general result. Let R be Noetherian and reduced with minimal prime ideals $\mathfrak{p}_1, \ldots, \mathfrak{p}_r$ and let $\overline{R}$ denote the integral closure of R in $Q(R)$. Then $\overline{R} = \overline{R_1} \times \cdots \times \overline{R_r}$, $R_i := R/\mathfrak{p}_i$, $i = 1, \ldots, r$. In particular, R is integrally closed in $Q(R)$ if and only if $R \cong R_1 \times \cdots \times R_r$, where each R_i is a normal Noetherian domain.

Now we will prove an important criterion for the normality of a local Noetherian ring, due to Krull and Serre.

6.B.4. Theorem (Criterion of Krull–Serre for Normality) *For a Noetherian local ring R the following conditions are equivalent:*

(1) *R is a normal domain.*

(2) a) *For every prime ideal $\mathfrak{p}$ of height 1 in R, the ring $R_\mathfrak{p}$ is regular (i. e. $R_\mathfrak{p}$ is a discrete valuation ring).*
b) *R is reduced and for every non-zero divisor $f \in R$ the associated primes of the R-module R/Rf are of height 1 in R.*

Before going into the proof of the theorem we refer to Exercise 6.A.11 for the concept of associated primes. We note that the associated primes of R/Rf are sometimes also called the associated primes of Rf. More generally, for an ideal $\mathfrak{a} \in R$ the associate primes of $R/\mathfrak{a}$ are also called the associated primes of $\mathfrak{a}$.

PROOF. To prove the implication $(1) \Rightarrow (2)$ let R be a normal domain and let $\mathfrak{p}$ be a prime ideal in R of height 1. Then, $R_{\mathfrak{p}}$ is a normal local domain of dimension 1 and hence by Theorem 6.B.1 $R_{\mathfrak{p}}$ is regular. This proves condition (2) a). To prove condition (2) b) we have to show that for $0 \neq f \in R$ and $\mathfrak{p} \in \mathrm{Ass}(R/Rf)$ the ideal $\mathfrak{p}R_{\mathfrak{p}}$ is principal. For this we may replace R by $R_{\mathfrak{p}}$ and hence assume that R is local with maximal ideal $\mathfrak{m} = \mathfrak{p} \in \mathrm{Ass}(R/Rf)$. By definition of $\mathrm{Ass}(R/Rf)$ there exists a $g \in R$, such that $\mathfrak{m}g \subseteq Rf$ and $g \notin Rf$. Let $x := f/g \in Q(R)$. Then, $x^{-1} = g/f \notin R$ and $x^{-1}\mathfrak{m} \subseteq R$. In case $x^{-1}\mathfrak{m} \subseteq \mathfrak{m}$ the element x^{-1} would be integral over R (cf. Proposition 1.E.2). Therefore, $x^{-1}\mathfrak{m} = R$ and so $\mathfrak{m} = Rx$, because R is normal and $x^{-1} \notin R$.

To prove $(2) \Rightarrow (1)$ let R be reduced and let $\mathfrak{p}_1, \dots, \mathfrak{p}_r$ be the minimal primes in R. Then, as in the proof of Proposition 6.B.2 (see also Exercise 6.B.3) we have, $R \subseteq R_1 \times \cdots \times R_r$, where $R_i := R/\mathfrak{p}_i$, and $Q(R) = Q(R_1) \times \cdots \times Q(R_r)$ as well as $\overline{R} = \overline{R_1} \times \cdots \times \overline{R_r}$, where $\overline{R_i}$ is the integral closure of R_i in $Q(R_i)$. We have to show that $\overline{R} = R$. Let $c = f/g \in Q(R)$ be integral over R. Then, $c/1 \in Q(R_{\mathfrak{p}})$ is integral over $R_{\mathfrak{p}}$ and hence $c/1 \in R_{\mathfrak{p}}$ for all prime ideals $\mathfrak{p} \in \mathrm{Ass}(R/Rg)$, since $R_{\mathfrak{p}}$ is normal for such $\mathfrak{p}$ by hypothesis. Therefore, $f \in R_{\mathfrak{p}}g$ for all $\mathfrak{p} \in \mathrm{Ass}(R/Rg)$ and it follows from the following lemma that $f \in Rg$, hence $c \in R$.   •

6.B.5. Lemma *Let f, g be elements of a Noetherian ring R with $f/1 \in R_{\mathfrak{p}}g$ for all $\mathfrak{p} \in \mathrm{Ass}(R/Rg)$. Then $f \in Rg$.*

PROOF. Suppose that $(Rf + Rg)/Rg \neq 0$. Let $\mathfrak{p} \in \mathrm{Ass}((Rf + Rg)/Rg) \subseteq \mathrm{Ass}(R/Rg)$. Then $\mathfrak{p}R_{\mathfrak{p}} \in \mathrm{Ass}((Rf + Rg)/Rg)_{\mathfrak{p}}$, but $((Rf + Rg)/Rg)_{\mathfrak{p}} = 0$ by hypothesis, a contradiction.   •

6.B.6. Exercise Show that for a Noetherian integral domain R and a subset $Z \subseteq \mathrm{Spec}\, R$ the following conditions are equivalent: (1) $R = \bigcap_{\mathfrak{p} \in Z} R_{\mathfrak{p}}$. (2) For any $g \in R \setminus \{0\}$ and any $\mathfrak{q} \in \mathrm{Ass}(R/Rg)$ there exists a $\mathfrak{p} \in Z$ with $\mathfrak{q} \subseteq \mathfrak{p}$. (**Hint:** Imitate the proof of Lemma 6.B.5.)

6.B.7. Exercise Show that for a Noetherian normal domain R one has $R = \bigcap_{\mathrm{ht}\,\mathfrak{p}=1} R_{\mathfrak{p}}$. (**Hint:** Use Exercise 6.B.6 and Theorem 6.B.4.)

6.B.8. Exercise Let R be a Noetherian local ring, M a finite R-module and $f \in \mathfrak{m}_R$ a non-zero divisor of M.

(1) If g is a non-zero-divisor of M/fM, then g is a non-zero divisor of M and f is a non-zero divisor of M/gM. (**Hint:** Apply the Snake Lemma to the following diagram. From $\mathrm{Ker}\, g = f \, \mathrm{Ker}\, g$ it follows that $\mathrm{Ker}\, g = 0$ by Nakayama's Lemma.)

$$
\begin{array}{ccccccccc}
0 & \longrightarrow & M & \xrightarrow{\ f\ } & M & \longrightarrow & M/fM & \longrightarrow & 0 \\
& & {\scriptstyle g}\downarrow & & {\scriptstyle g}\downarrow & & {\scriptstyle g}\downarrow & & \\
0 & \longrightarrow & M & \xrightarrow{\ f\ } & M & \longrightarrow & M/fM & \longrightarrow & 0
\end{array}
$$

(2) For every $\mathfrak{p} \in \mathrm{Ass}\, M$ there exists a $\mathfrak{q} \in \mathrm{Ass}(M/fM)$ with $\mathfrak{p} \subset \mathfrak{q}$. (**Hint**: By part (1) $\mathfrak{p} \subseteq \bigcup_{\mathfrak{q} \in \mathrm{Ass}(M/Mf)} \mathfrak{q}$. Obviously, the converse is true: If $\mathfrak{q} \in \mathrm{Ass}(M/fM)$, then there exists a $\mathfrak{p} \in \mathrm{Ass}\, M$ with $\mathfrak{p} \subseteq \mathfrak{q}$.)

6.B.9. Exercise Let R be a Noetherian local ring and $p \in \mathfrak{m}_R$ a non-zero divisor. Show that if R/Rp is reduced or an integral domain or normal or regular, then R has the corresponding property. (**Hint**: For the case of normal rings one checks the condition (2) in Theorem 6.B.4: Let $f \in R \setminus Rp$. If $\mathfrak{p} \in \mathrm{Ass}(R/Rf)$, then there exists a $\mathfrak{q} \in \mathrm{Ass}(R/(Rf + Rp))$ with $\mathfrak{p} \subseteq \mathfrak{q}$ (cf. Exercise 6.B.8(2)). By the normality of R/Rp the ring $R_\mathfrak{q}/R_\mathfrak{q}p$ is a discrete valuation ring. It follows that $\mathfrak{p}R_\mathfrak{q} + pR_\mathfrak{q} = gR_\mathfrak{q} + pR_\mathfrak{q}$ for some $g \in \mathfrak{p}$. Then $\mathfrak{p}R_\mathfrak{q} = gR_\mathfrak{q} + p\mathfrak{p}R_\mathfrak{q}$ (since $p \notin \mathfrak{p}R_\mathfrak{q}$) and $\mathfrak{p}R_\mathfrak{q} = gR_\mathfrak{q}$ by Nakayama's Lemma and hence $\mathfrak{p}R_\mathfrak{p} = gR_\mathfrak{p}$.)

6.B.10. Exercise Let R be a Noetherian local ring. If R is normal (resp. regular), then the power series ring $R[[T_1, \ldots, T_n]]$ is also normal (resp. regular). (**Hint**: Use Exercise 6.B.9.)

6.B.11. Exercise A Noetherian regular local ring R is normal. (**Hint**: Use induction on dim R together with Exercise 6.B.9. With the method described in the hint for Exercise 6.B.9 one proves also very elementarily that a regular local ring of dimension 2 is factorial.)

6.B.12. Remark (R e g u l a r   s e q u e n c e s   a n d   d e p t h) The criterion for normality in Theorem 6.B.4 can be conveniently formulated in terms of the depth of a local ring. The following exercises elaborate these concepts. In this remark unless otherwise specified, R denotes a Noetherian local ring with maximal ideal $\mathfrak{m} = \mathfrak{m}_R$ and M a finite non-zero R-module. We start with a definition:

6.B.13. Definition A sequence $f_1, \ldots, f_r$ of elements in R is called a r e g u l a r   s e q u e n c e for M or an M - s e q u e n c e   of   l e n g t h r, if $f_1, \ldots, f_r \in \mathfrak{m}$ and if f_{i+1} is a non-zero divisor of $M/(f_1, \ldots, f_i)M$ for $i = 0, \ldots, r - 1$. – For an arbitrary (commutative) ring R and an arbitrary module M over R, a sequence $f_1, \ldots, f_r$ of elements of R is called a s t r o n g l y   r e g u l a r   s e q u e n c e (for M), if f_{i+1} is a non-zero divisor of $M/(f_1, \ldots, f_i)M$ for $i = 0, \ldots, r - 1$.

6.B.14. Exercise (1) Let $0 \le s \le r$. Show that $f_1, \ldots, f_r \in R$ is an M-sequence if and only if $f_1, \ldots, f_s$ is an M-sequence and $f_{s+1}, \ldots, f_r$ is an $(M/(f_1, \ldots, f_s)M)$-sequence.

(2) Show that if $f_1, \ldots, f_r$ is an M-sequence, then $f_{\sigma 1}, \ldots, f_{\sigma r}$ is also an M-sequence for any permutation $\sigma \in \mathfrak{S}_r$. (**Hint**: By Exercise 6.B.8 (1) this is true for any transposition of the indices $1, \ldots, r$.)

(3) If $f_1, \ldots, f_r$ is an M-sequence, then $r \le \dim R/\mathfrak{p}$ for all $\mathfrak{p} \in \mathrm{Ass}\, M$. (**Hint**: Use induction on r. If $r > 0$ and $\mathfrak{p} \in \mathrm{Ass}\, M$ with minimal dimension $\dim R/\mathfrak{p}$, then there is a $\mathfrak{q} \in \mathrm{Ass}\, M/f_1M$ with $\mathfrak{p} \subset \mathfrak{q}$ by Exercise 6.B.8 (2).) In particular, $r \le \dim M$.[2])

(4) If $f_1, \ldots, f_r$ is an M-sequence, then ht $\mathfrak{p} = r$ for every minimal prime ideal $\mathfrak{p}$ containing $f_1 R + \cdots + f_r R$. (**Hint**: One can assume that Ann $M = 0$. Now use induction on r. $f_1 \notin \mathfrak{p}$

---

[2]) The d i m e n s i o n of a finite module N over a Noetherian ring S is by definition the dimension $\dim(S/\mathrm{Ann}_S N)$ of $\mathrm{Supp}\, N = V(\mathrm{Ann}_S N) \subseteq \mathrm{Spec}\, S$. It is the supremum of the dimensions of the cyclic submodules of N. For an arbitrary S-module the dimension is defined as the supremum of the dimensions of its finite submodules or as the supremum of the dimensions $\dim (A/\mathfrak{p})$, $\mathfrak{p} \in \mathrm{Ass}_S N$.

for all minimal primes $\mathfrak{p} \subseteq R$ (cf. Exercise 6.A.10 (9)). The inequality $\operatorname{ht} \mathfrak{p} \leq r$ holds by Theorem 3.B.29.)

By Exercise 6.B.14 (3) (or (4)) the length of any M-sequence is $\leq \dim M \leq \dim R$. An M-sequence $f_1, \ldots, f_r$ of length r is called m a x i m a l, if it cannot be extended to an M-sequence of length $r + 1$. By Exercise 6.A.10 (7) this condition is equivalent with $\mathfrak{m} \in \operatorname{Ass}(M/(f_1, \ldots, f_r)M)$.

6.B.15. Theorem *Let R be a Noetherian local ring and M a finite R-module $\neq 0$. Then any two maximal M-sequences have the same length.*

PROOF. We use induction on the length r of a maximal M-sequence. The case $r = 0$ is trivial and the case $r = 1$ is covered by the following Exercise 6.B.17. Let $f_1, \ldots, f_r$ and $g_1, \ldots, g_s$ be maximal M-sequences with $2 \leq r \leq s$. Then $\mathfrak{m} \nsubseteq \bigcup \mathfrak{p}$, where $\mathfrak{p}$ runs through the union $\operatorname{Ass}(M/(f_1, \ldots, f_{r-1})M) \cup \operatorname{Ass}(M/(g_1, \ldots, g_{s-1})M)$. Therefore, there exists $h \in \mathfrak{m}$ such that $f_1, \ldots, f_{r-1}, h$ and $g_1, \ldots, g_{s-1}, h$ and hence $h, f_1, \ldots, f_{r-1}$ and $h, g_1, \ldots, g_{s-1}$ are M-sequences (cf. Exercise 6.B.14 (2)). By induction hypothesis $r - 1 = s - 1$. •

6.B.16. Definition The d e p t h
$$\operatorname{depth} M = \operatorname{depth}_R M$$
of a finite non-zero module M over a Noetherian local ring R is the length of a maximal M-sequence. M is called a C o h e n – M a c a u l a y m o d u l e if depth $M = \dim M$. The ring R is called a C o h e n – M a c a u l a y r i n g if it is a Cohen–Macaulay R-module. A Cohen–Macaulay R-module M with depth $M = \dim M = \dim R$ is called a m a x i m a l C o h e n – M a c a u l a y m o d u l e over R.

A local Noetherian ring R is a Cohen–Macaulay ring if and only if there exists a parameter system which is an R-sequence. In this case, any parameter system is an R-sequence (cf. Theorem 6.B.15 and Exercise 6.B.14 (3)). If $f_1, \ldots, f_r$ is an M-sequence for a Cohen–Macaulay module M, then $M/(f_1, \ldots, f_r)M$ is also a Cohen–Macaulay module. If $f_1, \ldots, f_d$ is a regular parameter system of the regular local ring R, then $(f_1, \ldots, f_r)$, $r = 0, \ldots, d$, are prime ideals by Lemma 6.A.5 and Proposition 6.A.3. Hence $f_1, \ldots, f_d$ is an R-sequence. It follows that *regular local rings are Cohen–Macaulay rings*. Moreover, if $f_1, \ldots, f_r$ is part of a (not necessarily regular) parameter system, then $R/(f_1, \ldots, f_r)$ is a Cohen–Macaulay ring. Rings of this type are called l o c a l c o m p l e t e i n t e r s e c t i o n s. Every non-zero R-module M with $\dim M = 0$ is a Cohen–Macaulay module over R; it is a maximal Cohen–Macaulay module over R if and only if R itself is 0-dimensional. Every Cohen–Macaulay module M with annihilator $\mathfrak{a} := \operatorname{Ann}_R M$ is a maximal Cohen–Macaulay module over $R/\mathfrak{a}$.

For a non-zero R-module M the condition depth $M = 0$ is equivalent with $\mathfrak{m} \in \operatorname{Ass} M$ or with $\operatorname{Soc} M \neq 0$, where the so called s o c l e $\operatorname{Soc} M \subseteq M$ is the R-submodule of M containing all the elements of M which are annihilated by the maximal ideal $\mathfrak{m}$ of R :
$$\operatorname{Soc} M = \operatorname{Soc}_R M := 0 :_M \mathfrak{m} = \{x \in M \mid \mathfrak{m}x = 0\} = \operatorname{Hom}_R(R/\mathfrak{m}, M) .$$
The socle $\operatorname{Soc} M$ is an $(R/\mathfrak{m})$-vector space.

For a finite module M over an arbitrary Noetherian ring A, by Exercise 6.A.11 (8) one has $\operatorname{Ass}_A M = \{\mathfrak{p} \in \operatorname{Spec} A \mid \operatorname{depth}_{A_\mathfrak{p}} M_\mathfrak{p} = 0\}$.

6.B.17. Exercise (1) Let $f, g \in \mathfrak{m}$ be non-zero divisors of (the finite non-zero modu- le) M over (the Noetherian local ring) R. Show that there is a canonical isomorphism

$\mathrm{Soc}\,(M/fM) = (fM : \mathfrak{m})/fM \cong \mathrm{Soc}\,(M/gM) = (gM : \mathfrak{m})/gM$. In particular, $\mathfrak{m} \in \mathrm{Ass}(M/fM)$ if and only if $\mathfrak{m} \in \mathrm{Ass}(M/gM)$. (**Hint:** For an element $x \in (fM : \mathfrak{m})\ (= \{x \in M \,|\, \mathfrak{m}x \subseteq fM\})$ there exists a (unique) element $y \in M$ with $gx = fy$. The isomorphism is given by $\overline{x} \mapsto \overline{y}$.)

(2) (T y p e o f a l o c a l r i n g a n d a m o d u l e) Let $\mathrm{depth}_R M = r$ and $k := R/\mathfrak{m}$. Show that

$$\mathrm{Dim}_k \mathrm{Soc}\,(M/(f_1, \ldots, f_r)M) = \mathrm{Dim}_k \mathrm{Soc}\,(M/(g_1, \ldots, g_r)M)$$

for any two (maximal) regular sequences $f_1, \ldots, f_r$ and $g_1, \ldots, g_r$ of M. (**Hint:** Induction on r. If $r > 0$, then there exists $h \in \mathfrak{m}$ such that $f_1, \ldots, f_{r-1}, h$ and $g_1, \ldots, g_{r-1}, h$ are regular sequences for M and by induction hypothesis $\mathrm{Dim}_k \mathrm{Soc}\,(M/(h, f_1, \ldots, f_{r-1})M) = \mathrm{Dim}_k \mathrm{Soc}\,(M/(h, g_1, \ldots, g_{r-1})M)$, further by part (1), $\mathrm{Dim}_k \mathrm{Soc}\,(M/(f_1, \ldots, f_{r-1}, h)M) = \mathrm{Dim}_k \mathrm{Soc}\,(M/(f_1, \ldots, f_r)M)$ and similarly for $g_1, \ldots, g_{r-1}, h$ and $g_1, \ldots, g_r$.)

If M is a Cohen–Macaulay module, i.e. if $\mathrm{depth}\,M = \dim M = r$, then the dimension $\mathrm{Dim}_k \mathrm{Soc}\,(M/(f_1, \ldots, f_r)M)$ which is independent of the chosen maximal regular sequence $f_1, \ldots, f_r$ for M is called the t y p e o f M and is usually denoted by $\tau(M)$. If M is a Cohen–Macaulay module and if $f_1, \ldots, f_s$ is an *arbitrary* regular sequence for M, then $\tau(M) = \tau(M/(f_1, \ldots, f_s)M)$. A Cohen–Macaulay local ring R of type 1 is called a G o r e n s t e i n r i n g. *Every regular local ring and hence every local complete intersection is Gorenstein.*

6.B.18. Exercise Show that for a Noetherian local ring R of dimension d the following conditions are equivalent: a) R is a Cohen–Macaulay ring. b) If $f_1, \ldots, f_r \in \mathfrak{m}$ are elements with $\mathrm{ht}(f_1, \ldots, f_r) = r$, then $f_1, \ldots, f_r$ is an R-sequence. c) If $f_1, \ldots, f_r \in \mathfrak{m}$ are elements with $\mathrm{ht}(f_1, \ldots, f_r) = r$, then $\mathrm{ht}\,\mathfrak{p} = s$ for all $s \leq r$ and $\mathfrak{p} \in \mathrm{Ass}(R/(f_1, \ldots, f_s))$. d) There is a parameter system $f_1, \ldots, f_d \in \mathfrak{m}$ such that $\mathrm{ht}\,\mathfrak{p} = s$ for all $s = 0, \ldots, d - 1$ and all $\mathfrak{p} \in \mathrm{Ass}(R/(f_1, \ldots, f_s))$.

6.B.19. Exercise If R is a Cohen–Macaulay ring, then $R_\mathfrak{p}$ is a Cohen–Macaulay ring for all $\mathfrak{p} \in \mathrm{Spec}\,R$. (**Hint:** If $\mathrm{ht}\,\mathfrak{p} = r$, then there are elements $f_1, \ldots, f_r \in \mathfrak{p}$ with $\mathrm{ht}(f_1, \ldots, f_r) = r$. Now use Exercise 6.B.18.)

6.B.20. Exercise For a Noetherian local ring R show that the following conditions are equivalent: a) R is reduced. b) If $\mathfrak{p} \in \mathrm{Spec}\,R$ and $\mathrm{depth}\,R_\mathfrak{p} = 0$ (i.e. if $\mathfrak{p} \in \mathrm{Ass}\,R$), then $R_\mathfrak{p}$ is regular (i.e. a field).

6.B.21. Exercise Let R be a Noetherian local ring. Show that:

(1) If R is normal, then $\mathrm{depth}\,R \geq \min\,(2, \dim R)$. (**Hint:** Use Theorem 6.B.4.)

(2) The following conditions are equivalent: a) R is normal. b) If $\mathfrak{p} \in \mathrm{Spec}\,R$ and $\mathrm{depth}\,R_\mathfrak{p} \leq 1$, then $R_\mathfrak{p}$ is regular. (**Hint:** Use Theorem 6.B.4.)

(3) A Cohen–Macaulay ring R is normal if and only if $R_\mathfrak{p}$ is regular for all $\mathfrak{p} \in \mathrm{Spec}\,R$ with $\mathrm{ht}\,\mathfrak{p} \leq 1$.

A very important formula for the depth is the following:

6.B.22. Theorem (Auslander–Buchsbaum) *Let M be a finite non-zero module of finite projective dimension over the Noetherian local ring R. Then*

$$\mathrm{depth}\,M + \mathrm{pd}\,M = \mathrm{depth}\,R\,.$$

PROOF. We use induction on depth R. If depth $R = 0$, then M is free. Otherwise, from a minimal free resolution $0 \to F_r \to \cdots \to F_0 \to M \to 0$ with $r > 0$ we get the inclusion $0 \neq F_r \subseteq \mathfrak{m}_R F_{r-1}$ which is impossible.

Now assume depth $R > 0$. If depth $M > 0$, then there exists an element $f \in \mathfrak{m}$ which is a non-zero divisor for both M and R. In this case $\mathrm{pd}_{R/Rf} M/fM = \mathrm{pd}_R M$, since a free minimal resolution $0 \to F_s \to \cdots \to F_0 \to M \to 0$ of M over R gives the free minimal resolution $0 \to F_s/fF_s \to \cdots \to F_0/fF_0 \to M/fM \to 0$ of M/fM over R/Rf. By induction hypothesis, depth $M + \mathrm{pd}\, M = 1 + \mathrm{depth}\, M/fM + \mathrm{pd}\, M/fM = 1 + \mathrm{depth}\, R/Rf = \mathrm{depth}\, R$. If depth $M = 0$, then M is not free and we consider an exact sequence $0 \to N \to F \to M \to 0$ with a free R-module F. Then depth $N > 0$ and depth $N + \mathrm{pd}\, N = \mathrm{depth}\, N + \mathrm{pd}\, M - 1 = \mathrm{depth}\, R$. So we have to show that depth $N = 1$. But, let $f \in \mathfrak{m}$ be an arbitrary non-zero divisor of R. Then from the commutative diagram

$$
\begin{array}{ccccccccc}
0 & \longrightarrow & N & \longrightarrow & F & \longrightarrow & M & \longrightarrow & 0 \\
& & \downarrow{\scriptstyle f} & & \downarrow{\scriptstyle f} & & \downarrow{\scriptstyle f} & & \\
0 & \longrightarrow & N & \longrightarrow & F & \longrightarrow & M & \longrightarrow & 0
\end{array}
$$

and the Snake Lemma we get $0 \neq \{x \in M \mid \mathfrak{m}x = 0\} \subseteq \{x \in M \mid fx = 0\} \subseteq N/fN$.   •

Now let us come back to the language of schemes again.

6.B.23. Definition Let X be a locally Noetherian scheme. A point $x \in X$ is called a normal point of X, if the stalk $\mathcal{O}_{X,x}$ is a normal domain. X is called normal, if all its points are normal.

Let X be a locally Noetherian scheme of dimension 0 or 1. Then it follows from Theorem 6.B.1 that X is regular if and only if X is normal. An affine Noetherian scheme is normal if and only if it is the spectrum of a (finite) product of Noetherian normal domains. Noetherian normal domains of dimension 1, i. e. Noetherian regular domains of dimension 1, are called Dedekind domains. The principal ideal domains are special cases of such domains.

6.B.24. Exercise Let X be a locally Noetherian normal scheme, Z a closed subset of X of codimension ≥ 2 and $U := X \setminus Z$ its open complement. Then the restriction map $\Gamma(X, \mathcal{O}_X) \to \Gamma(U, \mathcal{O}_X)$ is bijective. (**Hint:** Without loss of generality, one can assume that X is affine. Then use Exercise 6.B.7. – The result of this exercise is sometimes called Riemann's extension theorem (for functions). It shows in particular the following: If X is an affine Noetherian normal scheme and if U is an open affine subset of X, then its complement $X \setminus U$ is either empty or of pure codimension 1.)

6.B.25. Exercise Let X be a normal locally Noetherian scheme and $n \in \mathbb{N}$. Then $\mathbb{A}^n_X$ and $\mathbb{P}^n_X$ are also normal locally Noetherian schemes. (**Hint:** Use Exercise 1.E.10 (2).)

6.B.26. Exercise If R is a positively graded Noetherian normal integral domain, then Proj R is a normal Noetherian scheme. (**Hint:** For $d > 0$ and $f \in R_d$, $f \neq 0$, the Veronese algebra $R^{[d]}$ and the algebra $R_f^{[d]} = \left(R_f^{[d]}\right)_0[f, f^{-1}]$ are also normal integral domains (cf. Exercise 1.E.11 (1)).) More generally, Proj R is normal, if the open subscheme $D(R_+) \subseteq \operatorname{Spec} R$ of the affine cone is normal. Conversely, if R is standardly graded, then if Proj R is normal, then $D(R_+)$ is an open normal subscheme of $\operatorname{Spec} R$. Give an example of a positively graded Noetherian non-normal domain R for which $\operatorname{Proj} R$ is normal. (**Remark:** If R is normal, then one says that $\operatorname{Proj} R$ is a r i t h m e t i c a l l y n o r m a l with respect to the coordinate ring R.)

6.C. Normalization of a Scheme

Let R be a Noetherian reduced ring and let $\mathfrak{p}_1, \ldots, \mathfrak{p}_r$ be the minimal primes in R. Then R is canonically embedded in the product $R_1 \times \cdots \times R_r$, $R_\rho := R/\mathfrak{p}_\rho$, and both rings have the same total quotient ring $Q(R) = Q(R_1) \times \cdots \times Q(R_r)$. It follows that the integral closure $\overline{R}$ of R in $Q(R)$ coincides with the product $\overline{R_1} \times \cdots \times \overline{R_r}$, where $\overline{R_\rho}$ is the integral closure of the integral domain R_ρ in its quotient field $Q(R_\rho)$. We call $\overline{R}$ the n o r m a l i z a t i o n of R. In general $\overline{R}$ is not Noetherian, but if $\overline{R}$ is finite over R or, equivalently, if all $\overline{R_\rho}$ are finite over R_ρ, then $\overline{R}$ is Noetherian and $\operatorname{Spec} \overline{R}$ an affine Noetherian normal scheme. In this case we say that R has a f i n i t e n o r m a l i z a t i o n. In any case the affine R-scheme $\overline{X} := \operatorname{Spec} \overline{R}$ is called the n o r m a l i z a t i o n of $X := \operatorname{Spec} R$. It is finite over X if and only if R has finite normalization. Note that $\overline{X} = \bigcup_{\rho=1}^r \overline{X_\rho}$ is the sum of the normalizations $\overline{X_\rho}$ of the irreducible components $X_\rho := \operatorname{Spec} R_\rho$ of X. So, in particular, *the normalization process separates the irreducible components of the base scheme.*

For an arbitrary $f \in R$ obviously $\overline{R_f} = (\overline{R})_f$. That is, the normalization of the open subscheme $X_f = D(f) \subseteq X := \operatorname{Spec} R$ is the preimage $\overline{X}_f = D(f) \subseteq \overline{X}$ of X with respect to the normalization morphism $\overline{X} \to X$. This allows us to define the n o r m a l i z a t i o n $\overline{X}$ of an arbitrary locally Noetherian reduced scheme X by gluing together the normalizations $\overline{U_i}$ of the elements of an open affine cover U_i, $i \in I$, of X. The normalization $\overline{X} \to X$ is finite if and only if for an affine open cover U_i, $i \in I$, of X the rings $\Gamma(U_i)$ have finite normalizations. In this case, every ring $\Gamma(U)$, $U \subseteq X$ open and affine, has finite normalization and we say that X has f i n i t e n o r m a l i z a t i o n. In any case $\dim \overline{X} = \dim X$, since $\dim R = \dim S$ for every integral ring extension $R \subseteq S$ (cf. Theorem 3.B.8).

For a reduced curve X, i. e. a locally Noetherian reduced scheme of pure dimension 1, the normalization $\overline{X}$ is again locally Noetherian and, therefore (cf. Theorem 6.B.1), *a regular scheme of dimension* 1. This is a special case of the theorem of Krull–Akizuki and is described in Exercise 6.C.1(3). So in dimension 1 the normalization of X yields a desingularization (= regularization) of X.[3]

6.C.1. Exercise (T h e o r e m o f K r u l l – A k i z u k i) A module M over a non-zero (commutative) ring R is called a m o d u l e w i t h r a n k, if $Q(R) \otimes_R M$ is a free $Q(R)$-

---

[3] In higher dimensions the normalization is only regular in codimension 1; to construct a full desingularization is, in general, much more complicated and done only for certain special cases, especially in characteristic 0.

module. M is a module of rank $n \in \mathbb{N}$, if the $Q(R)$-module $Q(R) \otimes_R M$ is free of rank n. This is precisely the case when there exist R-linearly independent elements $v_1, \ldots, v_n \in M$ such that M/F is an R-torsion module, $F := Rv_1 + \cdots + Rv_n$.

(1) Let R be a Noetherian ring of dimension 1, M a (not necessarily finite) R-module of rank n and $f \in R$ a non-zero divisor for R and M. Show that $\operatorname{length}(M/fM) \leq n \operatorname{length}(R/Rf)$. If, moreover, M is finite over R, then $\operatorname{length}(M/fM) = n \operatorname{length}(R/Rf)$. (**Hint:** It suffices to prove the last statement (one considers all the finite submodules of M of rank n). Let $F \subseteq M$ be a free submodule of M such that M/F is a (finite) torsion module and hence of finite length. Then from the chains $fF \subseteq fM \subseteq M$ and $fF \subseteq F \subseteq M$ we get

$$\operatorname{length}(M/fF) = \operatorname{length}(fM/fF) + \operatorname{length}(M/fM) = \operatorname{length}(F/fF) + \operatorname{length}(M/F).$$

Hence $\operatorname{length}(M/fM) = \operatorname{length}(F/fF) = n \operatorname{length}(R/Rf)$, since $fM/fF \cong M/F$.)

(2) Prove the following T h e o r e m o f K r u l l – A k i z u k i : Let R be a Noetherian integral domain of dimension 1, L a finite field extension of $K := Q(R)$ and let $S \subseteq L$ be an extension of R with quotient field L. Then S is a Noetherian ring of dimension ≤ 1. In particular, the integral closure of R in L is a Dedekind domain. (**Hint:** S is an R-module of rank $[L:K]$ (since $K \otimes_R S = Q(S) = L$). It is sufficient to show that for an arbitrary non-zero element $y \in S$ the ring S/Sy is Artinian. Let $f \in R \cap Sy$, $f \neq 0$ (cf. Lemma 1.F.5 (1)). By Part (1) S/Sf is an Artinian R-module and, in particular, an Artinian S-module.)

(3) Let R be a reduced Noetherian ring of dimension 1. Show that the normalization $\overline{R}$ of R is (as any ring S with $R \subseteq S \subseteq Q(R)$) Noetherian, hence $\overline{R}$ is a (finite) product of Dedekind domains.

(4) Let R be a Noetherian integral domain with an algebraically closed quotient field K. Show that $R = K$. (**Hint:** By (2) one may assume that R is a field or a discrete valuation ring.)

6.C.2. Exercise In this exercise we describe an important finiteness theorem for integral extensions. *Let R be a Noetherian normal domain and let L be a finite field extension of $K := Q(R)$. If L is separable over K, then the integral closure S of R in L is finite over R. In particular, S is finite over R, if the characteristic of R is 0.* (**Hint:** For every $z \in S$ the characteristic polynomial of z (which is a power of the minimal polynomial of z) has coefficients in R (cf. Lemma 1.E.9). In particular, the trace defines a non-degenerate symmetric bilinear form $S \times S \to R$, $(x, y) \to \operatorname{tr}(xy)$. Now apply the following lemma.)

6.C.3. Lemma *Let A be a Noetherian ring, M a torsion-free A-module of finite rank (in the sense of Exercise 6.C.1) and $\Phi : M \times M \to A$ a non-degenerate bilinear form. Then M is a finite A-module.*

PROOF. Since Φ is non-degenerate, its extension to the free $Q(A)$-module $Q(A) \otimes_A M$ defines a perfect duality. Now let $F \subseteq M$ be a finite free A-submodule such that M/F is an A-torsion module. Then $v \mapsto \Phi(-, v)$ is an injective A-module homomorphism $M \to \operatorname{Hom}_A(M, A) \subseteq \operatorname{Hom}_A(F, A)$.   •

For the proof of the following important theorem we use the last exercise.

6.C.4. Theorem *Let k be a field and X a reduced algebraic k-scheme. Then the normalization $\overline{X} \to X$ is finite. In particular, $\overline{X}$ is also an algebraic k-scheme.*

PROOF. We have to show that a k-algebra of finite type which is an integral domain has finite normalization. This is a consequence of the following more general lemma. •

6.C.5. Lemma *Let R be a k-algebra of finite type which is an integral domain and let L be a finite field extension of $K := Q(R)$. Then the integral closure of R in L is finite.*

PROOF. Using Noether's Normalization Lemma 1.F.2 we may replace R by a polynomial algebra $k[X_1, \ldots, X_n]$ (which is a normal domain). If char $k = 0$, then the lemma follows from Exercise 6.C.2. In the general case we proceed as follows: We may replace L by its splitting field over K (using the fact that a subalgebra of a finite algebra over a Noetherian ring is also a finite algebra) and hence assume that L is a normal extension of K. Then L is a separable extension of the subfield L' of the purely inseparable elements over K. Furthermore, by Exercise 6.C.2 we may even assume that $L = L'$ is purely inseparable over $K = k(X_1, \ldots, X_n)$. In this case L is a subfield of a field $L'' = k(\alpha_1^{1/q}, \ldots, \alpha_r^{1/q}, X_1^{1/q}, \ldots, X_n^{1/q})$ obtained from k by adjoining q-th roots of elements $\alpha_1, \ldots, \alpha_r \in k$ and the q-th roots of the indeterminates $X_1, \ldots, X_n$, where $q = p^e$ is a power of $p := \operatorname{char} k > 0$. Again we can assume that $L = L''$, but then the integral closure of $R = k[X_1, \ldots, X_n]$ in L is the finite $k[X_1, \ldots, X_n]$-algebra $k(\alpha_1^{1/q}, \ldots, \alpha_r^{1/q})[X_1^{1/q}, \ldots, X_n^{1/q}]$. •

The finiteness of the normalization of a reduced locally Noetherian scheme has an important consequence.

6.C.6. Proposition *Let X be a reduced locally Noetherian scheme with a finite normalization $\overline{X}$. Then the set*

$$\operatorname{Nor} X := \{x \in X \mid \mathcal{O}_{X,x} \text{ is normal}\}$$

is open and dense in X.

PROOF. We may assume that $X = \operatorname{Spec} R$ and $\overline{X} = \operatorname{Spec} \overline{R}$ are affine, where $\overline{R}$ is a finite normalization of R. Now for $\mathfrak{p} \in \operatorname{Spec} R$ the normalization of $R_{\mathfrak{p}}$ coincides with $(\overline{R})_{\mathfrak{p}}$. Hence $R_{\mathfrak{p}}$ is normal if and only if $(\overline{R}/R)_{\mathfrak{p}} = 0$, i.e. $\mathfrak{p} \notin \operatorname{Supp} \overline{R}/R = V(\operatorname{Ann}_R \overline{R}/R)$. Hence $\operatorname{Nor} X = D(\operatorname{Ann}_R \overline{R}/R)$ is open.

For a minimal prime $\mathfrak{p}$ in R the stalk $R_{\mathfrak{p}} = Q(R/\mathfrak{p})$ is a field and hence normal. This proves that $\operatorname{Nor} X$ is dense in X. •

6.C.7. Remark Note that in the last proof the annihilator $\operatorname{Ann}_R \overline{R}/R$ is the c o n d u c t o r $\mathfrak{C}_R = \mathfrak{C}_{\overline{R}|R}$ of R (in $\overline{R}$) (cf. Example 1.E.8 (4)). *In the situation of Proposition 6.C.6 the conductors $\mathfrak{C}_{\Gamma(U, \mathcal{O}_X)}$, $U \subseteq X$ open and affine, define a closed subscheme $C_X = C_{\overline{X}|X}$ of X with support $X \setminus \operatorname{Nor} X$.*

6.C.8. Example Let $m_1, \ldots, m_n$ be positive integers with GCD $(m_1, \ldots, m_n) = 1$ and let M be the submonoid of $\mathbb{N} = (\mathbb{N}, +)$ generated by $m_1, \ldots, m_n$. For a field k the normalization of the affine monomial curve $X := \operatorname{Spec} R$, $R := k[T^{m_1}, \ldots, T^{m_n}] = k[M] = \bigoplus_{m \in M} kT^m$, is the affine line $\mathbb{A}^1_k = \operatorname{Spec} k[T]$ (cf. Example 1.E.8 (6) and Exercise 2.B.14 (3)). The conductor $\mathfrak{C}_R = \operatorname{Ann}_{k[M]} k[T]/k[M]$ is the ideal $T^f k[T]$, where f is the least non-negative integer with $f + \mathbb{N} \subseteq M$. This integer $f = f_M$ is also called the c o n d u c t o r of M and $g = g_M := f - 1 \notin M$ is called the F r o b e n i u s n u m b e r of M. The (vector space) dimension $\operatorname{Dim}_k k[T]/k[M] = \operatorname{Card}(\mathbb{N} \setminus M)$ is the number of gaps and is called the d e g r e e o f s i n g u l a r i t y $\delta = \delta_R = \delta_M$ of R (or of M). If $m \in M$, $0 \le m < f$, then $f - 1 - m = g - m \notin M$ and hence $2\delta_M \ge f_M$. If $M \neq \mathbb{N}$, then the origin $0 \in X$ is the only non-normal (= singular) point of the curve X. For the simplest case $n = 2$ one has $f_M = (m_1 - 1)(m_2 - 1)$ and $\delta_M = f_M/2 = (m_1 - 1)(m_2 - 1)/2$ which was first proved by Sylvester. (**Hint:** If $m_1 < m_2$ show that $\delta_M = \delta_{M'} + \binom{m_1}{2}$ where $M' = \mathbb{N}m_1 + \mathbb{N}(m_2 - m_1)$.)

In general, if $f_M = 2\delta_M$, then the monoid M is called s y m m e t r i c or G o r e n s t e i n.[4])

The d e g r e e o f s i n g u l a r i t y δ_x is defined for every closed point x of a reduced algebraic k-scheme X of dimension 1, i. e. of a reduced algebraic curve X over k. By definition

$$\delta_x := \operatorname{Dim}_k \overline{\mathcal{O}}_x/\mathcal{O}_x = [\kappa(x) : k] \operatorname{length}_{\mathcal{O}_x}(\overline{\mathcal{O}}_x/\mathcal{O}_x),$$

(where $\overline{\mathcal{O}}_x$ is the normalization of $\mathcal{O}_x$). If $\mathfrak{p}_1, \ldots, \mathfrak{p}_r \subseteq \mathcal{O}_x$ are the minimal primes of $\mathcal{O}_x$, then $\delta_x = \operatorname{Dim}_k((\prod_{\rho=1}^r \mathcal{O}_x/\mathfrak{p}_\rho)/\mathcal{O}_x) + \sum_{\rho=1}^r \delta_\rho$, where δ_ρ is the degree of singularity of $\mathcal{O}_x/\mathfrak{p}_\rho$, i. e. of x considered as a point of the irreducible component X_ρ belonging to the minimal prime $\mathfrak{p}_\rho$, $\rho = 1, \ldots, r$, see also Exercise 7.E.7.

6.C.9. Exercise Let $R = \bigoplus_{m \in \mathbb{N}} R_m$ be a positively graded reduced Noetherian ring and $\overline{R}$ its normalization. By Exercise 1.E.10 (3) $\overline{R} = \bigoplus_{m \in \mathbb{N}} \overline{R}_m$ is also positively graded. Show that the canonical inclusion $R \to \overline{R}$ induces an everywhere-defined morphism $\operatorname{Proj} \overline{R} \to \operatorname{Proj} R$ which is the normalization of $\operatorname{Proj} R$. (**Hint:** For $d \in \mathbb{N}^*$ the Veronese transform $\overline{R}^{[d]}$ is the normalization of $R^{[d]}$ and for a homogeneous element $f \in R_d$ the normalization of $(R_f)_0[f/1, 1/f] = \left(R_f^{[d]}\right)_0[f/1, 1/f] = R_f^{[d]}$ is $(\overline{R}_f)_0[f/1, 1/f] = \left(\overline{R}_f^{[d]}\right)_0[f/1, 1/f] = \overline{R}_f^{[d]}$ and hence the normalization of $(R_f)_0$ is $(\overline{R}_f)_0$.) In particular, if $R_0 = k$ is a field, then the normalization of the projective algebraic k-scheme $X := \operatorname{Proj} R$ is the projective algebraic k-scheme $\overline{X} := \operatorname{Proj} \overline{R}$ (which is finite over X, but, in general, $\overline{R}_0 \neq k$).

6.D. The Module of Kähler Differentials

We mentioned already at the end of Example 6.A.13 that for an algebraic k-scheme X, k a field, and a k-rational point $x \in X(k)$, i. e. a point $x \in X$ with $\kappa(x) = k$, there is a close connection between regularity and smoothness described by the Jacobian criterion. In this section we want to explore this connection more carefully and start with algebraic aspects of derivations.

Let A be a ring, R a (commutative) A-algebra and M an R-module. A map $D : R \to M$ is called an A - d e r i v a t i o n if it is A-linear and satisfies the p r o d u c t

---

[4]) This is the case if and only if the local ring $R_{\mathfrak{m}_0} = \mathcal{O}_{X,0}$, $\mathfrak{m}_0 = RT^{m_1} + \cdots + RT^{m_n}$, in the origin $0 \in X$ is Gorenstein (in the sense of Exercise 6.B.17 (2), see also Exercise 7.E.22 (3)).

r u l e :
$$D(ab) = aD(b) + bD(a), \quad a, b \in R.$$

By the product rule we have $D(1) = D(1) + D(1)$, so that $D(1) = 0$, hence: *Any derivation $D : R \to M$ vanishes on $A \cdot 1_R$.* More generally, the c o n s t a n t s of D, i. e. the kernel of D, is an A-subalgebra of R. Furthermore, we mention the p o w e r r u l e : $D(a^n) = na^{n-1}D(a)$ for $a \in R$, $n \in \mathbb{N}^*$, which is proved easily using induction on n. Hence, if R is a ring of characteristic $m \in \mathbb{N}^*$, then $D(a^m) = 0$ for all $a \in R$. A derivation $D : R \to R$ satisfies for any $n \in \mathbb{N}$ the L e i b n i z f o r m u l a :

$$D^n(ab) = \sum_{i=0}^{n} \binom{n}{i} D^i(a) D^{n-i}(b), \quad a, b \in R.$$

In particular, if R is of prime characteristic $p > 0$, then $D^p = D \circ \cdots \circ D$ (p times) is also an A-derivation of R.

We denote by
$$\mathrm{Der}_A(R, M)$$

the set of all A-derivations of R into M. In the particular case $M = R$ we also use the short notation $\mathrm{Der}_A(R)$. $\mathrm{Der}_A(R, M)$ is an R-submodule of the module M^R of all maps $R \to M$. If $D : R \to M$ is an A-derivation and if $f : M \to N$ is an R-module homomorphism, then the composition $f \circ D$ is again an A-derivation.

The m o d u l e o f K ä h l e r d i f f e r e n t i a l s $\Omega_R = \Omega_{R|A}$ is an R-module together with a u n i v e r s a l A-d e r i v a t i o n

$$d_R = d_{R|A} : R \to \Omega_R = \Omega_{R|A}$$

satisfying the following property: Given any R-module M and an A-derivation $D : R \to M$, there exists a unique R-module homomorphism $h : \Omega_R \to M$, such that $D = h \circ d_R$, i. e. for any R-module M the canonical R-module homomorphism

$$\mathrm{Hom}_R(\Omega_R, M) \to \mathrm{Der}_A(R, M), \quad h \mapsto h \circ d_R,$$

is an isomorphism. In particular, $\mathrm{Der}_A(R)$ can be identified with the dual Ω_R^* of the module Ω_R of Kähler differentials. If it exists, the pair (Ω_R, d_R) is unique up to unique isomorphism.

A simple way to prove the existence of the universal derivation is the following: Let $R^{(R)}$ be the free R-module with the standard basis e_a, $a \in R$, and let $U \subseteq R^{(R)}$ be the submodule generated by the elements

$$e_{ab} - ae_b - be_a, \quad a, b \in R; \qquad e_{\alpha 1_R}, \quad \alpha \in A.$$

Obviously, $a \mapsto [e_a]$ is the universal A-derivation $d_R : R \to \Omega_R := R^{(R)}/U$. This description of the module of Kähler differentials gives less insight into its structure. From the point of view that the algebra R is represented as a homomorphic image of a polynomial algebra $A[X_i \mid i \in I]$ one wants to construct the module Ω_R in two steps: First construct the universal derivation for a polynomial algebra and second construct the universal derivation for a homomorphic image of an A-algebra with given universal A-derivation. Both these steps can be carried out quite easily.

6.D.1. Proposition *Let I be a set and let P be the polynomial algebra $A[X_i \mid i \in I]$. For $i \in I$, let $\partial_i = \partial/\partial X_i : P \to P$ be the partial derivative with respect to the indeterminate X_i (which is an A-derivation). Then*

$$d_P : P \to \Omega_P := P^{(I)}, \quad F \mapsto dF = d_P F := (\partial_i F)_{i \in I} = \sum_{i \in I} \frac{\partial F}{\partial X_i} d_P X_i \,,$$

is the universal A-derivation of P. In particular, Ω_P is a free P-module with P-basis $dX_i = d_P X_i$, $i \in I$.

PROOF. Let $D : P \to M$ be an A-derivation and let $m_i := DX_i$, $i \in I$. By the product rule $DF = \sum_{i \in I} (\partial_i F) DX_i = \sum_{i \in I} (\partial_i F) m_i$. Hence $D = h \circ d_P$, where $h : P^{(I)} \to M$ is the P-module homomorphism defined by $e_i = d_P X_i \mapsto m_i$.   ●

6.D.2. Exercise Prove the following generalization of Proposition 6.D.1: Let I be a set and let P be the polynomial algebra $R[X_i \mid i \in I]$, where R is a (commutative) A-algebra. Then

$$P \to (\Omega_{R|A} \otimes_R P) \oplus \left(\bigoplus_{i \in I} P dX_i \right), \quad F = \sum_\nu f_\nu X^\nu \mapsto \sum_\nu d_{R|A} f_\nu \otimes X^\nu + \sum_i \frac{\partial F}{\partial X_i} dX_i \,,$$

is the universal A-derivation of P and the differentials dX_i, $i \in I$, are linearly independent over P.

6.D.3. Proposition *Let R be an A-algebra with universal A-derivation $d_R : R \to \Omega_R$ and let $\mathfrak{a} \subseteq R$ be an ideal. Then d_R induces an $(R/\mathfrak{a})$-module homomorphism*

$$\mathfrak{a}/\mathfrak{a}^2 \to \Omega_R/\mathfrak{a}\Omega_R \,, \quad [a] \mapsto [d_R a] \,, \quad a \in \mathfrak{a} \,.$$

Its cokernel is the module $\Omega_{R/\mathfrak{a}}$ of Kähler differentials of $R/\mathfrak{a}$ and the universal A-derivation is $d_{R/\mathfrak{a}} : R/\mathfrak{a} \to \Omega_{R/\mathfrak{a}}$ induced by d_R. In other words: There is a canonical exact sequence

$$\mathfrak{a}/\mathfrak{a}^2 \to \Omega_R/\mathfrak{a}\,\Omega_R \to \Omega_{R/\mathfrak{a}} \to 0$$

of $(R/\mathfrak{a})$-modules and a commutative diagram

$$
\begin{array}{ccc}
R & \xrightarrow{\ d_R\ } & \Omega_R \\
\downarrow & & \downarrow \\
R/\mathfrak{a} & \xrightarrow{\ d_{R/\mathfrak{a}}\ } & \Omega_{R/\mathfrak{a}} = \Omega_R/(R\,d_R\,\mathfrak{a} + \mathfrak{a}\,\Omega_R) \,.
\end{array}
$$

The simple proof of the last proposition is left to the reader. Note that because of $a\,d_R b = d_R(ab) - b\,d_R a$, $a \in \mathfrak{a}$, $b \in R$, we have $\mathfrak{a}\Omega_R \subseteq R\,d_R\,\mathfrak{a}$, i.e. $\Omega_{R/\mathfrak{a}} = \Omega_R/R\,d_R\,\mathfrak{a}$. Furthermore, if the ideal $\mathfrak{a}$ is generated by the elements f_j, $j \in J$, then

$$R\,d_R\,\mathfrak{a} = \sum_{j \in J} R\,d_R\,f_j + \mathfrak{a}\,\Omega_R \,.$$

In view of 6.D.1 and 6.D.3 a representation $R = A[X_i \mid i \in I]/(F_j ; j \in J) = A[x_i \mid i \in I]$ of an A-algebra R by algebra generators x_i, $i \in I$, and relations

$F_j \in A[X_i \,|\, i \in I]$, $j \in J$, yields the following representation of the module
$\Omega_R = \Omega_{R\,|\,A}$ of Kähler differentials by generators and relations:

$$\Omega_R = R^{(I)} \Big/ \sum_{j \in J} R\,(\overline{\partial_i F_j})_{i \in I}$$

(where $\overline{\partial_i F_j}$ is the residue class of the partial derivative $\partial_i F_j = \partial F_j / \partial X_i \in R$).
The universal derivation d_R maps the element $F(x) \in R$ to the residue class
$\sum_{i \in I} \partial_i F(x) d_R x_i$ of $(\partial_i F(x))_{i \in I} \in R^{(I)}$ in Ω_R. In particular, if the elements
x_i, $i \in I$, generate the A-algebra R, then the differentials $d_R x_i$, $i \in I$, generate
the module $\Omega_R = \Omega_{R\,|\,A}$, and Ω_R is a finite R-module if R is an A-algebra of finite
type.

The next exercise describes another important way to prove the existence of the
universal derivation of an A-algebra R.

6.D.4. Exercise Let R be an A-algebra, $\mu : R \otimes_A R \to R$ the multiplication homomorphism
$a \otimes b \mapsto ab$, and $I = I_R := \operatorname{Ker} \mu$ (cf. Lemma 4.D.11). Then I/I^2 is an R-module. Show
that

$$d : R \to I/I^2, \quad a \mapsto (1 \otimes a - a \otimes 1) + I^2,$$

is the universal A-derivation of R. (**Hint:** One checks immediately that d is a derivation.
Let $D : R \to M$ be an arbitrary A-derivation. Then the A-bilinear map $(a, b) \mapsto a\,Db$
induces an A-linear map $R \otimes_A R \to M$ and hence an A-linear map $H : I \to M$. Since
H vanishes on all the elements $(c \otimes d)(1 \otimes a - a \otimes 1)(1 \otimes b - b \otimes 1)$, $a, b, c, d \in R$, it
induces an A-linear map $h : I/I^2 \to M$ which is even R-linear. Obviously, $D = h \circ d$.
Since I/I^2 is generated as an R-module by the residue classes of $1 \otimes a - a \otimes 1$, $a \in R$, h
is the only R-homomorphism $I/I^2 \to M$ with this property.)

6.D.5. Exercise Let R be an A-algebra and let S be a multiplicatively closed subset of R.
Show that if $D : R \to M$ is an A-derivation of R, then $a/s \mapsto (s\,Da - a\,Ds)/s^2$ is a
(well-defined) A-derivation $\widetilde{D} : S^{-1}R \to S^{-1}M$ of $S^{-1}R$. One has $\widetilde{d}_{R\,|\,A} = d_{S^{-1}R\,|\,A}$ and,
in particular, $\Omega_{S^{-1}R\,|\,A} = S^{-1}\Omega_{R\,|\,A}$.

6.D.6. Exercise Let R and A' be two A-algebras and let $R' := A' \otimes_A R$. If $D : R \to M$
is an A-derivation, then $A' \otimes_A D : R' \to M' = A' \otimes_A M$ is an A'-derivation. One has
$A' \otimes_A d_{R\,|\,A} = d_{R'\,|\,A'}$ and, in particular, $\Omega_{R'\,|\,A'} = A' \otimes_A \Omega_{R\,|\,A} = R' \otimes_R \Omega_{R\,|\,A}$.

6.D.7. Exercise Let R and S be A-algebras. Show that:

(1) $\Omega_{(R \times S)\,|\,A} = \Omega_{R\,|\,A} \times \Omega_{S\,|\,A}$ and the universal derivation $d_{(R \times S)\,|\,A}$ is given by $d_{(R \times S)\,|\,A} = d_{R\,|\,A} \times d_{S\,|\,A}$.

(2) $\Omega_{(R \otimes_A S)\,|\,A} = (\Omega_{R\,|\,A} \otimes_A S) \oplus (R \otimes_A \Omega_{S\,|\,A})$ and the universal derivation $d_{(R \otimes_A S)\,|\,A}$ is
given by $d_{(R \otimes_A S)\,|\,A}(x \otimes y) = (d_{R\,|\,A}\,x) \otimes y + x \otimes (d_{S\,|\,A}\,y)$ for $x \in R$, $y \in S$.

6.D.8. Exercise Let $k \to A \to R$ be ring homomorphisms. Then $d_{R\,|\,k}\,|\,A : A \to \Omega_{R\,|\,k}$ de-
fines a canonical A-homomorphism $\Omega_{A\,|\,k} \to \Omega_{R\,|\,k}$ and hence a canonical R-homomorphism
$R \otimes_A \Omega_{A\,|\,k} \to \Omega_{R\,|\,k}$. Since every A-derivation is also a k-derivation $d_{R\,|\,A} : R \to \Omega_{R\,|\,A}$ in-
duces a canonical R-homomorphism $\Omega_{R\,|\,k} \to \Omega_{R\,|\,A}$. Show that the sequence

$$R \otimes_A \Omega_{A\,|\,k} \to \Omega_{R\,|\,k} \to \Omega_{R\,|\,A} \to 0$$

is exact.

We continue with some results on the derivation modules of field extensions. We start with a general lemma.

6.D.9. Lemma *Let K be a field and R a finite (commutative) K-algebra, such that all points in $\operatorname{Spec} R$ are K-rational. Then, $\Omega_{R|K} = 0$ if and only if R is reduced, i. e. $R \cong K^n$ for some integer n.*

PROOF. One has $R = R_1 \times \cdots \times R_n$, where R_i are local K-algebras, $i = 1, \ldots, n$. (Indeed, R_i are the stalks of the structure sheaf of $\operatorname{Spec} R$.) So, by Exercise 6.D.7 (1), we may assume that R is local with maximal ideal $\mathfrak{m} = \mathfrak{m}_R$ and with residue field K. Then $R = K \oplus \mathfrak{m}$.

If $R = K$, then $\Omega_{R|K} = 0$. Conversely, let $\Omega_{R|K} = 0$. We have to show that $\mathfrak{m} = 0$, which is equivalent to $\mathfrak{m}/\mathfrak{m}^2 = 0$ by Nakayama's Lemma. Since also $\Omega_{(R/\mathfrak{m}^2)|K} = 0$, by Proposition 6.D.3 we may replace R by $R/\mathfrak{m}^2$ and assume that $\mathfrak{m}^2 = 0$. Then the result is a consequence of the following simple exercise. •

6.D.10. Exercise Let M be an A-module and R the A-algebra $A \oplus M$ with $M^2 = 0$. Show that $(a, m) \mapsto m$ is an A-derivation of R (different from 0 if $M \neq 0$).

6.D.11. Corollary *Let K be a field and R a finite K-algebra. Then $\Omega_{R|K} = 0$ if and only if R is a separable K-algebra, i. e. isomorphic to a finite product of finite separable field extensions of K.*

PROOF. A finite K-algebra R is separable if and only if $L \otimes_K R$ is reduced for all field extensions L of K. Since there is a (finite) field extension L of K such that all the residue fields of $L \otimes_K R$ coincide with L and since $\Omega_{(L \otimes_K R)|L} = L \otimes_K \Omega_{R|K}$, one has $\Omega_{R|K} = 0$ if R is a separable field extension of K. Conversely, let $\Omega_{R|K} = 0$. If L is an arbitrary field extension of K, there exists a field extension L' of L such that the residue fields of $L' \otimes_K R$ coincide with L'. Hence $L' \otimes_K R$ is reduced by Lemma 6.D.9, and $L \otimes_K R \subseteq L' \otimes_K R$ is reduced too. •

The following theorem is a partial generalization of the last corollary.

6.D.12. Theorem (Schmidt–MacLane) *Let $k \subseteq K$ be a finitely generated field extension. Then:*

(1) $\operatorname{Dim}_K \Omega_{K|k} \geq \operatorname{trdeg}_k K$.

(2) $\operatorname{Dim}_K \Omega_{K|k} = \operatorname{trdeg}_k K$ *if and only if $k \subseteq K$ is separably generated.* [5]

(3) *If k is a perfect field, then $\operatorname{Dim}_K \Omega_{K|k} = \operatorname{trdeg}_k K$, i. e. K is separably generated over k.*

In particular:

(4) $\Omega_{K|k} = 0$ *if and only if K is a finite separable field extension of k.*

---

[5]) Let $k \subseteq K$ be an arbitrary field extension. Then K is said to be s e p a r a b l y g e n e r a t e d over k if there exists a transcendence basis $y_i, i \in I$, of K over k such that the extension $k(y_i, i \in I) \subseteq K$ is a separable algebraic extension. Such a transcendence basis is called a s e p a r a t i n g t r a n s c e n d e n c e b a s i s of K over k.

PROOF. K is the quotient field of an integral k-algebra of finite type. Hence $\mathrm{Dim}_K\,\Omega_{K\,|\,k}$ is finite (cf. Exercise 6.D.5).

There is a chain $k = L_0 \subseteq L_1 \subseteq \cdots \subseteq L_n = K$ of subfields of K such that every field extension $L_i \subseteq L_{i+1}$, $i = 0,\ldots, n-1$, is a simple extension $L \subseteq L(\alpha)$ of one of the following types: a) α is transcendental over L. b) α is algebraic over L. In case b) we may even assume that α is separably algebraic over L or purely inseparable of degree p, i.e. $\beta := \alpha^p \in L \setminus L^p$, $p := \mathrm{char}\,k > 0$.

Let $\rho(L) := \mathrm{Dim}_L\,\Omega_{L\,|\,k}$. We shall use the following lemma which we shall prove later.

Lemma *Let $L \subseteq L(\alpha)$ be as above. Then:*

a) *If α is transcendental over L, then $\rho(L(\alpha)) = \rho(L) + 1$.*

b) *If α is algebraic over L, then $\rho(L) \le \rho(L(\alpha)) \le \rho(L) + 1$.*

c) *If α is separably algebraic over L, then $\rho(L) = \rho(L(\alpha))$.*

d) *If k is a perfect field of characteristic $p > 0$ and if $\beta := \alpha^p \in L \setminus L^p$, then $\rho(L) = \rho(L(\alpha))$.*

The chain of field extensions constructed above together with parts a) and b) of the lemma yield part (1) of the theorem, whereas parts a), c) and d) yield part (3). To prove part (2) let $\alpha_1,\ldots,\alpha_m \in K$ be such that $d_{K\,|\,k}(\alpha_\mu)$, $\mu = 1,\ldots,m$, form a K-basis of $\Omega_{K\,|\,k}$, and let $L = k(\alpha_1,\ldots,\alpha_m) \subseteq K$ be the subfield generated by $\alpha_1,\ldots,\alpha_m$. In the exact sequence

$$K \otimes_L \Omega_{L\,|\,k} \to \Omega_{K\,|\,k} \to \Omega_{K\,|\,L} \to 0$$

of Exercise 6.D.8 the first homomorphism is surjective by construction. Hence $\Omega_{K\,|\,L} = 0$. By part (1) the field extension $L \subseteq K$ is algebraic and even separable algebraic by Corollary 6.D.11. So if $m = \mathrm{trdeg}_k\,K$, then $\alpha_1,\ldots,\alpha_m$ form a separating transcendence basis of K over k. Conversely, if $\beta_1,\ldots,\beta_m$ is a separating transcendence basis of K over k and if $L = k(\beta_1,\ldots,\beta_m)$, then $\Omega_{K\,|\,L} = 0$ and $\mathrm{Dim}_L\,\Omega_{L\,|\,k} = m$ and the above exact sequence yields $\mathrm{Dim}_K\,\Omega_{K\,|\,k} \le m$. The converse inequality follows from part (1).

We now provide the p r o o f of t h e l e m m a. For the polynomial algebra $L[X] = L \otimes_k k[X]$ we get by Exercise 6.D.7 (2)

$$\Omega_{L[X]\,|\,k} = \big(L[X] \otimes_L \Omega_{L\,|\,k}\big) \oplus \Omega_{L[X]\,|\,L}.$$

In particular, $\Omega_{L[X]\,|\,k}$ is a free $L[X]$-module of rank $\mathrm{Dim}_L\,\Omega_{L\,|\,k} + 1 = \rho(L) + 1$. With Exercise 6.D.5 this implies $\mathrm{Dim}_{L(X)}\,\Omega_{L(X)\,|\,k} = \rho(L) + 1$ and proves part a) of the lemma.

Now suppose that α is algebraic. Then, $L(\alpha) = L[\alpha] = L[X]/(F)$, where F is the minimal polynomial of α over L. Part b) of the lemma now follows from the exact sequence

$$(F)/(F^2) \to L[\alpha] \otimes_{L[X]} \Omega_{L[X]\,|\,k} \to \Omega_{L[\alpha]\,|\,k} \to 0$$

of Proposition 6.D.3 and from $\mathrm{Dim}_{L[\alpha]}(L[\alpha] \otimes_{L[X]} \Omega_{L[X]|k}) = \mathrm{rank}_{L[X]} \Omega_{L[X]|k} = \rho(L) + 1$.

To prove parts c) and d) of the lemma we use the exact sequence

$$L[\alpha] \otimes_L \Omega_{L|k} \overset{h}{\to} \Omega_{L[\alpha]|k} \to \Omega_{L[\alpha]|L} \to 0.$$

If α is separably algebraic over k, then $\Omega_{L[\alpha]|L} = 0$. Hence the $L(\alpha)$-homomorphism h is surjective and even bijective because of $\rho(L) \leq \rho(L[\alpha])$. This proves c).
In the case d) we have $L[\alpha] \cong L[X]/(F)$, $F := X^p - \beta$, and hence $\Omega_{L[\alpha]|L} \cong L[\alpha]/F'(\alpha)L[\alpha] = L[\alpha]$, i.e. $\mathrm{Dim}_{L[\alpha]} \Omega_{L[\alpha]|L} = 1$. Because of $\rho(L) \leq \rho(L[\alpha])$ it is sufficient to show that the homomorphism h is not injective. But $k = k^p \subseteq L^p \subseteq L$. Let k' be a maximal subfield of L not containing β but containing L^p. Because of the maximality of k' we have $L = k'[\beta] \cong k'[X]/(X^p - \gamma)$, $\gamma := \beta^p \in L^p \subseteq k'$, hence $d_{L|k'}\beta \neq 0$ and, in particular, $d_{L|k}\beta \neq 0$. But $\beta = \alpha^p$, so that $d_{L[\alpha]|k}\beta = p\alpha^{p-1}d_{L[\alpha]|k}\alpha = 0$ and $1 \otimes d_{L|k}\beta \neq 0$ belongs to Ker h.   •

6.D.13. Exercise Let $k \subseteq K$ be a (not necessarily finitely generated) field extension with a separating transcendence basis x_i, $i \in I$, over k. Show that $d_{K|k}x_i \in \Omega_{K|k}$, $i \in I$, is a K-basis of $\Omega_{K|k}$. In particular, in this case $\mathrm{Dim}_K \Omega_{K|k} = \mathrm{trdeg}_k K$.

We now want to derive some results on the Kähler module of differentials for the stalks of an algebraic k-scheme, k a field, i.e. for the localizations of k-algebras of finite type. Such localizations are sometimes called local algebras e s s e n t i a l l y of finite type over k. For an arbitrary local k-algebra R with maximal ideal $\mathrm{m} = \mathrm{m}_R$ and residue field $K = R/\mathrm{m}$ we use the exact sequence of Proposition 6.D.3

$$\mathrm{m}/\mathrm{m}^2 \to \Omega_{R|k}/\mathrm{m}\,\Omega_{R|k} \to \Omega_{K|k} \to 0.$$

For $\overline{R} := R/\mathrm{m}^2$ and $\overline{\mathrm{m}} := \mathrm{m}_{\overline{R}} = \mathrm{m}/\mathrm{m}^2$ we get the commutative diagram

$$
\begin{array}{ccccccc}
\mathrm{m}/\mathrm{m}^2 & \to & \Omega_{R|k}/\mathrm{m}\,\Omega_{R|k} & \to & \Omega_{K|k} & \to & 0 \\
\downarrow{=} & & \downarrow{\cong} & & \downarrow{=} & & \\
\overline{\mathrm{m}} & \to & \Omega_{\overline{R}|k}/\overline{\mathrm{m}}\,\Omega_{\overline{R}|k} & \to & \Omega_{K|k} & \to & 0.
\end{array}
$$

6.D.14. Lemma *Let k be a field and R a local k-algebra such that the residue field $K = R/\mathrm{m}$ has a separating transcendence basis over k. Then the sequence*

$$0 \to \mathrm{m}/\mathrm{m}^2 \to \Omega_{R|k}/\mathrm{m}\,\Omega_{R|k} \to \Omega_{K|k} \to 0$$

is exact.

PROOF. By the last remark we may assume that $\mathrm{m}^2 = 0$. Let x_i, $i \in I$, be a family of elements in R, the residue classes of which form a separating transcendence basis of K over k. Then $L := k(x_i \,|\, i \in I) \subseteq R$ is a subfield of R which is purely transcendental over k with transcendence basis x_i, $i \in I$, and K is separably algebraic over L. Then there exists a (unique) field extension L' of L in R which is isomorphic to K.[6] To construct L' let $x \in R$ and let $\mu \in L[X]$ be the separable

---

[6] Such a field L' is called a c o e f f i c i e n t f i e l d of R.

minimal polynomial of x over L. Then $\mu(x) \in \mathfrak{m}$, $\mu'(x) \in R^\times = R \setminus \mathfrak{m}$ and by Newton's approximation formula,

$$x - \frac{\mu(x)}{\mu'(x)} \in R$$

is the only element in R with minimal polynomial μ and the same residue class as x in K.

Now $R = L' \oplus \mathfrak{m}$ and the projection $R \to \mathfrak{m}$ is an L'- and hence a k-derivation of R (see Exercise 6.D.10) which induces a homomorphism $\Omega_{R|k}/\mathfrak{m}\,\Omega_{R|k} \to \mathfrak{m}$ such that the composition $\mathfrak{m} \to \Omega_{R|k}/\mathfrak{m}\,\Omega_{R|k} \to \mathfrak{m}$ is the identity of $\mathfrak{m}$. Hence $\mathfrak{m} \to \Omega_{R|k}/\mathfrak{m}\,\Omega_{R|k}$ is injective. $\qquad\bullet$

If the local ring R in Lemma 6.D.14 is Noetherian and if $\Omega_{R|k}$ is a finitely generated R-module, then the exactness of the sequence in this lemma is equivalent to the formula $\mu_R(\Omega_{R|k}) = \operatorname{emdim} R + \operatorname{trdeg}_k K$, where $\mu_R(\cdot)$ denotes the minimal number of generators of an R-module. In general, we have the inequality $\mu_R(\Omega_{R|k}) \geq \operatorname{emdim} R + \operatorname{trdeg}_k K$.

6.D.15. Exercise Let A be a local ring with infinite residue field and B a finite A-algebra. If $\Omega_{B|A}$ is generated by one element, then B is a cyclic A-algebra, i.e. generated by one element as an A-algebra. (**Hint:** By the Lemma of Nakayama we may replace A by $k = A/\mathfrak{m}_A$ and B by $B/\mathfrak{m}_A B$, i.e. B is a finite algebra over the field $A = k$. Furthermore, we may extend k to a (finite) field extension K such that all the residue fields of $K \otimes_k B$ coincide with K. If $K \otimes_k B$ is a cyclic K-algebra, then B itself is a cyclic k-algebra. (The reader should prove this!) Therefore we may assume that the residue fields of the local components $B_1, \ldots, B_r$ of $B = B_1 \times \cdots \times B_r$ coincide with k. Now, by Exercise 6.D.7 (1) and Lemma 6.D.14 the maximal ideal $\mathfrak{m}_\rho$ of the local component B_ρ is generated by one element x_ρ. Then $B_\rho = k[x_\rho] \cong k[X]/(X^{m_\rho})$, where $m_\rho = \operatorname{Dim}_k B_\rho$ and hence $B \cong k[X]/(f)$ is cyclic, where $f := \prod_{\rho=1}^r (X - a_\rho)^{m_\rho}$, $a_1, \ldots, a_r \in k$ pairwise distinct. — More generally, one can prove if $\Omega_{B|A}$ is generated by n elements, then B is generated by $\max\{1, n\}$ elements as an A-algebra.)

Now we prove:

6.D.16. Theorem *Let k be a field, A a k-algebra of finite type and $\mathfrak{p} \subseteq A$ a prime ideal such that $K = \kappa(\mathfrak{p}) = Q(A/\mathfrak{p})$ is separably generated over k. For the localization $R := A_\mathfrak{p}$ with maximal ideal $\mathfrak{m} = \mathfrak{p}A_\mathfrak{p}$ and residue field K the following conditions are equivalent:*

(1) *R is regular.*

(2) *$\Omega_{R|k} = (\Omega_{A|k})_\mathfrak{p}$ is a free R-module of rank*

$$\dim R + \operatorname{trdeg}_k K = \dim R + \operatorname{Dim}_K \Omega_{K|k} .$$

PROOF. Let R be regular and hence an integral domain. Replacing A by $A/\mathfrak{q}$ where $\mathfrak{q}$ is the (unique) minimal prime ideal of A contained in $\mathfrak{p}$ we may assume that A is also an integral domain. By Lemma 6.D.14 $\mu_R(\Omega_{R|k}) = \operatorname{emdim} R + \operatorname{trdeg}_k K =$

$\dim R + \operatorname{trdeg}_k K$. But by Theorem 6.D.12 (1) $\operatorname{rank}_R \Omega_{R|k} = \operatorname{Dim}_L \Omega_{L|k} \geq \operatorname{trdeg}_k L$ $= \dim A$, where $L := Q(R) = Q(A)$ is the quotient field of A. By Corollary 3.B.25 $\dim A = \dim R + \dim A/\mathfrak{p} = \dim R + \operatorname{trdeg}_k K$. Combining these we get $\mu_R(\Omega_{R|k}) = \operatorname{rank}_R \Omega_{R|k}$ which, of course, is equivalent to the freeness of $\Omega_{R|k}$.

Conversely, let $\Omega_{R|k}$ be free of rank $\dim R + \operatorname{Dim}_K \Omega_{K|k}$. Then, by Lemma 6.D.14, $\dim R + \operatorname{Dim}_K \Omega_{K|k} = \mu_R(\Omega_{R|k}) = \operatorname{emdim} R + \operatorname{Dim}_K \Omega_{K|k}$ and hence $\dim R = \operatorname{emdim} R$, i.e. R is regular.   •

6.D.17. Corollary *Let A be an algebra of finite type over the field k and let $\mathfrak{m} \in k$- $\operatorname{Spec} A \subseteq \operatorname{Spec} A$ (i. e. $\mathfrak{m}$ is a maximal ideal in A with $A/\mathfrak{m} = k$). The following conditions are equivalent*:

(1) $A_\mathfrak{m}$ *is regular.*

(2) *The $A_\mathfrak{m}$-module $\Omega_{A_\mathfrak{m}|k} = (\Omega_{A|k})_\mathfrak{m}$ is free of rank $\dim A_\mathfrak{m}$.*

6.D.18. Corollary *Let A be an integral domain of finite type over the field k and $\mathfrak{p} \in A$ a prime ideal. If $K := Q(A/\mathfrak{p})$ is a separably generated field extension of k and if the localization $A_\mathfrak{p}$ is regular, then $L := Q(A)$ is also separably generated over k.*

PROOF. By Theorem 6.D.16 $\operatorname{Dim}_L \Omega_{L|k} = \operatorname{rank}_A \Omega_{A|k} = \dim A_\mathfrak{p} + \dim(A/\mathfrak{p}) = \dim A = \operatorname{trdeg}_k L$. Now Theorem 6.D.12 (2) implies that the field L is separably generated over k.   •

6.D.19. Corollary *Let k be a perfect field, A a k-algebra of finite type, $\mathfrak{p} \subseteq A$ a prime ideal and $R := A_\mathfrak{p}$. The following conditions are equivalent*:

(1) *R is regular.*

(2) *$\Omega_{R|k} = (\Omega_{A|k})_\mathfrak{p}$ is a free R-module of rank*

$$\dim R + \operatorname{trdeg}_k K = \dim R + \operatorname{Dim}_K \Omega_{K|k}.$$

(3) *R is reduced and the R-module $\Omega_{R|k} = (\Omega_{A|k})_\mathfrak{p}$ is free.*

(4) *$R_\mathfrak{q}$ is reduced for all minimal prime ideals $\mathfrak{q} \subseteq R$ and the R-module $\Omega_{R|k} = (\Omega_{A|k})_\mathfrak{p}$ is free.*

PROOF. By Theorem 6.D.16 we have to show that condition (4) implies that $\Omega_{R|k}$ is free of rank $\dim R + \operatorname{trdeg}_k K$, $K := Q(A/\mathfrak{p})$. Let $\tilde{\mathfrak{q}} \subseteq \mathfrak{p}$ be a minimal prime of A, such that $\dim R = \dim(R/\mathfrak{q})$, $\mathfrak{q} := \tilde{\mathfrak{q}} A_\mathfrak{p}$. Since $L := R_\mathfrak{q}$ is reduced, $L = Q(A/\tilde{\mathfrak{q}})$ and by Theorem 6.D.12 (3) $\operatorname{rank}_R \Omega_{R|k} = \operatorname{Dim}_L \Omega_{L|k} = \operatorname{trdeg}_k L = \dim(A/\tilde{\mathfrak{q}}) = \dim A_\mathfrak{p} + \operatorname{trdeg}_k K$.   •

6.D.20. Corollary *Let k be a field of characteristic 0, A a k-algebra of finite type, $\mathfrak{p} \subseteq A$ a prime ideal and $R := A_\mathfrak{p}$. The following conditions are equivalent*:

(1) *R is regular.*

(2) *The R-module $\Omega_{R|k} = (\Omega_{A|k})_\mathfrak{p}$ is free.*

PROOF. By Corollary 6.D.19 it is enough to show that R is reduced if $\dim R = 0$ and if $\Omega_{R|k}$ is a free R-module. By Lemma 6.D.14 the differentials $d f_i = d_{R|k} f_i$, $i = 1, \ldots, r$, belong to an R-basis of $\Omega_{R|k}$, if $f_1, \ldots, f_r$ minimally generate $\mathfrak{m} = \mathfrak{m}_R$. Assume $r \geq 1$ and let $n_1 \geq 2$ be the minimal exponent with $f_1^{n_1} = 0$. Then $0 = d f_1^{n_1} = n_1 f_1^{n_1-1} d f_1$, hence $f_1^{n_1-1} = 0$, a contradiction.   •

6.D.21. Example Let k be a field of characteristic $p > 0$, A a k-algebra of finite type, $\mathfrak{p} \in \operatorname{Spec} A$ and $R := A_{\mathfrak{p}}$.

(1) $\Omega_{R|k}$ can be free without R being regular. Take for instance $R := k[X]/(X^p)$. Even there exist non-regular integral domains R for which $\Omega_{R|k}$ is free, for instance, $R := k[X, Y]_{(X,Y)}/(X^p + aY^p)$, if $a \in k$ is not a p-th power in k.

(2) R can be regular without $\Omega_{R|k}$ being free. To construct an example let $p > 2$ and $a \in k$ not a p-th power in k and take $A := k[X, Y]/(X^2 - Y^p + a) = k[x, y]$, $\mathfrak{m} := (x) \in \operatorname{Spm} A$. Then $R = A_{\mathfrak{m}}$ is a discrete valuation ring, but $\Omega_{R|k} \cong R \oplus (R/\mathfrak{m})$ is not free.

In view of Theorem 6.D.16 we give the following definition:

6.D.22. Definition The localization $A_{\mathfrak{p}}$ of an algebra A of finite type over a field k with respect to a prime ideal $\mathfrak{p} \subseteq A$ is called s m o o t h if $\Omega_{A_{\mathfrak{p}}|k} = (\Omega_{A|k})_{\mathfrak{p}}$ is a free module of rank $\dim A_{\mathfrak{p}} + \operatorname{trdeg}_k \kappa(\mathfrak{p})$. – A point x in a scheme X locally of finite type over a field k is called s m o o t h if its local ring $\mathcal{O}_{X,x}$ is smooth.

Non-smooth points are also called s i n g u l a r points, but the reader should take notice of the remarks behind Definition 6.A.12. By Corollary 6.D.19, *if k is a perfect field, the smooth points of* $\operatorname{Spec} A$ *coincide with the regular points of* $\operatorname{Spec} A$. Over a non-perfect field k there exist regular points which are not smooth, for instance, the spectrum of every finite inseparable field extension of k. On the other hand, there exist smooth points $\mathfrak{p} \in \operatorname{Spec} A$ for which the residue field $\kappa(\mathfrak{p})$ is not separably generated over k. Every closed point $\mathfrak{m} \in \mathbb{A}_k^n = \operatorname{Spec} k[X_1, \ldots, X_n]$ for which the field $\kappa(\mathfrak{m}) = k[X_1, \ldots, X_n]/\mathfrak{m}$ is not separable over k is such an example. The relationship between smoothness and regularity is described in the following proposition.

6.D.23. Proposition *Let k be a field and A a k-algebra of finite type. For a point $\mathfrak{p} \in \operatorname{Spec} A$ the following conditions are equivalent:*

(1) *$\mathfrak{p}$ is a smooth point.*

(2) *For every field extension $k \subseteq k'$ every point $\mathfrak{p}' \in \operatorname{Spec} A'$ lying over $\mathfrak{p}$ is a smooth point of* $\operatorname{Spec} A'$, $A' := k' \otimes_k A$.

(3) *There exist a field extension $k \subseteq k'$ and a point $\mathfrak{p}' \in \operatorname{Spec} A'$ lying over $\mathfrak{p}$ such that $\mathfrak{p}'$ is a smooth point of* $\operatorname{Spec} A'$, $A' := k' \otimes_k A$.

(4) *For every field extension $k \subseteq k'$ every point $\mathfrak{p}' \in \operatorname{Spec} A'$ lying over $\mathfrak{p}$ is a regular point of* $\operatorname{Spec} A'$, $A' := k' \otimes_k A$.

(5) *There exist a field extension $k \subseteq k'$ with a perfect field k' and a point $\mathfrak{p}' \in \operatorname{Spec} A'$ lying over $\mathfrak{p}$ such that $\mathfrak{p}'$ is a regular point of* $\operatorname{Spec} A'$, $A' := k' \otimes_k A$.

For the proof we use two lemmas:

6.D.24. Lemma *Let* A *be an algebra of finite type over a field* k, $k \subseteq k'$ *a field extension and let* $A' := k' \otimes_k A$. *Let* $\mathfrak{p}' \in \operatorname{Spec} A'$ *and* $\mathfrak{p} = \mathfrak{p}' \cap A$. *Then*

(1) *If* $\mathfrak{p}'$ *is minimal in* A', *then* $\mathfrak{p}$ *is minimal in* A *and* $\dim A'/\mathfrak{p}' = \dim A/\mathfrak{p}$. *In particular,* $\dim A'/\mathfrak{p}' \leq \dim A/\mathfrak{p}$ *in any case.*

(2) $\dim A'_{\mathfrak{p}'} + \dim A'/\mathfrak{p}' = \dim A_{\mathfrak{p}} + \dim A/\mathfrak{p}$.

PROOF. (1) Let $\mathfrak{p}' \subseteq A'$ be minimal. Then $\mathfrak{p}$ is minimal in A by the comments to the Going-down Theorem 3.B.12. Now, replacing A by $A/\mathfrak{p}$ (and A' by $A'/\mathfrak{p}A' = (A/\mathfrak{p})'$) we may assume that A is an integral domain and $\mathfrak{p} = 0$. Let $T \subseteq k'$ be a transcendence basis of k' over k, and let $k'' := k(T)$. Then $A'' := k'' \otimes_k A = k'' \otimes_k \left(k[T] \otimes_{k[T]} A \right) = S^{-1} \left(k[T] \otimes_k A \right)$ is again an integral domain and $\dim A'' = \operatorname{trdeg}_{k''} A'' = \operatorname{trdeg}_k A = \dim A$. Therefore, we may assume that k' is algebraic over k. Then $A'/\mathfrak{q}'$ is integral over A and $\dim A'/\mathfrak{q}' = \dim A$ by Theorem 3.B.8.

(2) Let $\mathfrak{q} \subseteq \mathfrak{p}$ be a prime ideal with $\dim A/\mathfrak{q} = \dim A_{\mathfrak{p}} + \dim A/\mathfrak{p}$. There exists a prime ideal $\mathfrak{q}' \subseteq \mathfrak{p}'$ with $\mathfrak{q}' \cap A = \mathfrak{q}$ (see the comments to the Going-down Theorem 3.B.12). Among these choose $\mathfrak{q}'$ minimal. Then $\mathfrak{q}'/\mathfrak{q}A'$ is a minimal prime in $A'/\mathfrak{q}A' = k' \otimes_k (A/\mathfrak{q})$. Now, by (1) $\dim A/\mathfrak{q} = \dim A'/\mathfrak{q}'$ and hence $\dim A_{\mathfrak{p}} + \dim A/\mathfrak{p} = \dim A/\mathfrak{q} = \dim A'/\mathfrak{q}' \leq \dim A'_{\mathfrak{p}'} + \dim A'/\mathfrak{p}'$. Conversely, if $\mathfrak{q}' \in \operatorname{Spec} A'$ with $\mathfrak{q}' \subseteq \mathfrak{p}'$ and $\dim A'/\mathfrak{q}' = \dim A'_{\mathfrak{p}'} + \dim A'/\mathfrak{p}'$, then $\mathfrak{q} := \mathfrak{q}' \cap A \subseteq \mathfrak{p}$ and $\mathfrak{q}'/\mathfrak{q}A'$ is a minimal prime in $A'/\mathfrak{q}A'$ and hence $\dim A/\mathfrak{q} = \dim A'/\mathfrak{q}'$ again by (1) and $\dim A/\mathfrak{q} \leq \dim A/\mathfrak{p} + \dim A/\mathfrak{p}$.   •

6.D.25. Lemma *Let* A *be an algebra of finite type over a field* k, $k \subseteq k'$ *a field extension and let* $A' := k' \otimes_k A$. *Let* $\mathfrak{p}' \in \operatorname{Spec} A'$ *lie over* $\mathfrak{p} \in \operatorname{Spec} A$. *If* $A'_{\mathfrak{p}'}$ *is regular, then* $A_{\mathfrak{p}}$ *is also regular.*

Since the extension $A_{\mathfrak{p}} \to A'_{\mathfrak{p}'}$ is faithfully flat, Lemma 6.D.25 is a special case of the general result discussed in Exercise 6.A.9. But there exists an elementary proof using the simple fact (see Example 6.A.13) that the localizations of a polynomial ring of finite type over a field are regular local. The details are left to the reader.

Now we prove Proposition 6.D.23. By Exercises 6.D.5 and 6.D.6 we have

$$\Omega_{A'_{\mathfrak{p}'} \mid k'} = (\Omega_{A' \mid k'})_{\mathfrak{p}'} = (k' \otimes_k \Omega_{A \mid k})_{\mathfrak{p}'} = A'_{\mathfrak{p}'} \otimes_{A_{\mathfrak{p}}} (\Omega_{A \mid k})_{\mathfrak{p}} = A'_{\mathfrak{p}'} \otimes_{A_{\mathfrak{p}}} \Omega_{A_{\mathfrak{p}} \mid k}.$$

It follows that $\Omega_{A'_{\mathfrak{p}'} \mid k'}$ is free if and only if $\Omega_{A_{\mathfrak{p}} \mid k}$ is free. In this case both these modules have the same rank. Now the equivalence of conditions (1), (2) and (3) follows from Lemma 6.D.24. The equivalence with conditions (4) and (5) follows from Lemma 6.D.25 and the fact that over a perfect field the regular points coincide with the smooth points. (Note that every field k has an extension k' which is perfect, for example, its algebraic closure.)   •

Because of Proposition 6.D.23 smooth points are also called g e o m e t r i c a l l y or u n i v e r s a l l y r e g u l a r p o i n t s. In particular, they are regular points.

6.D.26. Exercise In the situation of Noether's normalization lemma let $A = k[x_1, \ldots, x_n]$ be an algebra of finite type over the field k which is an integral domain such that the quotient field $K := Q(A)$ is separably generated over k. Then there exist $z_1, \ldots, z_m \in A$ which are algebraically independent over k such that A is integral over $k[z_1, \ldots, z_m]$ and K is (finite and) separable over $Q(k[z_1, \ldots, z_m]) = k(z_1, \ldots, z_m)$. (For the proof follow the lines of the proof of 1.F.2. If Char $k = p > 0$ and if $x_1, \ldots, x_n$ are algebraically dependent over k one may assume that $K dx_1 + \cdots + K dx_{n-1} = \Omega_{K|k}$. For appropriate sufficiently big multiples $\gamma_1, \ldots, \gamma_{n-1}$ of p the element x_n is integral over $A' := k[y_1, \ldots, y_{n-1}]$, $y_i := x_i - x_n^{\gamma_i}$, $i = 1, \ldots, n-1$, see the proof of Lemma 1.F.1. Then $dy_i = dx_i$ and $\Omega_{K|K'} = \Omega_{K|k}/(K dx_1 + \cdots + K dx_{n-1}) = 0$, hence K is separable over $K' := Q(A')$. Now apply the induction hypothesis on A'. – For an application of this result cf. Example 1.F.4.)

6.E. Quasi-coherent Sheaves and the Sheaf of Kähler Differentials

Let $(X, \mathcal{O}_X)$ be a ringed space. Recall from Section 4.A that a sheaf $\mathcal{F}$ on X is called a **sheaf of modules over** $\mathcal{O}_X$ or an $\mathcal{O}_X$-**module**, if for each open set $U \subseteq X$ the set of sections $\mathcal{F}(U) = \Gamma(\mathcal{F}, U)$ is provided with an $\mathcal{O}_X(U)$-module structure, such that for each inclusion of open sets $V \subseteq U$ in X the restriction map $\rho_V^U : \mathcal{F}(U) \to \mathcal{F}(V)$ respects the module structures, i.e. ρ_V^U is additive and $\rho_V^U(as) = \rho_V^U(a) \cdot \rho_V^U(s)$ for every $a \in \mathcal{O}_X(U)$ and $s \in \mathcal{F}(U)$. In a similar way, presheaves of modules are defined (over presheaves of rings). It is easy to check that the sheafification of a presheaf of modules over a presheaf $\mathcal{A}$ of rings is a sheaf of modules over the sheafification of $\mathcal{A}$. Morphisms of sheaves (or presheaves) of modules are defined in an obvious way as already indicated in Section 4.A.

6.E.1. Exercise Let X be a topological space, A a ring and M an A-module. Show that the constant sheaf M_X is (in a canonical way) a sheaf of modules over the constant sheaf A_X. (See Exercise 4.A.11.) Show also that any homomorphism $M \to N$ of A-modules induces an A_X-morphism $M_X \to N_X$.

The definition of exact sequences of sheaves of modules is not straightforward. First we define the kernel and the cokernel of a morphism $f : \mathcal{F} \to \mathcal{G}$ of sheaves of $\mathcal{O}_X$-modules given by the compatible family of $\mathcal{O}_X(U)$-module homomorphisms $f_U : \mathcal{F}(U) \to \mathcal{G}(U)$, U open in X. It is easy to verify that the family of kernels Ker f_U with the restriction maps $\rho_V^U \mid$ Ker f_U is a sheaf of modules over $\mathcal{O}_X$. By definition this is the **kernel** Ker f of f. It is a sub-$\mathcal{O}_X$-module of $\mathcal{F}$. The family of the images Im f_U and also the family of the cokernels coker $f_U = \mathcal{G}(U)/\text{Im } f_U$, form, in general, only presheaves of modules over $\mathcal{O}_X$ (see Exercise 4.A.16 (2) for an example). By definition, the **image** Im f and the **cokernel** coker f of f are the sheafifications of these presheaves. Note that Im f can be identified with the following subsheaf $\mathcal{G}'$ of $\mathcal{G}$: A section $s \in \mathcal{G}(U)$ belongs to $\mathcal{G}'(U)$ if and only if there is an open cover $U_i, i \in I$, of U such that $s \mid U_i \in \text{Im } f_{U_i}$ for every $i \in I$. For a sub-$\mathcal{O}_X$-module $\mathcal{F}' \subseteq \mathcal{F}$ the **quotient** (sheaf) $\mathcal{F}/\mathcal{F}'$ is defined as the cokernel of the inclusion $\mathcal{F}' \to \mathcal{F}$. i.e. the sheafification of the presheaf $\mathcal{F}(U)/\mathcal{F}'(U)$, U open in X. The proof of the following proposition is left to the reader as an exercise.

6.E.2. Proposition *Let* $(X, \mathcal{O}_X)$ *be a ringed space and* $f : \mathcal{F} \to \mathcal{G}$ *a morphism of* $\mathcal{O}_X$-*modules.*

(1) *The following conditions are equivalent*:

a) $\operatorname{Ker} f = 0$.

b) *The homomorphism* $f_x : \mathcal{F}_x \to \mathcal{G}_x$ *of the stalks is injective for every* $x \in X$.

c) *The homomorphism* $f_U : \mathcal{F}(U) \to \mathcal{G}(U)$ *is injective for every open set* $U \subseteq X$.

(2) *The following conditions are equivalent*:

a) $\operatorname{coker} f = 0$. a') $\operatorname{Im} f = \mathcal{G}$.

b) *The homomorphism* $f_x : \mathcal{F}_x \to \mathcal{G}_x$ *of the stalks is surjective for every* $x \in X$.

Note that in part (2) of the proposition there is no condition analogous to (c) of part (1) (cf. Exercise 4.A.16).

The last proposition motivates the definition of exact sequences of $\mathcal{O}_X$-modules.

6.E.3. Definition Let $(X, \mathcal{O}_X)$ be a ringed space. A sequence

$$\mathcal{F}' \xrightarrow{\;f'\;} \mathcal{F} \xrightarrow{\;f\;} \mathcal{F}''$$

of $\mathcal{O}_X$-modules and $\mathcal{O}_X$-module homomorphisms is called e x a c t, if the sequence

$$\mathcal{F}'_x \xrightarrow{\;f'_x\;} \mathcal{F}_x \xrightarrow{\;f_x\;} \mathcal{F}''_x$$

of homomorphisms of stalks is exact for every $x \in X$.

A sequence as in Definition 6.E.3 is exact if and only if $\operatorname{Im} f' = \operatorname{Ker} f$. The inclusion $\operatorname{Im} f' \subseteq \operatorname{Ker} f$ is equivalent to $f \circ f' = 0$. In this case the sequence is called a c o m p l e x and the sheaf $\operatorname{Ker} f / \operatorname{Im} f'$ is called its h o m o l o g y or its c o h o m o l o g y (s h e a f) – depending on the point of view.

Now let $X = \operatorname{Spec} A$ be the spectrum of a (commutative) ring A with its structure sheaf $\mathcal{O}_X = \widetilde{A}$ and let M be an A-module. In the same way as the sheaf $\widetilde{A}$ is constructed in Section 4.B a sheaf $\widetilde{M}$ of modules over $\widetilde{A}$ can be constructed with

$$\Gamma(D(f), \widetilde{M}) := M_f$$

for $f \in A$ and with $\Gamma(U, \widetilde{M}) = \varprojlim_{D(f) \subseteq U} \Gamma(D(f), \widetilde{M})$ for an arbitrary open subset U of X. The proof that this indeed defines a *sheaf* of modules is identical to that of Theorem 4.B.1. In particular, we have

$$\Gamma(X, \widetilde{M}) = M.$$

6.E.4. Exercise Let $(X, \mathcal{O}_X) = (\operatorname{Spec} A, \widetilde{A})$ be an affine scheme and M an A-module. Show that for an arbitrary $\mathcal{O}_X$-module $\mathcal{F}$ the canonical group homomorphism

$$\operatorname{Hom}_{\mathcal{O}_X}(\widetilde{M}, \mathcal{F}) \to \operatorname{Hom}_A(M, \Gamma(X, \mathcal{F}))$$

is bijective.

The stalk of $\widetilde{M}$ at a point $x \in X$ is the localization $M_{\mathfrak{p}_x}$ of M with respect to the prime ideal $\mathfrak{p}_x$ corresponding to x. Since a sequence $M' \to M \to M''$ of A-modules and A-module homomorphisms is exact if and only if all the localized sequences $M'_{\mathfrak{p}} \to M_{\mathfrak{p}} \to M''_{\mathfrak{p}}$, $\mathfrak{p} \in \mathrm{Spec}\, A$, are exact, we get the following proposition:

6.E.5. Proposition *Let $M' \to M \to M''$ be a sequence of A-module homomorphisms. The sequence $\widetilde{M}' \to \widetilde{M} \to \widetilde{M}''$ is a complex if and only if the sequence $M' \to M \to M''$ of global sections is a complex. In this case if $\mathcal{H}$ is the homology of this complex, then $\mathcal{H}$ is in a canonical way isomorphic to the homology sheaf of $\widetilde{M}' \to \widetilde{M} \to \widetilde{M}''$. In particular, the sequence $\widetilde{M}' \to \widetilde{M} \to \widetilde{M}''$ of $\widetilde{A}$-modules is exact if and only if the sequence $M' \to M \to M''$ of global sections is exact.*

An $\mathcal{O}_X$-module $\mathcal{F}$ over an affine scheme $(X, \mathcal{O}_X) = (\mathrm{Spec}\, A, \widetilde{A})$ is called q u a s i - c o h e r e n t, if $\mathcal{F} \cong \widetilde{M}$ for an A-module M. In this case, necessarily $M \cong \Gamma(X, \mathcal{F})$ and for a standard open set $\mathrm{D}(f)$ the restriction $\mathcal{F}\,|\,\mathrm{D}(f) \cong \widetilde{M}\,|\,\mathrm{D}(f) = \widetilde{M_f}$ is also quasi-coherent. Furthermore, by Proposition 6.E.5 kernels and cokernels of homomorphisms of quasi-coherent $\mathcal{O}_X$-modules are again quasi-coherent. If $\mathcal{F}$ is quasi-coherent, then $\mathcal{F}\,|\,U$ is quasi-coherent for *every* open affine subset $U \subseteq \mathrm{Spec}\, A$ (and not only for the standard open sets $\mathrm{D}(f)$). This is a consequence of the following more general result.

6.E.6. Proposition *Let $(X, \mathcal{O}_X) = (\mathrm{Spec}\, A, \widetilde{A})$ be an affine scheme and $\mathcal{F}$ an $\mathcal{O}_X$-module. Then the following conditions are equivalent:*

(1) $\mathcal{F}$ is quasi-coherent.

(2) There is an open affine cover U_i, $i \in I$, of X such that $\mathcal{F}\,|\,U_i$ is quasi-coherent for every $i \in I$.

(3) There is a finite open cover $\mathrm{D}(f_1), \ldots, \mathrm{D}(f_m)$ of X by standard open sets such that $\mathcal{F}\,|\,\mathrm{D}(f_i)$ is quasi-coherent for every $i = 1, \ldots, m$.

PROOF. The implication "(1) $\Rightarrow$ (2)" is trivial. In order to prove "(2) $\Rightarrow$ (3)" let $f \in A$ be an element with $\mathrm{D}(f) \subseteq U_i$ for some $i \in I$ and let $f\,|\,U_i \cong \widetilde{M_i}$ for some $\Gamma(U_i)$-module M_i. Then $\mathcal{F}\,|\,\mathrm{D}(f) \cong \widetilde{M_i}\,|\,\mathrm{D}(f) = \widetilde{(M_i)_{f\,|\,U_i}}$ is also quasi-coherent. Since X is quasi-compact, (3) follows from (2).

"(3) $\Rightarrow$ (1)": For $i = 1, \ldots, m$ let $U_i := \mathrm{D}(f_i)$ and $M_i := \Gamma(U_i, \mathcal{F})$. Hence $\mathcal{F}\,|\,U_i \cong \widetilde{M_i}$. By Serre's sheaf conditions (S1) and (S2) for an arbitrary open subset $U \subseteq X$ the sequence

$$0 \to \mathcal{F}(U) \to \bigoplus_i \Gamma(U \cap U_i, \mathcal{F}) \to \bigoplus_{i,j} \Gamma(U \cap U_i \cap U_j, \mathcal{F})$$

is exact, where the last homomorphism is given by

$$(s_i)_i \mapsto \left(s_j\,|\,U \cap U_i \cap U_j - s_i\,|\,U \cap U_i \cap U_j\right)_{i,j}.$$

But $\bigoplus_i \Gamma(U \cap U_i, \mathcal{F})$ and $\bigoplus_{i,j} \Gamma(U \cap U_i \cap U_j, \mathcal{F})$ are the sections of the quasi-coherent $\mathcal{O}_X$-modules $\bigoplus_i \widetilde{M_i}$ and $\bigoplus_{i,j} \widetilde{M_{ij}}$ respectively, where $M_{ij} := (M_i)_{f_j} =$

$(M_j)_{f_i} = \Gamma(U_i \cap U_j, \mathcal{F})$. So $\mathcal{F}$ is the kernel of a homomorphism of quasi-coherent $\mathcal{O}_X$-modules and hence quasi-coherent too.   •

The last proposition allows the following general definition.

6.E.7. Definition Let X be an (arbitrary) scheme and $\mathcal{F}$ an $\mathcal{O}_X$-module. Then $\mathcal{F}$ is called q u a s i - c o h e r e n t, if one of the following equivalent conditions holds:

(1) For every open affine subset $U \cong \operatorname{Spec} A$ of X the restriction $\mathcal{F}|U$ is quasi-coherent, i. e. isomorphic to $\widetilde{M}$ for some A-module M.

(2) There is a cover of X by open affine subsets U_i, $i \in I$, such that $\mathcal{F}|U_i$ is quasi-coherent for every $i \in I$, i. e. isomorphic to $\widetilde{M}_i$ for a $\Gamma(U_i, \mathcal{O}_X)$-module M_i.

That these conditions are equivalent is an immediate consequence of Proposition 6.E.6. The restriction of a quasi-coherent $\mathcal{O}_X$-module to an arbitrary open subset of X is again quasi-coherent. Furthermore, direct sums of quasi-coherent $\mathcal{O}_X$-modules are quasi-coherent and so are the kernels and cokernels of homomorphisms of quasi-coherent $\mathcal{O}_X$-modules. The tensor product of quasi-coherent $\mathcal{O}_X$-modules is quasi-coherent. For an affine scheme $X = \operatorname{Spec} A$, we have $\widetilde{M} \otimes_{\widetilde{A}} \widetilde{N} = (M \otimes N)^{\sim}$ for A-modules M, N, this follows from the canonical isomorphisms $(M \otimes_A N)_f \cong M_f \otimes_{A_f} N_f$ for every $f \in A$. The sheaf $\mathcal{H}om_{\mathcal{O}_X}(\mathcal{F}, \mathcal{G})$ of quasi-coherent $\mathcal{O}_X$- modules is, in general, not quasi-coherent; however, if $\Gamma(U, \mathcal{F})$ is a finitely represented $\mathcal{O}(U)$-module for every affine open subset U of X, then it is. This follows from the canonical isomorphisms $\operatorname{Hom}_A(M, N)_f \cong \operatorname{Hom}_{A_f}(M_f, N_f)$ for every $f \in A$, if M is a finitely represented A-module. These isomorphisms imply $\mathcal{H}om_{\widetilde{A}}(\widetilde{M}, \widetilde{N}) = (\operatorname{Hom}_A(M, N))^{\sim}$ in this case.

The s u p p o r t $\operatorname{Supp} \mathcal{F}$ of an $\mathcal{O}_X$-module $\mathcal{F}$ is, by definition, the set of points $x \in X$ with $\mathcal{F}_x \neq 0$. If $\mathcal{F} = \widetilde{M}$ is a quasi-coherent $\mathcal{O}_X$-module over an affine scheme $X = \operatorname{Spec} A$, then the $\operatorname{Supp} \mathcal{F}$ coincides with the support of the A-module $M = \Gamma(X, \mathcal{F})$. In particular, $\operatorname{Supp} \mathcal{F} = \operatorname{Supp} M = V(\operatorname{Ann}_A M)$ is a closed subset of X if M is a finite A-module. It follows that for an arbitrary scheme X and a quasi-coherent $\mathcal{O}_X$-module $\mathcal{F}$ the support $\operatorname{Supp} \mathcal{F}$ is closed if $\Gamma(U, \mathcal{F})$ is a finite $\Gamma(U, \mathcal{O}_X)$-module for all open affine subset $U \subseteq X$. The last condition is easily seen to hold if $\Gamma(U_i, \mathcal{F})$ is a finite $\Gamma(U_i, \mathcal{O}_X)$-module for an open affine cover U_i, $i \in I$, of X. For locally Noetherian schemes these quasi-coherent $\mathcal{O}_X$-modules are called c o h e r e n t.

6.E.8. Definition Let $X = (X, \mathcal{O}_X)$ be a locally Noetherian scheme. An $\mathcal{O}_X$-module $\mathcal{F}$ is called c o h e r e n t, if it is quasi-coherent and if one of the following equivalent conditions holds:

(1) For every open affine subset $U \subseteq X$ the module $\Gamma(U, \mathcal{F})$ is Noetherian, i. e. finite, over $\Gamma(U, \mathcal{O}_X)$.

(2) There is a cover of X by open affine subsets U_i, $i \in I$, such that $\Gamma(U_i, \mathcal{F})$ is a finite $\Gamma(U_i, \mathcal{O}_X)$-module for every $i \in I$.

By the remark preceding Definition 6.E.8, *the support of a coherent $\mathcal{O}_X$-module is always a closed subset of X*. Furthermore, $\mathcal{F} \otimes_{\mathcal{O}_X} \mathcal{G}$ and $\mathcal{H}om_{\mathcal{O}_X}(\mathcal{F}, \mathcal{G})$ are coherent $\mathcal{O}_X$- modules if $\mathcal{F}$ and $\mathcal{G}$ are coherent. Furthermore, one sets $\mathrm{Ass}\, \mathcal{F} := \{x \in X \mid \mathrm{depth}\, \mathcal{F}_x = 0\}$ for any coherent $\mathcal{O}_X$-module $\mathcal{F}$, cf. Exercise 6.A.11 (8).

6.E.9. Example (Closed subschemes and closed embeddings) Let $X = (X, \mathcal{O}_X)$ be a scheme. The quasi-coherent $\mathcal{O}_X$-submodules of the structure sheaf $\mathcal{O}_X$ are the q u a s i - coherent ideal sheaves of X. Let $\mathcal{I}$ be such a quasi-coherent ideal sheaf. The quotient $\mathcal{O}_X/\mathcal{I}$ is a sheaf of rings. For an open affine subset $U = \mathrm{Spec}\, A$ in X we have $\mathcal{O}_X \mid U \cong \widetilde{A}$ and $\mathcal{I} \mid U = \widetilde{\mathfrak{a}}$ for some ideal $\mathfrak{a} \subseteq A$ and hence $(\mathcal{O}_X/\mathcal{I}) \mid U = \widetilde{A/\mathfrak{a}}$. It follows that $\mathcal{O}_X/\mathcal{I}$ defines a scheme structure on the closed support Y of $\mathcal{O}_X/\mathcal{I}$. The schemes obtained in this way are called c l o s e d s u b s c h e m e s of X. In this case the canonical projection $\mathcal{O}_X \to \mathcal{O}_X/\mathcal{I}$ defines a c l o s e d e m b e d d i n g $Y \hookrightarrow X$ of schemes in the sense of Definition 4.E.6. An arbitrary closed embedding $Z \to X$ of schemes is a composition of an isomorphism $Z \overset{\sim}{\to} Y$ and the canonical closed embedding $Y \hookrightarrow X$, where Y is a closed subscheme of X.

6.E.10. Exercise Let $(X, \mathcal{O}_X)$ be a scheme, $\mathcal{A} \subseteq \mathcal{O}_X$ an ideal sheaf and $Y := \mathrm{Supp}\, \mathcal{O}_X/\mathcal{A}$. Show that if the sheaf of rings $\mathcal{O}_X/\mathcal{A}$ defines a scheme structure on Y, then $\mathcal{A}$ is quasi-coherent. (This exercise though somewhat involved is not really difficult.)

6.E.11. Example (Affine morphisms of schemes) Let $f = (f, \Theta) : X \to Y$ be a morphism of schemes. In the Definition 4.E.1 of affine morphisms we used the following result: *If there exists an open affine cover V_i, $i \in I$, of Y such that for every $i \in I$ the preimage $f^{-1}(V_i)$ is affine, then for every open affine subset $V \subseteq Y$ the preimage $f^{-1}(V)$ is also affine*. For the p r o o f note first that we may assume without loss of generality that the open affine subsets V_i, $i \in I$, form a basis of the topology of X. Furthermore, by definition Θ is a morphism $\Theta : \mathcal{O}_Y \to f_*\mathcal{O}_X$ of sheaves of rings, where $f_*\mathcal{O}_X$ is the direct image of the structure sheaf $\mathcal{O}_X$ (see 4.A.14). In particular, Θ defines an $\mathcal{O}_Y$-module structure (even an $\mathcal{O}_Y$-algebra structure) on $f_*\mathcal{O}_X$. By assumption $(f_*\mathcal{O}_X) \mid V_i \cong \Gamma(f^{-1}(V_i), \mathcal{O}_X)^\sim$ for every $i \in I$. It follows that $f_*\mathcal{O}_X$ is a quasi-coherent $\mathcal{O}_Y$-module. Now let V be an arbitrary open affine subset of Y and $U := f^{-1}(V)$. We have to show that the canonical scheme morphism $U \to \mathrm{Spec}\, \Gamma(U, \mathcal{O}_X)$ belonging to the identity of $\Gamma(U, \mathcal{O}_X)$ is an isomorphism. But for an arbitrary $s \in \Gamma(V, \mathcal{O}_Y)$ with $V_s = \{y \in V \mid s(y) \neq 0\} \subseteq V_i \subseteq V$ for some $i \in I$ the canonical morphism $f^{-1}(V_s) \to \mathrm{Spec}\, \Gamma(f^{-1}(V_s), \mathcal{O}_X) = \mathrm{Spec}\, \Gamma(U, \mathcal{O}_X)_s$ is an isomorphism because of $V_s = (V_i)_{s \mid V_i}$. Finally, the fact that these sets V_s form an open cover of V completes the proof.   •

If $\mathcal{F}$ is a quasi-coherent $\mathcal{O}_X$-module, then its direct image $f_*\mathcal{F}$ under an affine morphism $f : X \to Y$ is a quasi-coherent $\mathcal{O}_Y$-module. Indeed, if $V \subseteq Y$ is an affine open subset, then $f_*(\mathcal{F}|V) = \Gamma(f^{-1}(V), \mathcal{F})^\sim$, where $\Gamma(f^{-1}(V), \mathcal{F})$ has to be considered as a $\Gamma(V, \mathcal{O}_Y)$-module with respect to the homomorphism $\Theta_V : \Gamma(V, \mathcal{O}_Y) \to \Gamma(f^{-1}(V), \mathcal{O}_X)$.

Moreover, *if $f : X \to Y$ is a finite morphism of locally Noetherian schemes and if $\mathcal{F}$ is coherent, then the direct image $f_*\mathcal{F}$ is also coherent*. Namely, by definition, for any affine open subset V in Y, $\Gamma(f^{-1}(V), \mathcal{O}_X)$ is a finite $\Gamma(V, \mathcal{O}_Y)$-algebra with respect to Θ_V and $\Gamma(f^{-1}(V), \mathcal{F})$ is a finite $\Gamma(f^{-1}(V), \mathcal{O}_X)$-module.

Give examples that, in general, for a morphism $f : X \to Y$ of schemes, $f_*\mathcal{O}_X$ is not a quasi-coherent $\mathcal{O}_Y$-module. Such examples are rather pathological. Show, for example, the following general result: *If $f : X \to Y$ is a quasi-compact morphism of schemes then $f_*\mathcal{F}$*

is a quasi-coherent $\mathcal{O}_Y$*-module for every quasi-coherent* $\mathcal{O}_X$*-module* $\mathcal{F}$. (For the proof one may assume that Y is affine. Then use the following fact: If X is quasi-compact and if $\mathcal{F}$ is a quasi-coherent $\mathcal{O}_X$-module then $\Gamma(X_f, \mathcal{F}) = \Gamma(X, \mathcal{F})_f$ for every $f \in \Gamma(X, \mathcal{O}_X)$, cf. the proof of Proposition 6.E.6.)

6.E.12. Exercise (Pull back of a module) Let $f : X \to Y$ be a morphism of ringed spaces and let $\mathcal{G}$ be an $\mathcal{O}_Y$-module. Define an $\mathcal{O}_X$-module $f^*\mathcal{G}$ and an $\mathcal{O}_X$-homomorphism $\mathcal{G} \to f_*(f^*\mathcal{G})$ with the following universal property: For every $\mathcal{O}_X$-module $\mathcal{F}$ and every $\mathcal{O}_Y$-homomorphism $\mathcal{G} \to f_*\mathcal{F}$ there exists a unique $\mathcal{O}_X$-homomorphism $f^*\mathcal{G} \to \mathcal{F}$ such that the composition $\mathcal{G} \to f_*(f^*\mathcal{G}) \to f_*\mathcal{F}$ coincides with the given homomorphism $\mathcal{G} \to f_*\mathcal{F}$. For every $x \in X$ one has $(f^*\mathcal{G})_x = \mathcal{O}_x \otimes_{\mathcal{O}_{f(x)}} \mathcal{G}_{f(x)}$. (This is a hint for the construction of $f^*\mathcal{G}$.) $f^*\mathcal{G}$ is called the pull back of $\mathcal{G}$ with respect to $f : X \to Y$.

Now, let $f : X \to Y$ be a morphism of schemes. Show that $f^*\mathcal{G}$ is quasi-coherent if $\mathcal{G}$ is quasi-coherent and that, in case X and Y are locally Noetherian, $f^*\mathcal{G}$ is coherent if $\mathcal{G}$ is coherent. (**Hint:** If $A \to B$ is a ring homomorphism and $f :$ Spec $B \to$ Spec A the corresponding morphism of affine schemes, then $f^*\widetilde{M} = (B \otimes_A M)^\sim$ for every A-module M.)

6.E.13. Exercise (Sheaf of rational functions for a scheme) Exercise 4.B.8 (2) suggests to construct a sheaf $\mathcal{R} = \mathcal{R}_X$ of rational functions on an arbitrary scheme X.

(1) For an *affine* open set $U \subseteq X$ let

$$\mathcal{R}'(U) = \mathcal{R}'_X(U) := Q\big(\mathcal{O}_X(U)\big) = Q\big(\Gamma(U, \mathcal{O}_X)\big).$$

Show that for affine open subsets $V \subseteq U$ the restriction $\mathcal{O}(U) \to \mathcal{O}(V)$ extends uniquely to a homomorphism $\mathcal{R}'(U) \to \mathcal{R}'(V)$ and $\mathcal{R}'$ is a presheaf of rings (with $\mathcal{R}'(U) = \lim_{U' \subseteq U \text{ affine}} \mathcal{R}'(U')$ for an arbitrary open subset $U \subseteq X$, cf. Section 4.A). The associated sheaf $\mathcal{R} = \mathcal{R}_X$ is called the sheaf of rational functions on X. If X is an integral scheme, then $\mathcal{R} = \mathcal{R}'$ is the constant sheaf $\mathcal{R}(X)$ and the field $\mathcal{R}(X)$ is called the field of rational functions on X, see Exercise 4.B.8 (2).

(2) The presheaf $\mathcal{R}'$ fulfills Serre's condition (S1) of Definition 4.A.2. Hence, for every $x \in X$, the canonical homomorphism

$$\mathcal{R}_x = \lim_{x \in U, U \text{ affine}} Q\big(\mathcal{O}(U)\big) \to Q(\mathcal{O}_x)$$

is injective and $\mathcal{R}'(U) \subseteq \mathcal{R}(U)$ for all open sets $U \subseteq X$. For an open $U \subseteq X$, the element $(r_x)_{x \in U} \in \prod_{x \in U} Q(\mathcal{O}_x)$ belongs to $\mathcal{R}(U)$ if and only if for every $x \in U$ there exists an affine open neighbourhood $U_x \subseteq U$ of x and an element $a/b \in Q\big(\mathcal{O}(U_x)\big)$ with $a/b = r_y$ in $Q(\mathcal{O}_y)$ for all $y \in U_x$.

(3) Let $r \in \Gamma(X, \mathcal{R}_X)$ be a rational function on X. The sheaf $\mathcal{D}_r$ of denominators of r is the subsheaf of $\mathcal{O}_X$ defined by

$$\mathcal{D}_r(U) := \{ s \in \Gamma(U, \mathcal{O}_X) \mid s_x r_x \in \mathcal{O}_x \text{ for all } x \in U \}, \quad U \text{ open in } X .$$

Show that $\mathcal{D}_r \subseteq \mathcal{O}_X$ is a quasi-coherent ideal sheaf on X. (Since $r_x \in \mathcal{O}_x$ is equivalent to $\mathcal{D}_{r,x} = \mathcal{O}_x$, the closed support of $\mathcal{O}_X/\mathcal{D}_r$ is called the locus of indeterminacy of r. Its complement, sometimes called the domain of definition of r, is dense in X, and the restriction of r to this open subset of X is a section in $\mathcal{O}_X$.) If $U \subseteq X$ is open and affine and if $\mathcal{D}_r(U)$ contains a non-zero divisor $b \in \Gamma(U, \mathcal{O}_X)$, then $a := b \cdot (r|U) \in \Gamma(U, \mathcal{O}_X)$ and $r|U = a/b \in Q\big(\Gamma(U, \mathcal{O}_X)\big) = \mathcal{R}'(U) \subseteq \mathcal{R}(U)$.

Determine the domain of definition for the rational function $a/b \in Q(A) = \mathcal{R}_X(X)$, $a, b \in A$, $b \neq 0$, $X := \operatorname{Spec} A$, in the following cases: a) A is a factorial domain. b) A is a Noetherian normal domain, for example $A := k[X, Y, U, V]/(XU - YV) = k[x, y, u, v]$, k a field, $a/b := v/x = u/y$. c) $A := k[T, TU, U(U-1), U^2(U-1)]$ $\left(\subseteq k[T, U] \right)$, k a field, $a/b := U = TU/T \in Q(A) = k(T, U)$. (The last example is from R. Hartshorne, Am. J. Math. **84**, 497–508 (1962). Note that $k[T, U]$ is the normalization of A. Study the normalization map $\mathbb{A}_k^2 = \operatorname{Spec} k[T, U] \to \operatorname{Spec} A$.)

(4) Let X be Noetherian and affine. Then $U \mapsto \operatorname{NZD}(\Gamma(U, \mathcal{O}_X)) =$ monoid of the non-zero-divisors of $\Gamma(U, \mathcal{O}_X)$ and $U \mapsto Q(\Gamma(U, \mathcal{O}_X))$, U open in X, are sheaves on X. (Use (3) and Exercise 6.A.11 and show that a section $s \in \Gamma(U, \mathcal{O}_X)$ is a non-zero divisor if and only if $s_x \in \mathcal{O}_x^\times$ for all $x \in U$ with depth $\mathcal{O}_x = 0$.) – For an arbitrary Noetherian scheme X, $U \mapsto \operatorname{NZD}(\Gamma(U, \mathcal{O}_X))$ is not necessarily a presheaf. For example, let $X := X_1 \cup X_2 \setminus \{P\}$ where $X_1 \cup X_2$ is the union of two lines in $\mathbb{P}_k^2$ (with its reduced structure), k a field, and $P \in X_1 \cup X_2$ is a closed point different from the intersection point $X_1 \cap X_2$.

(5) Let X be locally Noetherian. Then $\mathcal{R}'(U) = \mathcal{R}(U)$ for all affine open sets $U \subseteq X$, i.e. the presheaf $\mathcal{R}'$ coincides already with the sheaf $\mathcal{R}$ of rational functions on X. Moreover, for every (not necessarily affine) open set V contained in an affine open set $U \subseteq X$, $\Gamma(V, \mathcal{R}(X)) = Q(\Gamma(V, \mathcal{O}_X))$. (Use (4).) Furthermore, $\mathcal{R}_x = Q(\mathcal{O}_x)$ for all $x \in X$. (If b is a non-zero divisor in $\mathcal{O}_x$, then there exists an open affine neighbourhood U of x and a non-zero divisor $f \in \Gamma(U, \mathcal{O}_X)$ with $f_x = b$.)

(6) Let X be locally Noetherian. Give examples that $\mathcal{R}_X$ ($= \mathcal{R}_X'$) is not necessarily a quasi-coherent $\mathcal{O}_X$-module. But, if the associated prime ideals of all stalks $\mathcal{O}_x$, $x \in X$, are minimal prime ideals (i.e. if $\dim \mathcal{O}_x > 0$ implies depth $\mathcal{O}_x > 0$ for all $x \in X$), then $\mathcal{R}_X$ is quasi-coherent, more precisely: $\Gamma(U, \mathcal{R}_X) = \prod_{\xi \in U, \, \xi \text{ generic}} \mathcal{O}_\xi$ for every open subset $U \subseteq X$. In particular, $\mathcal{R}_X$ is quasi-coherent if X is reduced.

6.E.14. Example (Quasi-coherent modules over projective schemes) Let $R = \bigoplus_{m \in \mathbb{N}} R_m$ be a positively graded ring and $X = \operatorname{Proj} R$ the corresponding projective scheme. A graded R-module $M = \bigoplus_{m \in \mathbb{Z}} M_m$ defines in a canonical way a quasi-coherent $\mathcal{O}_X$-module. As in the affine case this module is denoted by

$$\widetilde{M}.$$

For a homogeneous element f of positive degree the module of fractions M_f is a graded module over R_f and by definition

$$\Gamma(D_+(f), \widetilde{M}) := (M_f)_0 \, .$$

It is easy to check that this defines a sheaf of modules over $\mathcal{O}_X$ and that its restriction to $D_+(f) = \operatorname{Spec}(R_f)_0$ coincides with $\widetilde{(M_f)_0}$. So $\widetilde{M}$ is indeed quasi-coherent. If $M' \to M \to M''$ is an exact sequence of homogeneous homomorphisms of degree 0 of graded R-modules, then the induced sequence $\widetilde{M'} \to \widetilde{M} \to \widetilde{M''}$ of $\mathcal{O}_X$-modules is also exact, but not conversely. The sheaf $\widetilde{M}$ may be 0 without M being 0.

For $n \in \mathbb{Z}$ the shift

$$M(n)$$

of the graded R-module M is defined by $M(n)_m := M_{m+n}$. In general, the corresponding $\mathcal{O}_X$-modules

$$\widetilde{M(n)}$$

are not isomorphic to $\widetilde{M}$. We have $\widetilde{M(n)} \mid D_+(f) = \widetilde{(M_f)_n}$. If $f \in R_1$ is a homogeneous element of degree 1, then $M_f \cong (M_f)_0 \otimes_{(R_f)_0} R_f$ as R_f-modules and $(M_f)_0 \cong (M_f)_n$ as $(R_f)_0$-modules and hence $\widetilde{M(n)} \mid D_+(f) \cong \widetilde{M} \mid D_+(f)$. It follows that if R is generated as an R_0-algebra by R_1, i. e. if R_+ is generated as an ideal by R_1, then the modules $\widetilde{M(n)}$ are locally isomorphic to $\widetilde{M}$ for all $n \in \mathbb{Z}$. In particular, in this case the $\mathcal{O}_X$-modules

$$\mathcal{O}_X(n) := \widetilde{R(n)}$$

are locally isomorphic to the structure sheaf $\mathcal{O}_X$, i. e. locally free of rank 1, furthermore

$$\widetilde{M(n)} \cong \widetilde{M} \otimes_{\mathcal{O}_X} \mathcal{O}_X(n).$$

The tensor product $M \otimes_R N$ of graded R-modules M, N is again graded with

$$\deg(x \otimes y) = \deg x + \deg y$$

for homogeneous elements $x \in M$ and $y \in N$. There is a canonical homomorphism $\widetilde{M} \otimes_{\mathcal{O}_X} \widetilde{N} \to (M \otimes_R N)^\sim$ given by the canonical homomorphisms $(M_f)_0 \otimes_{(R_f)_0} (N_f)_0 \to (M_f \otimes_{R_f} N_f)_0$ for homogeneous elements f of positive degrees. *This is an isomorphism if R is standardly graded*, since for $\deg f = 1$ the homomorphism above is obviously an isomorphism. In general, this is *not* an isomorphism. In the same way, the canonical isomorphism $M(n) \otimes_R N(p) \cong (M \otimes_R N)(n + p)$ of graded R-modules, $n, p \in \mathbb{Z}$, defines a canonical homomorphism $\widetilde{M(n)} \otimes_{\mathcal{O}_X} \widetilde{N(p)} \to (M \otimes_R N)(n + p)^\sim$ which is an isomorphism if R is standardly graded. But, in general, even the canonical homomorphisms $\widetilde{M} \otimes_{\mathcal{O}_X} \mathcal{O}_X(p) \to \widetilde{M(p)}$ are *not* isomorphisms.

For arbitrary graded R-modules M, N the R-module $\mathrm{Hom}_R(M, N)$ contains the R-submodule $\bigoplus_{n \in \mathbb{Z}} \mathrm{Hom}_R(M, N)_n$ which is graded, where $\mathrm{Hom}_R(M, N)_n$ is the group of homogeneous R-homomorphisms $M \to N$ of degree $n \in \mathbb{Z}$. If R is Noetherian and if M is finitely generated, then $\mathrm{Hom}_R(M, N)$ and $\bigoplus_{n \in \mathbb{Z}} \mathrm{Hom}_R(M, N)_n$ coincide. The canonical homomorphisms

$$\left(\left(\bigoplus_{n \in \mathbb{Z}} \mathrm{Hom}_R(M, N)_n \right)_f \right)_0 \longrightarrow \mathrm{Hom}_{(R_f)_0}((M_f)_0, (N_f)_0),$$

f homogeneous of positive degree, define a homomorphism

$$\left(\bigoplus_{n \in \mathbb{Z}} \mathrm{Hom}_R(M, N)_n \right)^\sim \longrightarrow \mathcal{H}om_{\mathcal{O}_X}(\widetilde{M}, \widetilde{N}),$$

which is an isomorphism if R is Noetherian and standardly graded and if M is finitely generated.

If the graded module M is generated by the family x_i, $i \in I$, of homogeneous elements of degrees $n_i \in \mathbb{Z}$, the canonical homomorphism of degree 0

$$\bigoplus_{i \in I} R(-n_i) \to M$$

which maps the i-th canonical basis element e_i (which is homogeneous of degree n_i) to x_i is surjective and thereby defines a surjective homomorphism

$$\bigoplus_{i \in I} \mathcal{O}_X(-n_i) \to \widetilde{M}.$$

This representation of $\widetilde{M}$ shows the importance of the modules $\mathcal{O}_X(n)$, $n \in \mathbb{Z}$.

Let R be a Noetherian graded ring. Then the scheme $X = \mathrm{Proj}\, R$ is Noetherian and for every finite graded R-module M the $\mathcal{O}_X$-module $\widetilde{M}$ is coherent and has a finite representation

$$\mathcal{O}_X(-n_1) \oplus \cdots \oplus \mathcal{O}_X(-n_r) \longrightarrow \widetilde{M}$$

for certain integers $n_1, \ldots, n_r$. In the next chapter we will show that every coherent $\mathcal{O}_X$-module is of the form $\widetilde{M}$ for some finite graded R-module M (cf. Corollary 7.A.2).

Let A be a (commutative) ring, X an A-scheme and $\mathcal{F}$ an $\mathcal{O}_X$-module. By definition, an A-d e r i v a t i o n $D : \mathcal{O}_X \to \mathcal{F}$ is a sheaf homomorphism such that $D_U : \mathcal{O}_X(U) \to \mathcal{F}(U)$ is an A-derivation for every open set $U \subseteq X$. D is called u n i v e r s a l, if for every A-derivation $\delta : \mathcal{O}_X \to \mathcal{G}$ there exists a unique $\mathcal{O}_X$-homomorphism $h : \mathcal{F} \to \mathcal{G}$ such that $\delta = h \circ D$. In this case for every open set $U \subseteq X$ the restriction $D \,|\, U : \mathcal{O}_X \,|\, U \to \mathcal{F} \,|\, U$ is also universal for the A-scheme $(U, \mathcal{O}_X \,|\, U)$. To prove this let $\delta : \mathcal{O}_X \,|\, U \to \mathcal{G}$ be an A-derivation on U and then apply the universal property of D to the A-derivation $\delta' : \mathcal{O}_X \to \iota_* \mathcal{G}$ defined by $\delta'_{U'} := \delta_{U \cap U'} \circ \rho^{U'}_{U \cap U'} : \mathcal{O}_X(U') \to (\iota_* \mathcal{G})(U') = \mathcal{G}(U \cap U'), U'$ open in X. It follows that universal A-derivations $d_i : \mathcal{O}_X \,|\, U_i \to \Omega_{U_i \,|\, A}, U_i, i \in I$, an open cover of X, can be glued together to a universal A-derivation $d : \mathcal{O}_X \to \Omega_{X \,|\, A}$. If X is the spectrum of an A-algebra B, then the A-derivation $d : \mathcal{O}_X = \widetilde{B} \to \Omega_{X \,|\, A} := \widetilde{\Omega}_{B \,|\, A}$ given by the universal A-derivations $d_f : B_f \to (\Omega_{B \,|\, A})_f = \Omega_{B_f \,|\, A}, f \in B$, is obviously universal. This proves the following theorem:

6.E.15. Theorem *For every A-scheme X there exists the universal A-derivation*

$$d = d_{X \,|\, A} : \mathcal{O}_X \to \Omega_{X \,|\, A} .$$

For an open affine subset $U \subseteq X$ one has

$$(\Omega_{X \,|\, A}) \,|\, U = \widetilde{\Omega}_{\mathcal{O}_X(U) \,|\, A} .$$

In particular, the $\mathcal{O}_X$-module $\Omega_{X \,|\, A}$ is quasi-coherent and even coherent, if, in addition, A is Noetherian and X of finite type over A.

The $\mathcal{O}_X$-module $\Omega_{X \,|\, A}$ is called the s h e a f or m o d u l e of K ä h l e r d i f f e r e n t i a l s on X over A.

6.E.16. Theorem *Let X be an algebraic scheme over a field k. Then the set of smooth points of X is an open subset of X.*

PROOF. By definition 6.D.22 a point $x \in X$ is smooth if and only if $\Omega_x := \Omega_{X \,|\, k, x}$ is free of rank $\dim \mathcal{O}_{X,x} + \operatorname{trdeg}_k \kappa(x)$. By Lemma 6.E.18 below the set of points $x \in X$ such that the stalk Ω_x is free over $\mathcal{O}_x$ is an open set X' in X and the function $x \mapsto \operatorname{rank}_{\mathcal{O}_x} \Omega_x$ is locally constant on X'. Therefore, it is sufficient to show that for a smooth point $x \in X$ there exists an open neighbourhood U such that the function $x \mapsto \dim \mathcal{O}_x + \operatorname{trdeg}_k \kappa(x)$ is constant on U. Since $\mathcal{O}_x$ is regular (cf. Proposition 6.D.23(4)), there is an open affine neighbourhood U such that $B := \Gamma(U, \mathcal{O}_X)$ is an integral domain. But then $\dim \mathcal{O}_x + \operatorname{trdeg}_k \kappa(x) = \dim B_{\mathfrak{p}_x} + \dim B/\mathfrak{p}_x = \dim B$ for all $x \in U$ (cf. Theorem 3.B.22).   ●

6.E.17. Corollary *Let X be a reduced algebraic scheme over a perfect field k. Then the set of regular points of X and the set of smooth points coincide; this set is open and dense in X.*

PROOF. By Corollary 6.D.19 a point $x \in X$ is regular if and only if it is smooth. Since the stalks of the generic points of the irreducible components of X are fields, they are regular. Since the set of smooth points of X is open by Theorem 6.E.16, it is also dense in X.   •

We mention without proof that for an arbitrary algebraic scheme X over an arbitrary field k the set of regular points is also open in X.

6.E.18. Lemma *Let X be a locally Noetherian scheme and $\mathcal{F}$ a coherent $\mathcal{O}_X$-module. Then the set* Free $\mathcal{F} := \{x \in X \mid \mathcal{F}_x$ *is free over* $\mathcal{O}_x\}$ *is open in X and the function $x \mapsto \mathrm{rank}_{\mathcal{O}_x} \mathcal{F}_x$ is locally constant on* Free $\mathcal{F}$.

PROOF. We may assume that $X = \mathrm{Spec}\, A$ is affine and $\mathcal{F} = \widetilde{M}$ with a finite A-module M. Let $\mathfrak{p} \in \mathrm{Spec}\, A$ be a prime ideal and $x_1, \ldots, x_n \in M$ be elements such that $x_1/1, \ldots, x_n/1 \in M_\mathfrak{p}$ is an $A_\mathfrak{p}$-base of $M_\mathfrak{p}$. Then the kernel and the cokernel of the A-module homomorphism $A^n \to M$, $e_i \mapsto x_i$, vanish at the point $\mathfrak{p}$ and hence so in an open neighbourhood U of $\mathfrak{p} \in \mathrm{Spec}\, A$, since the supports of the kernel and the cokernel are closed in $\mathrm{Spec}\, A$. This means that $M_\mathfrak{q}$ is free of rank n over $A_\mathfrak{q}$ for all $\mathfrak{q} \in U$.   •

6.E.19. Example (The sheaf of Kähler differentials of projective A-schemes) Let $R = \bigoplus_{m \in \mathbb{N}} R_m$ be a positively graded algebra over $A = R_0$ and $X := \mathrm{Proj}\, R$ the projective A-scheme defined by R. We want to describe the module $\Omega_{X|A}$ of Kähler differentials on X over A.

The differential module $\Omega := \Omega_{R|A}$ is graded in a canonical way such that $\deg d_{R|A} f = \deg f$ for every homogeneous element $f \in R$. This follows, for example, from the description of Ω in Propositions 6.D.1 and 6.D.3. Furthermore, the mapping $R \to R$ with $f \mapsto mf$ for every $f \in R_m$, $m \in \mathbb{N}$, is obviously an A-derivation of R called the Euler derivation of R. It is equivalent to a homogeneous R-linear map $e : \Omega \to R$ of degree 0 with $d_{R|A} f \mapsto mf$, $f \in R_m$, $m \in \mathbb{N}$. This Euler map defines the Euler homomorphism of $\mathcal{O}_X$-modules

$$\widetilde{e} : \widetilde{\Omega} = \widetilde{\Omega}_{R|A} \to \mathcal{O}_X .$$

The universal derivation $d = d_{R|A}$ defines a derivation $\widetilde{d} : \mathcal{O}_X \to \widetilde{\Omega}$ which coincides on $D_+(f) = \mathrm{Spec}(R_f)_0$, f homogeneous of positive degree, with the restriction of the universal derivation $d_f : R_f \to \Omega_{R_f|A} = \Omega_f$ to $(R_f)_0$. Obviously, the image of $(d_f)_0$ belongs to the kernel of the linear map $(e_f)_0 : (\Omega_f)_0 \to (R_f)_0$, i.e.

$$(d_f)_0 : (R_f)_0 \to \mathrm{Ker}(e_f)_0 = (\mathrm{Ker}\, e_f)_0 ,$$

or equivalently Im $\widetilde{d} \subseteq$ Ker $\widetilde{e}$. If $\deg f = 1$, *then* $(d_f)_0 : (R_f)_0 \to$ Ker$(e_f)_0$ *is the universal A-derivation of* $(R_f)_0$. Since $R_f = (R_f)_0[f/1, 1/f]$, where $f/1$ is transcendental over $(R_f)_0$, this follows from the description of $\Omega_f = \Omega_{R_f|A}$ in Exercise 6.D.2. Furthermore, $(e_f)_0(d\, f/f) = \deg f = 1$. Hence $(e_f)_0 : (\Omega_f)_0 \to (R_f)_0$ is surjective. Combining these observations we get the following theorem:

6.E.20. Theorem *Let R be a standardly graded A-algebra with $A = R_0$ and X the projective A-scheme* Proj R. *Then there exists a canonical exact sequence*

$$0 \to \Omega_{X|A} \to \tilde{\Omega}_{R|A} \xrightarrow{\tilde{e}} \mathcal{O}_X \to 0,$$

where $\tilde{e}$ is the Euler homomorphism derived from the Euler derivation $R \to R$.

For the projective space $\mathbb{P}_A^n = $ Proj R, $R := A[X_0, \ldots, X_n]$ with $\deg X_i = 1, i = 0, \ldots, n$, the graded universal differential module $\Omega_{R|A}$ is isomorphic to $R(-1)^{n+1}$ and we get the classical exact sequence

$$0 \to \Omega_{\mathbb{P}_A^n|A} \to \mathcal{O}_{\mathbb{P}_A^n}(-1)^{n+1} \to \mathcal{O}_{\mathbb{P}_A^n} \to 0$$

for the sheaf of Kähler differentials of the projective space $\mathbb{P}^n$. For later use we mention the following consequence:

$$\omega_{\mathbb{P}_A^n|A} := \Lambda^n \left(\Omega_{\mathbb{P}_A^n|A}\right) \cong \Lambda^{n+1} \left(\mathcal{O}_{\mathbb{P}_A^n}(-1)^{n+1}\right) = \mathcal{O}_{\mathbb{P}_A^n}(-1)^{\otimes(n+1)} = \mathcal{O}_{\mathbb{P}_A^n}(-(n+1)).$$

6.E.21. Exercise Let X be an algebraic scheme over a field k. Show that X is regular or smooth if (and only if) all closed points of X are regular or smooth points respectively.

6.E.22. Exercise (Strong topologies and analytic structures on complex and real algebraic schemes · Harnack's equality) Let $X = $ Spec $\mathbb{C}[x_1, \ldots, x_n] \subseteq \mathbb{A}_\mathbb{C}^n$ be an affine $\mathbb{C}$-scheme of finite type. Then, in the strong topology, $X(\mathbb{C}) = \mathbb{C}$-Spec $\mathbb{C}[x_1, \ldots, x_n]$ is a closed and in particular locally compact subset of $\mathbb{C}^n = \mathbb{A}_\mathbb{C}^n(\mathbb{C})$ (cf. Example 2.C.4). For an arbitrary algebraic $\mathbb{C}$-scheme X, the set $X(\mathbb{C}) \subseteq X$ of complex (= closed) points of X carries the strong topology which induces on each open affine subset the strong topology just described and which is, for $\dim X > 0$, bigger than the Zariski topology. With the strong topology, $X(\mathbb{C})$ is a locally compact Hausdorff space. (It is a Hausdorff space because the diagonal $\Delta_X(\mathbb{C}) \subseteq (X \times_\mathbb{C} X)(\mathbb{C}) = X(\mathbb{C}) \times X(\mathbb{C})$ is Zariski closed and in particular strongly closed.) Show that for a smooth algebraic $\mathbb{C}$-scheme X the space $X(\mathbb{C})$ carries a canonical structure as a complex(-analytic) manifold (usually denoted by X^{an}). (Start with the affine case. – For an arbitrary algebraic $\mathbb{C}$-scheme, the set $X(\mathbb{C})$ carries the structure of a so called complex space X^{an}.)

Let now X be an algebraic $\mathbb{R}$-scheme. Then the set X_0 of closed points of X can be identified with the orbit space $X(\mathbb{C})\backslash\{\mathrm{id}, \sigma\}$, where $\sigma : X(\mathbb{C}) \to X(\mathbb{C})$ is the involution of $X(\mathbb{C}) = X_{(\mathbb{C})}(\mathbb{C})$, $X_{(\mathbb{C})} = \mathbb{C} \otimes_\mathbb{R} X$, induced by the conjugation of $\mathbb{C}$. The set $X(\mathbb{R}) \subseteq X_0$ corresponds to the set of fixed points of σ (cf. Example 2.C.2). If X is smooth of (pure) dimension n, then $X_{(\mathbb{C})}$ is a smooth algebraic $\mathbb{C}$-scheme and $\sigma : X_{(\mathbb{C})}^{an} \to X_{(\mathbb{C})}^{an}$ preserves the canonical orientation of the complex manifold $X_{(\mathbb{C})}^{an}$ if and only if n is even. Furthermore, $X_0 \backslash X(\mathbb{R})$ and $X(\mathbb{R})$ carry the structure of real-analytic manifolds of dimension $2n$ and n respectively.

If $n = 1$ and if X is smooth, then X_0 is a 2-dimensional real-analytic manifold X_0^{an} with boundary $X(\mathbb{R})^{an}$. Assume furthermore that X is even a smooth *projective* scheme of dimension 1 over $\mathbb{R}$. Then $X_0^{an} = X_{(\mathbb{C})}^{an}\backslash\{\mathrm{id}, \sigma\}$ is a *compact* real-analytic surface with boundary $X(\mathbb{R})^{an}$ and $X_{(\mathbb{C})}^{an}$ is a compact complex-analytic manifold of dimension 1, i.e. a *compact* Riemann surface (not necessarily connected). The real manifold $X(\mathbb{R})^{an}$ is a compact real-analytic manifold of dimension 1, i.e. a finite disjoint union of, say, $r = r(X)$ circles, which are called the ovals of X. The compact Riemann surface $X(\mathbb{C})^{an}$ is a disjoint union of, say, $s = s(X)$ spheres with handles; we denote the number of these handles by $g_1, \ldots, g_s$. X_0^{an} is a disjoint union of, say, $t(X) = t = t_0 + t_1$ topological surfaces with holes, where t_0 is the number of orientable components and t_1 is the number of non-orientable components. We denote by $h_1, \ldots, h_{t_0}$ (≥ 0) the number of handles of the orientable components, by $c_1, \ldots, c_{t_1}$ (> 0) the number of cross-caps of the non-orientable components and by $r_1^0, \ldots, r_{t_0}^0, r_1^1, \ldots, r_{t_1}^1$ the number of holes of all these components. From topology, besides the simple relations $s/2 \leq t = t_0 + t_1 \leq s$, $r = r_1^0 + \cdots + r_{t_0}^0 + r_1^1 + \cdots + r_{t_1}^1$, we have the following equations (where χ denotes the Euler–Poincaré characteristic of a topological space):

(1) For each component X_i of $X_{(\mathbb{C})}^{an}$, $\chi(X_i) = 2 - 2g_i$;
(2) for each orientable component X_j^0 of X_0^{an}, $\chi(X_j^0) = 2 - 2h_j - r_j^0$;
(3) for each non-orientable component X_j^1 of X_0^{an}, $2 \cdot \chi(X_j^1) = 2 - c_j - r_j^1$;
(4) from the equality $X_0^{an} = X_{(\mathbb{C})}^{an} \backslash \{id, \sigma\}$, we get $2 \cdot \chi(X_0^{an}) = \chi(X_{(\mathbb{C})}^{an})$. (Note that the Euler–Poincaré characteristic of a circle is 0.)

Combining these equalities and defining $g := \sum_{i=1}^{s} g_i$, $h := \sum_{j=1}^{t_0} h_j$, $c := \sum_{j=1}^{t_1} c_j$ we get

$$2t - (2h + c) = r + s - g$$

which is sometimes called H a r n a c k ' s   e q u a l i t y. In the most important case $s = 1$ (which implies $t = 1$), i.e. $X_{(\mathbb{C})}^{an}$ or, by Theorem 6.E.24 below, $X_{(\mathbb{C})}$ is connected, we get

$$r = \begin{cases} 1 + g - 2h, & \text{if } X_0^{an} \text{ is orientable,} \\ 1 + g - c, & \text{if } X_0^{an} \text{ is non-orientable.} \end{cases}$$

In particular, $r \leq 1 + g$ (H a r n a c k ' s   i n e q u a l i t y). With these considerations in mind a reader should again look at the (smooth) curves in Example 2.B.6 and also discuss the example $X := \mathrm{Proj}\big(\mathbb{R}[X_0, X_1, X_2]/(X_0^2 + X_1^2 + X_2^2)\big)$ with $X(\mathbb{R}) = \emptyset$, cf. Exercise 1.B.4 (3). The above non-negative integer $g = g(X_{(\mathbb{C})})$ is called the g e n u s of the smooth complex projective curve $X_{(\mathbb{C})}$. The importance of this invariant will be discussed in the next chapter. See Section 7.E, in particular, Example 7.E.17.

6.E.23. Exercise Let $\mathfrak{p}$ be a prime ideal in $\mathbb{R}[X_1, \ldots, X_n]$ and $A := \mathbb{R}[X_1, \ldots, X_n]/\mathfrak{p}$. Show that $\mathfrak{J}(V_\mathbb{R}(\mathfrak{p})) = \mathfrak{p}$ if and only if $V_\mathbb{R}(\mathfrak{p}) = \mathbb{R}\text{-Spec}\, A = X(\mathbb{R})$ contains a smooth point of $X := \mathrm{Spec}\, A$. (**Hint:** If $X(\mathbb{R})$ is dense in X then $X(\mathbb{R})$ contains a smooth point, since the set of smooth points of X is open and dense in X by 6.E.17. Conversely, assume that $0 = (0, \ldots, 0) \in V_\mathbb{R}(\mathfrak{p}) \subseteq \mathbb{R}^n$ is a smooth point of X and let $\mathfrak{q} := \mathfrak{J}_\mathbb{R}(V_\mathbb{R}(\mathfrak{p})) \supseteq \mathfrak{p}$. Then the ideal $\mathfrak{p}\,\mathbb{R}[X_1, \ldots, X_n]_{\mathfrak{m}_0} \subseteq \mathbb{R}[X_1, \ldots, X_n]_{\mathfrak{m}_0}$ is generated by polynomials $F_1, \ldots, F_r \in \mathfrak{p}$ such that rank $\partial(F_1, \ldots, F_r)/\partial(X_1, \ldots, X_n)|_{X=0} = r := \mathrm{ht}\,\mathfrak{p}$ and there is an open neighbourhood U of 0 in the strong topology of $\mathbb{R}^n$ such that $\{F_1 = \cdots = F_r = 0\} \cap U = V_\mathbb{R}(\mathfrak{p}) \cap U$ is a real-analytic submanifold of U. – Any convergent power series $F \in \mathbb{R}\langle\!\langle X_1, \ldots, X_n \rangle\!\rangle$ which vanishes on $V_\mathbb{R}(\mathfrak{p})$ in some neighbourhood of 0 is of the form $F = G_1 F_1 + \cdots + G_r F_r$ with $G_1, \ldots, G_r \in \mathbb{R}\langle\!\langle X_1, \ldots, X_n \rangle\!\rangle$, in particular, $\mathfrak{q}\,\mathbb{R}[X_1, \ldots, X_n]_{\mathfrak{m}_0} \subseteq (F_1, \ldots, F_r)\,\mathbb{R}[\![X_1, \ldots, X_n]\!] \cap \mathbb{R}[X_1, \ldots, X_n]_{\mathfrak{m}_0} = (F_1, \ldots, F_r)\,\mathbb{R}[X_1, \ldots, X_n]_{\mathfrak{m}_0} = \mathfrak{p}\,\mathbb{R}[X_1, \ldots, X_n]_{\mathfrak{m}_0}$ which implies that $\mathfrak{q} = \mathfrak{p}$. The equality $(F_1, \ldots, F_r)\,\mathbb{R}[\![X_1, \ldots, X_n]\!] \cap \mathbb{R}[X_1, \ldots, X_n]_{\mathfrak{m}_0} = (F_1, \ldots, F_r)\,\mathbb{R}[X_1, \ldots, X_n]_{\mathfrak{m}_0}$ can be proved more generally: If k is a field and if $F_1, \ldots, F_r \in k[X_1, \ldots, X_n]_{\mathfrak{m}_0}$, $\mathfrak{m}_0 := (X_1, \ldots, X_n)$, is part of a regular system of parameters for $k[X_1, \ldots, X_n]_{\mathfrak{m}_0}$, then $\mathfrak{a} := (F_1, \ldots, F_r)\,k[X_1, \ldots, X_n]_{\mathfrak{m}_0} = (F_1, \ldots, F_r)\,k[\![X_1, \ldots, X_n]\!] \cap k[X_1, \ldots, X_n]_{\mathfrak{m}_0}$. To show this, look at the homomorphism $k[X_1, \ldots, X_n]_{\mathfrak{m}_0}/\mathfrak{a} \to k[\![X_1, \ldots, X_n]\!]/\mathfrak{a}k[\![X_1, \ldots, X_n]\!]$ of regular local rings of dimensions $d := n - r$ and use Exercise 3.B.43. Indeed, the equality $\mathfrak{a}k[\![X_1, \ldots, X_n]\!] \cap k[X_1, \ldots, X_n]_{\mathfrak{m}_0} = \mathfrak{a}$ holds for arbitrary ideals $\mathfrak{a} \subseteq k[X_1, \ldots, X_n]_{\mathfrak{m}_0}$. This follows from the fact that *the extension* $k[X_1, \ldots, X_n]_{\mathfrak{m}_0} \hookrightarrow k[\![X_1, \ldots, X_n]\!]$ *is faithfully flat*. $k[\![X_1, \ldots, X_n]\!]$ is the ($\mathfrak{m}_0$-adic) "completion" of the local ring $k[X_1, \ldots, X_n]_{\mathfrak{m}_0}$, a concept which we have not discussed in these lectures.)

We finish this section with some comparison theorems for the Zariski and the strong topology of an algebraic $\mathbb{C}$-scheme (cf. the beginning of Exercise 6.E.22 above). They are used very often (and occasionally even unconsciously). The basic result is the following:

6.E.24. Theorem *Let* U *be a non-empty* (*Zariski*) *open subset of an integral algebraic* $\mathbb{C}$-*scheme* X. *Then* $U(\mathbb{C}) \subseteq X(\mathbb{C})$ *is strongly connected and strongly dense in* $X(\mathbb{C})$.

PROOF. We may assume that $X = \operatorname{Spec} A$ is affine with an integral $\mathbb{C}$-algebra A of finite type of positive dimension m.

First we prove the density result. Let $\mathrm{D}(g) \subseteq U$ for some $g \in A$, $g \neq 0$, and let $x \in X \setminus U \subseteq \mathrm{V}(g)$ be a closed point corresponding to a maximal ideal $\mathfrak{m} \subseteq A$ with $g \in \mathfrak{m}$. We have to show that x belongs to the strong closure of $U(\mathbb{C})$ in $X(\mathbb{C})$. There is a prime ideal $\mathfrak{p} \subseteq A$ with $\mathfrak{p} \subseteq \mathfrak{m}$, $g \notin \mathfrak{p}$ and $\dim A/\mathfrak{p} = 1$. We may replace X by $\operatorname{Spec} A/\mathfrak{p}$ ($\subseteq X$) and U by $U \cap (\operatorname{Spec} A/\mathfrak{p})$ and hence assume that $m = 1$. Let $f : X' \to X$ be the (finite) normalization of X. $X'(\mathbb{C})$ is a complex manifold of dimension 1, and $f^{-1}(X(\mathbb{C}) \setminus U(\mathbb{C})) \subseteq X'(\mathbb{C})$ is finite. Hence $f^{-1}(U(\mathbb{C})) = X'(\mathbb{C}) \setminus f^{-1}(X(\mathbb{C}) \setminus U(\mathbb{C}))$ is strongly dense in $X'(\mathbb{C})$. Since f is surjective (and strongly continuous) $U(\mathbb{C})$ is also strongly dense in $X(\mathbb{C})$.

To prove the strong connectedness of $U(\mathbb{C})$ it suffices to show (using the first part of the proof) that U contains a non-empty Zariski open subset U' for which $U'(\mathbb{C})$ is strongly connected. Now we use a Noether normalization $P := \mathbb{C}[z_1, \ldots, z_m] \subseteq A$ (cf. 1.F.2) and the corresponding finite morphism $f : X \to \mathbb{A}_{\mathbb{C}}^m$. Let $y \in A$ generate the quotient field $L := \mathrm{Q}(A)$ over $K := \mathrm{Q}(P) = \mathbb{C}(z_1, \ldots, z_m)$ and let $F \in P[Y]$ be its minimal polynomial which is a monic prime polynomial of degree $n := [L : K]$, cf. Lemma 1.E.9. As the reader will check easily, there is a polynomial $h \in P$, $h \neq 0$, with (1) $A_h = P_h[y]$, (2) $U' := X_h = \operatorname{Spec} A_h = \operatorname{Spec} P_h[y] \subseteq U$, (3) $\Omega_{A_h|P_h} = (\Omega_{A|P})_h = 0$. It follows that the projection $(z; y) \mapsto y$ from $X_h(\mathbb{C}) = \{(z; y) \in \mathbb{C}^m \times \mathbb{C} \mid h(z) \neq 0, F(z; y) = 0\}$ onto $D := \{z \in \mathbb{C}^m \mid h(z) \neq 0\}$ is, with respect to the strong topologies, a proper regular (i. e. locally biholomorphic) analytic map of m-dimensional complex manifolds. Then this is a covering map and its sheet number is n. *We show that* $X_h(\mathbb{C})$ *is strongly connected.*

The base D of the covering is dense in $\mathbb{C}^m$ and connected: For convenience, we assume $0 \in D$, i. e. $h(0) \neq 0$. Then the restriction of h to any line $\mathbb{C}z_0$, $z_0 \in \mathbb{C}^m$, $z_0 \neq 0$, has only a finite number of zeros and $D \cap \mathbb{C}z_0$ is strongly connected. – Any strong connected component Z of $X_h(\mathbb{C})$ is also a covering of D with a well-defined sheet number $k \leq n$. We are to show: $k = n$.

To do this we consider the characteristic polynomial $G \in \mathrm{H}(D)[Y]$ of degree k of the function $y|Z$ with $G(z_0; Y) = \prod_{(z_0; y) \in Z}(Y - y)$ for a fixed $z_0 \in D$. $\mathrm{H}(D)$ denotes the algebra of complex-analytic (= holomorphic) functions on the domain D. *We want to show that* $G \in P[Y]$ *is a polynomial*. Since $Z \subseteq X_h(\mathbb{C})$ it follows then already that F is in the radical ideal of the principal ideal $G P_h[Y]$,

i. e. $F|G$ in $P_h[Y]$ because F is prime and hence $k = n$. Or: If $Z_1, \ldots, Z_r$ are all the strong connected components of $X_h(\mathbb{C})$ with the characteristic polynomials $G_1, \ldots, G_r \in P[Y]$ of $y|Z_1, \ldots, y|Z_r$ then $F = G_1 \cdots G_r$ in $P_h[Y]$, hence $r = 1$, since F is prime.

Finally, we prove that indeed $G = a_0 + \cdots + a_{k-1} Y^{k-1} + Y^k \in P[Y]$. First of all, the coefficients $b_\nu \in \mathbb{C}[z_1, \ldots, z_m]$ of F can be estimated by $b_\nu(z) = O(1 + \|z\|^\ell)$ for some $\ell \in \mathbb{N}^*$ (with a norm $\|-\|$ on $\mathbb{C}^m$). The zeros of $F(z; Y)$ are then also $= O(1 + \|z\|^\ell)$. Hence, the coeffizients a_κ of G are $O\big((1 + \|z\|^\ell)^k\big) = O(1 + \|z\|^{\ell k})$. Let $a_\kappa(z) = \sum_{i \in \mathbb{N}} H_{i,\kappa}(z)$ be the power series expansion of a_κ around 0 with homogeneous polynomials $H_{i,\kappa}$ of degree i. On the complex line $\mathbb{C} z_0$ with fixed $z_0 \in \mathbb{C}^m \setminus \{0\}$ we have $a_\kappa(t z_0) = \sum_i H_{i,\kappa}(z_0) t^i$. The functions $t \mapsto a_\kappa(t z_0)$ are complex-analytic on $\mathbb{C} \setminus \{t \mid h(t z_0) = 0\}$ and locally bounded on $\mathbb{C}$. By the Riemann's extension theorem they are complex-analytic on $\mathbb{C}$ and, because of $a_\kappa(t z_0) = O(1 + |t|^{\ell k})$, even polynomials of degree $\leq \ell k$ according to Liouville's theorem. It follows $H_{i,\kappa}(z_0) = 0$ for all z_0 and all $i > \ell k$, and hence all the a_κ are polynomials as asserted.   •

We mention the following direct consequences:

6.E.25. Corollary *Let X be an algebraic $\mathbb{C}$-scheme and $X_1, \ldots, X_r$ its (Zariski) connected components. Then the strong connected components of $X(\mathbb{C})$ are $X_1(\mathbb{C})$, $\ldots, X_r(\mathbb{C})$. In particular, $X(\mathbb{C})$ is strongly connected if and only if X is (Zariski) connected.*

6.E.26. Corollary *Let X be an algebraic $\mathbb{C}$-scheme and $Y \subseteq X$ a (Zariski) constructible subset. Then the strong closure of $Y(\mathbb{C})$ in $X(\mathbb{C})$ coincides with the Zariski closure $\overline{Y}(\mathbb{C}) = \overline{Y(\mathbb{C})}$ of $Y(\mathbb{C})$ in $X(\mathbb{C})$. In particular, $Y(\mathbb{C})$ is strongly closed in $X(\mathbb{C})$ if and only if Y is (Zariski) closed in X.*

Note that for algebraic $\mathbb{R}$-schemes X no analogous result hold with respect to the strong topologies of the spaces $X(\mathbb{R})$. For example, if S is a set of $n \in \mathbb{N}^*$ points in $\mathbb{P}^1_\mathbb{R}(\mathbb{R})$, then $\mathbb{P}^1_\mathbb{R}(\mathbb{R}) \setminus S$ is connected in the Zariski toplogy, but has n connected components in the strong topology. The number of the strong connected components of $X(\mathbb{R})$ is always finite. For other examples look at the curves in Example 2.B.6 or at the discussion of the ovals in Exercise 6.E.22.

6.E.27. Example Let $f : X \to Y$ be a morphism of algebraic $\mathbb{C}$-schemes. *If the restriction $f|X(\mathbb{C}) : X(\mathbb{C}) \to Y(\mathbb{C})$ is strongly closed, then f is (Zariski) closed.* This follows from 6.E.26, since the image $f(A)$ of a Zariski closed set $A \subseteq X$ is constructible in Y by the Mapping Theorem of Chevalley 5.C.3. It follows: *If $f|X(\mathbb{C}) : X(\mathbb{C}) \to Y(\mathbb{C})$ strongly proper, then $\mathrm{id}_Z \times_\mathbb{C} f : Z \times_\mathbb{C} X \to Z \times_\mathbb{C} Y$ is closed for all algebraic $\mathbb{C}$-schemes Z.* One can show that this implies the properness of f. (Formally, the condition for properness is a little bit stronger, cf. Defintion 5.B.6.) The converse is also true: If $f : X \to Y$ is proper, then $f|X(\mathbb{C}) : X(\mathbb{C}) \to Y(\mathbb{C})$ is strongly proper. In particular: X *is a proper $\mathbb{C}$-scheme if and only if $X(\mathbb{C})$ is compact.* Instructive examples are the projective algebraic $\mathbb{C}$-schemes. Cf. also [6], Chap. I, §10.

CHAPTER 7: Riemann–Roch Theorem

In this last chapter we prove one of the highlights of classical Algebraic Geometry, i.e. the theorem of Riemann–Roch for projective algebraic curves over arbitrary fields. Furthernore, we allow arbitrary singularities and formulate this theorem for arbitrary coherent sheaves. We do not use cohomology, but include in the Riemann–Roch formulae the duality theorem as it was done originally. In the beginning we collect some preliminaries.

7.A. Coherent Modules on Projective Schemes

Let $R = \bigoplus_{m \in \mathbb{N}} R_m$ be a positively graded algebra of finite type over the Noetherian ring $A = R_0$ and $X := \operatorname{Proj} R$ the corresponding Noetherian projective algebraic A-scheme (see Section 5.A). Any finite graded R-module M defines a coherent $\mathcal{O}_X$-module $\tilde{M}$ as described in Example 6.E.14. We want to show that all coherent $\mathcal{O}_X$-modules are obtained in this way.

Let $\mathcal{F}$ be an arbitrary $\mathcal{O}_X$-module. We set

$$\mathcal{F}(m) := \mathcal{F} \otimes_{\mathcal{O}_X} \mathcal{O}_X(m), \quad m \in \mathbb{Z},$$

where $\mathcal{O}_X(m) := R(m)^\sim$. If R is standardly graded, then $\mathcal{O}_X(m) = \mathcal{O}_X(1)^{\otimes m}$ for all $m \in \mathbb{Z}$ and the $\mathcal{O}_X(m)$ are locally free sheaves of rank 1, cf. Example 6.E.14. Every homogeneous element $f \in R_d$ defines $\mathcal{O}_X$-module homomorphisms

$$\mathcal{F}(m) \xrightarrow{\ f\ } \mathcal{F}(m+d), \quad m \in \mathbb{Z},$$

which are induced by multiplication $R(m) \xrightarrow{\ f\ } R(m+d)$ with f and the corresponding $\mathcal{O}_X$-homomorphisms $\mathcal{O}_X(m) \to \mathcal{O}_X(m+d)$ and which are isomorphisms on $D_+(f) \subseteq X$. These homomorphisms define on the direct sum

$$\Gamma_*(\mathcal{F}) := \bigoplus_{m \in \mathbb{Z}} \Gamma(X, \mathcal{F}(m))$$

a canonical structure as a graded R-module. Furthermore, for a homogeneous element $f \in R_d$ of positive degree $d > 0$ a canonical homomorphism

$$\left(\Gamma_*(\mathcal{F})_f \right)_0 \longrightarrow \Gamma(D_+(f), \mathcal{F})$$

is defined, since an element s/f^ν, $\nu \in \mathbb{N}$, $s \in \Gamma(X, \mathcal{F}(\nu d))$, defines an element in $\Gamma(D_+(f), \mathcal{F})$ as the preimage of $s|D_+(f)$ with respect to the isomorphism

$$f^\nu | D_+(f) : \mathcal{F}|D_+(f) \longrightarrow \mathcal{F}(\nu d)|D_+(f).$$

Therefore we get an $\mathcal{O}_X$-homomorphism

$$\beta_{\mathcal{F}} : \left(\Gamma_*(\mathcal{F}) \right)^\sim \longrightarrow \mathcal{F}$$

which is called the S e r r e   h o m o m o r p h i s m of $\mathcal{F}$. Note that $\left(\Gamma_*(\mathcal{F}) \right)^\sim$ depends only on the R-submodule $\Gamma_*(\mathcal{F})_{\geq 0} = \bigoplus_{m \in \mathbb{N}} \Gamma(X, \mathcal{F}(m)) \subseteq \Gamma_*(\mathcal{F})$.

7.A.1. Theorem *For any quasi-coherent* $\mathcal{O}_X$-*module* $\mathcal{F}$ *the Serre homomorphism*

$$\beta_{\mathcal{F}} : \left(\Gamma_*(\mathcal{F})\right)^{\sim} \longrightarrow \mathcal{F}$$

is an isomorphism.

PROOF. Let $f_0, \ldots, f_n$ be homogeneous elements in R of positive degrees $d_0, \ldots, d_n$ which generate the ideal R_+.

(1) $\beta_{\mathcal{F}}$ *is injective*: Let $f \in R_d$, $d > 0$. We have to show that $\left(\Gamma_*(\mathcal{F})_f\right)_0 \to \Gamma(D_+(f), \mathcal{F})$ is injective, that is, if $s \in \Gamma(X, \mathcal{F}(\nu d))$ and $s|D_+(f) = 0$, then $f^r s = 0$ for some $r \in \mathbb{N}$. Since $\mathcal{F}$ is quasi-coherent, from $s|D_+(ff_i) = 0$, it follows that there exists an $r_i \geq 0$ with $f^{r_i d_i} s / f_i^{r_i d} = 0$ on $D_+(f_i)$, i.e. $f^{r_i d_i} s = 0$ on $D_+(f_i)$. Then, for $r := \max(r_i d_i)$, $f^r s = 0$ on $D_+(f_i)$ for all i and hence $f^r s = 0$.

(2) $\beta_{\mathcal{F}}$ *is surjective*: Let $t \in \Gamma(D_+(f), \mathcal{F})$, $f \in R_d$, $d > 0$. We must show that there exists an $r \in \mathbb{N}$ and a section $s \in \Gamma(X, \mathcal{F}(rd))$ with $s|D_+(f) = f^r t$. Since $\mathcal{F}$ is quasi-coherent, there exist sections $t_i \in \Gamma(D_+(f_i), \mathcal{F})$ and $r_i \in \mathbb{N}$ such that $t|\Gamma(D_+(ff_i), \mathcal{F}) = f_i^{r_i d} t_i / f^{r_i d_i}$, i.e. $f^{r_i d_i} t = f_i^{r_i d} t_i$ on $\Gamma(D_+(ff_i), \mathcal{F}(r_i d_i d))$. Multiplying by appropriate powers of f we get sections $s_i \in \Gamma(D_+(f_i), \mathcal{F}(rd))$ with $f^r t|D_+(ff_i) = s_i|D_+(ff_i)$ for some $r \in \mathbb{N}$ and all i. But then, for some $\nu \in \mathbb{N}$, $f^\nu s_i|D_+(f_i f_j) = f^\nu s_j|D_+(f_i f_j)$ for all i, j, and these sections $f^\nu s_i$, $i = 0, \ldots, n$, define a section $s \in \Gamma(X, \mathcal{F}((r + \nu)d))$ with $f^{r+\nu} t = s|D_+(f)$.   •

7.A.2. Corollary *Let* R *be a positively graded Noetherian ring and let* $X := \operatorname{Proj} R$. *For any coherent* $\mathcal{O}_X$-*module* $\mathcal{F}$, *there exists a finite graded* R-*module* M *with* $\mathcal{F} \cong \tilde{M}$.

PROOF. Let $f_0, \ldots, f_n$ be homogeneous elements in R of positive degrees which generate the ideal R_+ and let M_j, $j \in J$, be the family of finite R-submodules of $\Gamma_*(\mathcal{F}) = \bigoplus_{m \in \mathbb{Z}} \Gamma(X, \mathcal{F}(m))$. By Theorem 7.A.1, $\mathcal{F} = \left(\Gamma_*(\mathcal{F})\right)^{\sim} = \bigcup_{j \in J} \tilde{M}_j$. Since $D_+(f_j) \subseteq X$ is Noetherian and affine there exists j_i with $\tilde{M}_{j_i}|D_+(f_i) = \mathcal{F}|D_+(f_i)$, $i = 0, \ldots, n$. Then obviously $\mathcal{F} \cong \tilde{M}$ with $M := \sum_{i=0}^n M_{j_i}$.   •

Now we consider for positively graded ring R and a graded A-module M the canonical homogeneous R-homomorphism

$$\alpha_M : M \longrightarrow \bigoplus_{m \in \mathbb{Z}} \Gamma\left(X, M(m)^{\sim}\right)$$

of degree 0 defined by the canonical homomorphisms $M_m \to \Gamma\left(X, M(m)^{\sim}\right)$, $m \in \mathbb{Z}$. (Observe the distinction between $M(m)^{\sim}$ and $\tilde{M}(m) = \tilde{M} \otimes_{\mathcal{O}_X} \mathcal{O}_X(m) = \tilde{M} \otimes_{\mathcal{O}_X} R(m)^{\sim}$ if R is not standardly graded.) In the next lemma a sequence $f_1, \ldots, f_r$ of elements of R is called a s t r o n g l y   r e g u l a r   s e q u e n c e for M if f_i is a non-zero divisor for $M/(f_1, \ldots, f_{i-1})M$, $i = 1, \ldots, r$. (See Definition 6.B.13. The definitions for regular sequences or M-sequences change in the literature.)

7.A.3. Lemma *Let R, M be as above and let $\alpha_M : M \longrightarrow \bigoplus_{m \in \mathbb{Z}} \Gamma\big(X, M(m)^{\sim}\big)$ be the canonical homomorphism.*

(1) *If R contains a homogeneous non-zero divisor f of positive degree for the module M, then α_M is injective.*

(2) *If R contains a strongly regular sequence f, g of homogeneous elements of positive degrees for the module M, then α_M is bijective.*

PROOF. (1) We have to show that $M_0 \to \Gamma(X, \widetilde{M})$ is injective. If $x_0 \in M_0$ defines the zero-section in $\widetilde{M}$, then, in particular, $x_0/1 = 0$ in $(M_f)_0$, i.e. $f^n x_0 = 0$ for some $n \in \mathbb{N}$ and hence $x_0 = 0$, since f is a non-zero divisor for M.

(2) By (1), we have to show that $M_0 \to \Gamma(X, \widetilde{M})$ is surjective. Let $s \in \Gamma(X, \widetilde{M})$ and $s|D_+(f) = x/f^m$, $s|D_+(g) = y/g^n$ with $m, n \in \mathbb{N}$ and homogeneous elements $x, y \in M$ of degrees $m \deg f$ and $n \deg g$, respectively. Because of $x/f^m = y/g^n$ in $\Gamma(D_+(fg), \widetilde{M})$ there is an $r \in \mathbb{N}$ with $x g^m (fg)^{n+r} = y f^n (fg)^{m+r}$ or $x g^{m+n+r} = y f^m g^{m+r}$ in M. Since g^{m+n+r} is a non-zero divisor for the module M/fM and hence for the module $M/f^m M$ there is $x_0 \in M_0$ with $x = f^m x_0$. Then s is *defined* by x_0. Namely, for an arbitrary homogeneous element $h \in R$ of positive degree, $s|D_+(hf) = x_0|D_+(hf)$ which implies $s|D_+(h) = x_0|D_+(h)$, since $M_h \to M_{hf}$ is injective. ●

7.A.4. Example Let A be an arbitrary (commutative) ring. Then, since $X_0, \ldots, X_n$ is a strongly regular sequence for $A[X_0, \ldots, X_n]$, by 7.A.3 (2) we have

$$\Gamma\big(\mathbb{P}^n_{\gamma, A}, \mathcal{O}_{\mathbb{P}^n_{\gamma, A}}(m)\big) \cong A[X_0, \ldots, X_n]_m$$

for all $n \geq 1$, all $m \in \mathbb{Z}$ and for arbitrary weights $\gamma = (\gamma_0, \ldots, \gamma_n) \in (\mathbb{N}^*)^{n+1}$ for the indeterminates $X_0, \ldots, X_n$. In particular,

$$\Gamma\left(\mathbb{P}^n_A, \mathcal{O}_{\mathbb{P}^n_A}(m)\right) \cong \begin{cases} A^{\binom{m+n}{n}}, & \text{if } m \geq 0, \\ 0, & \text{if } m < 0. \end{cases}$$

Of course, $\mathbb{P}^0_A \cong \operatorname{Spec} A$ and $\mathcal{O}_{\mathbb{P}^0_A}(m) = \widetilde{A}$ for all $m \in \mathbb{Z}$, hence $\Gamma\big(\mathbb{P}^0_A, \mathcal{O}_{\mathbb{P}^0_A}(m)\big) \cong A$ for all $m \in \mathbb{Z}$.

The next theorem is true for arbitrary positively graded Noetherian rings R, but we prove (and use) it only for the case that R_0 is a field.

7.A.5. Theorem *Let R be a positively graded Noetherian ring over a field $k = R_0$. Then for every coherent $\mathcal{O}_X$-module $\mathcal{F}$, $X := \operatorname{Proj}(R)$, the module $\Gamma(X, \mathcal{F})$ of sections is finite over $k = R_0$ (i.e. a finite-dimensional k-vector space).*

PROOF. By 7.A.2 there is a finite graded R-module M with $\mathcal{F} \cong \widetilde{M}$. By Exercise 6.A.11 (10) there is a chain $0 = M_0 \subset M_1 \subset \cdots \subset M_r = M$ of homogeneous submodules $M_i \subseteq M$ such that $M_i/M_{i-1} \cong (R/\mathfrak{p}_i)(-m_i)$ with homogeneous prime ideals $\mathfrak{p}_i \subseteq R$ and $m_i \in \mathbb{Z}$, $i = 1, \ldots, r$. By induction on r, $\Gamma(X, \widetilde{M})$ is finite over k if $\Gamma(X, \widetilde{M_i/M_{i-1}})$ is finite over k for all $i = 1, \ldots, r$. Because of

$\widetilde{M_i/M_{i-1}} \cong (M_i/M_{i-1})^\sim$, we may therefore assume that R is an integral domain and that $M = R(m)$ for some $m \in \mathbb{Z}$.

By Lemma 6.C.5 and Exercise 1.E.10 (3b) the normalization S of R is finite over R and moreover also positively graded. Then $R(m)^\sim$ is a subsheaf of $S(m)^\sim$. Since S_0 is a finite field extension of $R_0 = k$, we may replace R by S and then *assume that R is a normal domain.*

If $\dim R \leq 1$, then $R = R_0 = k$ or $R = k[f]$ with a homogeneous element of positive degree, i. e. $X = \emptyset$ or $X = \operatorname{Spec} k$ and the assertion is trivially true. If $\dim R \geq 2$, then $\operatorname{ht} R_+ \geq 2$ and, by condition (2b) in Theorem 6.B.4, there is a strongly regular sequence f, g in R of homogeneous elements of positive degrees. The canonical homomorphism $\alpha_R : R \longrightarrow \bigoplus_{m \in \mathbb{Z}} \Gamma(X, R(m)^\sim)$ is bijective by Lemma 7.A.3 (2) and, in particular, $\Gamma(X, R(m)^\sim) \cong R_m$ is finite-dimensional over $k = R_0$ for every $m \in \mathbb{Z}$.   •

7.A.6. Example Let k be a field and let X be a projective algebraic k-scheme. If $X_1, \ldots, X_r$ are the connected components of X, $X = X_1 \uplus \cdots \uplus X_r$, then, by Theorem 7.A.5, the k-algebra $\Gamma(X, \mathcal{O}_X)$ is the direct product of the finite k-algebras $\Gamma(X_\rho, \mathcal{O}_{X_\rho})$:

$$\Gamma(X, \mathcal{O}_X) = \Gamma(X_1, \mathcal{O}_{X_1}) \times \cdots \times \Gamma(X_r, \mathcal{O}_{X_r}).$$

By Exercise 4.A.17 these factors do not have non-trivial idempotent elements and are therefore local, hence *the $\Gamma(X_\rho, \mathcal{O}_{X_\rho})$, $\rho = 1, \ldots, r$, are the local components of the finite k-algebra* $\Gamma(X, \mathcal{O}_X)$. (Note that for an Artinian ring A with $\operatorname{Spec} A = \operatorname{Spm} A = \{x_1, \ldots, x_r\}$ one has always the unique decomposition $A = \Gamma(\operatorname{Spec} A, \tilde{A}) = \prod_{\rho=1}^{r} \tilde{A}_{x_\rho} = \prod_{\rho=1}^{r} A_{\mathfrak{m}_\rho}$, $\mathfrak{m}_\rho := \mathfrak{m}_{x_\rho}$, into its local components $A_\mathfrak{m}$, $\mathfrak{m} \in \operatorname{Spec} A$.) In particular, X *is connected if and only if* $\Gamma(X, \mathcal{O}_X)$ *is a (finite) local k-algebra.* This is an important criterion for connectedness.

If X is reduced, then $\Gamma(X, \mathcal{O}_X)$ is (also reduced and hence) a finite direct product of finite field extensions of k. It is isomorphic to the product algebra k^r if, in addition, k is algebraically closed, $r = \operatorname{Dim}_k \Gamma(X, \mathcal{O}_X) =$ number of connected components of X.

Let $X = \operatorname{Proj} R$, where R is a positively graded k-algebra of finite type with $\operatorname{Dim}_k R_0 < \infty$ (but not necessarily with $R_0 = k$). If R contains a strongly regular sequence of length 2 of homogeneous elements of positive degrees, then $R_0 = \Gamma(X, \mathcal{O}_X)$ by 7.A.3 (2) and X is connected if R_0 is local. For example, if $F_1, \ldots, F_m$ is a strongly regular sequence of homogeneous elements of positive degrees in the polynomial algebra $k[X_0, \ldots, X_n]$, $\deg X_i = \gamma_i > 0$, $i = 0, \ldots, n$, and if $m < n$, i. e. $\dim R \geq 2$, $R := k[X_0, \ldots, X_n]/(F_1, \ldots, F_m)$, then R contains such a sequence, since the localization $k[X_0, \ldots, X_n]_{(X_0, \ldots, X_n)}$ is regular and hence Cohen–Macaulay of dimension $n + 1$, cf. Exercise 6.B.17 (4). It follows that $X = \operatorname{Proj} R \subseteq \mathbb{P}_{\gamma, k}^n$ *is connected* and $\Gamma(X, \mathcal{O}_X) = k$. Such an X is called a c o m p l e t e i n t e r s e c t i o n of codimension m ($< n$) in $\mathbb{P}_{\gamma, k}^n$. The simplest non-trivial example of this kind is a hypersurface $X = \operatorname{Proj}(k[X_0, X_1, X_2]/(F))$ in $\mathbb{P}_{\gamma, k}^2$, where $F \neq 0$ is a homogeneous polynomial of positive degree.

We end this section with a simple formula which describes the behaviour of the spaces of global sections with respect to field extensions.

7.A.7. Proposition *Let* $k \subseteq K$ *be a field extension. Then, for an arbitrary algebraic k-scheme* X *and a quasi-coherent* $\mathcal{O}_X$*-module* $\mathcal{F}$*, there exists a canonical isomorphism*

$$K \otimes_k \Gamma(X, \mathcal{F}) \xrightarrow{\sim} \Gamma(X_{(K)}, \mathcal{F}_{(K)}),$$

in particular,

$$\text{Dim}_k \, \Gamma(X, \mathcal{F}) = \text{Dim}_K \, \Gamma(X_{(K)}, \mathcal{F}_{(K)}),$$

where $\mathcal{F}_{(K)}$ *denotes the pull back of* $\mathcal{F}$ *with respect to the canonical morphism* $X_{(K)} \to X$ *(cf. Exercise 6.E.12).*

The Proposition 7.A.7 is a particular case of the following more general result:

7.A.8. Lemma *Let* B *be a flat A-algebra. Then, for an arbitrary algebraic A-scheme* X *and a quasi-coherent* $\mathcal{O}_X$*-module* $\mathcal{F}$*, there exists a canonical isomorphism*

$$B \otimes_A \Gamma(X, \mathcal{F}) \xrightarrow{\sim} \Gamma(X_{(B)}, \mathcal{F}_{(B)}),$$

where $\mathcal{F}_{(B)}$ *denotes the pull back of* $\mathcal{F}$ *with respect to the canonical morphism* $X_{(B)} \to X$.

PROOF. Let U_i, $i \in I$, be a *finite* open affine covering of X. By Exercise 4.D.20 the intersections $U_i \cap U_j$, $i, j \in I$, are also affine. Then $(U_i)_{(B)}$, $i \in I$, is an open affine covering of $X_{(B)}$ and $(U_i)_{(B)} \cap (U_j)_{(B)}$ can be identified with the affine scheme $(U_i \cap U_j)_{(B)}$ $i, j \in I$.

By Serre's conditions (S1) and (S2) for a sheaf (cf. Definition 4.A.2) the canonical sequence of A-modules

$$0 \to \Gamma(X, \mathcal{F}) \longrightarrow \prod_{i \in I} \Gamma(U_i, \mathcal{F}) \longrightarrow \prod_{(i,j) \in I \times I} \Gamma(U_i \cap U_j, \mathcal{F})$$

is exact. Here the image of $(s_i) \in \prod_i \Gamma(U_i, \mathcal{F})$ is the tuple (s_{ij}) with $s_{ij} := s_j | U_i \cap U_j - s_i | U_i \cap U_j$. From this sequence we derive the exact sequence of B-modules

$$0 \to B \otimes_A \Gamma(X, \mathcal{F}) \longrightarrow \prod_i B \otimes_A \Gamma(U_i, \mathcal{F}) \longrightarrow \prod_{i,j} B \otimes_A \Gamma(U_i \cap U_j, \mathcal{F}).$$

But we have $B \otimes_A \Gamma(U_i, \mathcal{F}) = \Gamma((U_i)_{(B)}, \mathcal{F}_{(B)})$ and $B \otimes_A \Gamma(U_i \cap U_j, \mathcal{F}) = \Gamma((U_i \cap U_j)_{(B)}, \mathcal{F}_{(B)}) = \Gamma((U_i)_{(B)} \cap (U_j)_{(B)}, \mathcal{F}_{(B)})$ for all $i, j \in I$. This yields the canonical isomorphism $B \otimes_A \Gamma(X, \mathcal{F}) \xrightarrow{\sim} \Gamma(X_{(B)}, \mathcal{F}_{(B)})$.   ●

7.B. Projective Curves

Let k be an arbitrary field. An a l g e b r a i c   c u r v e   o v e r k is by definition an algebraic k-scheme X of pure dimension one, i. e. X is of finite type and separated over k, cf. Definition 4.D.12, and all irreducible components of X are one-dimensional, cf. 3.B. In particular, an affine algebraic curve over k is the spectrum of a pure one-dimensional k-algebra A of finite type. The local ring $\mathcal{O}_{X,x} = \mathcal{O}_x$ of a closed point x of an algebraic curve X over k is a one-dimensional Noetherian local k-algebra and its residue field $\kappa(x)$ is finite over k. There are only finitely many non-closed points, they are the generic points of the irreducible components of X (cf. Exercise 4.B.8 (1)). Their local rings are zero-dimensional with finitely generated residue fields of transcendence degree one over k (see Section 3.B for the dimension theory of k-algebras of finite type).

7.B.1. Cohen–Macaulayfication An algebraic curve X is called a C o h e n – M a c a u l a y curve if all the local rings $\mathcal{O}_x$, $x \in X$, are Cohen–Macaulay (cf. Definition 6.B.16). For the generic points of X this is always true, since 0-dimensional local rings are Cohen–Macaulay rings. For a closed point $x \in X$, the local ring $\mathcal{O}_x$ is Cohen–Macaulay if and only if the maximal ideal $\mathfrak{m}_x \subseteq \mathcal{O}_x$ contains a non-zero divisor or, equivalently, if $\mathfrak{m}_x \notin \mathrm{Ass}\, \mathcal{O}_x$. For each algebraic curve X, there is a canonical Cohen–Macaulay curve X_{CM} associated to X. More generally, for each coherent $\mathcal{O}_X$-module $\mathcal{F}$, we will construct a coherent Cohen–Macaulay sheaf $\mathcal{F}_{\mathrm{CM}}$ with $\dim \mathcal{F}_{\mathrm{CM},x} = \dim \mathcal{O}_x$ for all $x \in \mathrm{Supp}\, \mathcal{F}_{\mathrm{CM}}$. *In this whole chapter, by a* C o h e n – M a c a u l a y   m o d u l e *over an algebraic curve X, we mean a coherent $\mathcal{O}_X$-module $\mathcal{F}$ such that the stalks $\mathcal{F}_x$ for the closed points $x \in X$ of the support of $\mathcal{F}$ are Cohen–Macaulay $\mathcal{O}_x$-modules of dimension* 1, i. e. there is an element in the maximal ideal of $\mathcal{O}_x$ which is a non-zero divisor for $\mathcal{F}_x$ or, equivalently, $\mathfrak{m}_x \notin \mathrm{Ass}\, \mathcal{F}_x$.

First we construct the Cohen–Macaulayfication in the affine case. Let A be a pure one-dimensional k-algebra of finite type and let M be a finite A-module. By $\Gamma_0(M)$ we denote the biggest A-submodule of M of finite length, it contains precisely those elements of M with 0-dimensional support, i. e. which are annihilated by a 0-dimensional ideal in A. Then the following statements are equivalent: (1) *For all prime ideals $\mathfrak{p} \in \mathrm{Supp}\, M$, the localization $M_{\mathfrak{p}}$ is a Cohen–Macaulay module of dimension* $\dim A_{\mathfrak{p}}$. (2) $\Gamma_0(M) = 0$. (3) $\mathrm{Ass}_A\, M \cap \mathrm{Spm}\, A = \emptyset$. The equivalence of (2) and (3) is clear. For the implication "(1) $\Rightarrow$ (2)" assume $\Gamma_0(M) \neq 0$, i. e. there is an element $y \in M$ with 0-dimensional support. If $\mathfrak{m}$ belongs to this support, then $M_{\mathfrak{m}}$ is not Cohen–Macaulay of dimension $1 = \dim A_{\mathfrak{m}}$, since $\mathfrak{m} A_{\mathfrak{m}} \in \mathrm{Ass}\, M_{\mathfrak{m}}$. Conversely, if (3) is fullfiled and if $\mathfrak{m} \in \mathrm{Spm}\, A \cap \mathrm{Supp}\, M$, then $\mathfrak{m} A_{\mathfrak{m}} \notin \mathrm{Ass}\, M_{\mathfrak{m}}$, since $\mathfrak{m} \notin \mathrm{Ass}\, M$, i. e. $M_{\mathfrak{m}}$ is Cohen–Macauly of dimension $\dim A_{\mathfrak{m}}$.

The A-module $M_{\mathrm{CM}} := M / \Gamma_0(M)$ is called the C o h e n – M a c a u l a y f i c a t i o n of M. Obviously, $\Gamma_0(M_{\mathrm{CM}}) = 0$. Therefore M_{CM} is a Cohen–Macaulay module with $\dim M_{CM,\mathfrak{p}} = \dim A_{\mathfrak{p}}$ for all $\mathfrak{p} \in \operatorname{Supp} M_{CM}$. Note that $M_{CM} = 0$ if and only if $M = \Gamma_0(M)$ has finite length. Moreover, from the short exact sequence $0 \to \Gamma_0(M) \to M \to M_{CM} \to 0$ it is clear that $M_{\mathfrak{p}} = M_{CM,\mathfrak{p}}$ for all $\mathfrak{p} \in \operatorname{Spec} A$ except for the finitely many closed points in $\operatorname{Supp} \Gamma_0(M)$.

Now, let X be any algebraic curve over k and let $\mathcal{F}$ be a coherent $\mathcal{O}_X$-module. For an open affine subset $U \subseteq X$, we define $\Gamma_0(U, \mathcal{F}) := \Gamma_0(\Gamma(U, \mathcal{F})) \subseteq \Gamma(U, \mathcal{F})$. Then the modules $\Gamma_0(U, \mathcal{F})$ are the sections over U of a coherent subsheaf $\Gamma_0(\mathcal{F})$ of $\mathcal{F}$ on X and the coherent $\mathcal{O}_X$-module $\mathcal{F}_{CM} := \mathcal{F}/\Gamma_0(\mathcal{F})$ is called the C o h e n – M a c a u l a y f i c a t i o n of $\mathcal{F}$. In the affine case $X = \operatorname{Spec} A$ and $\mathcal{F} = \tilde{M}$, we have $\Gamma_0(\mathcal{F}) = \Gamma_0(M)\tilde{}$ and $\mathcal{F}_{CM} = \widetilde{M_{CM}}$.

From the exact sequence $0 \to \Gamma_0(\mathcal{F}) \to \mathcal{F} \to \mathcal{F}_{CM} \to 0$, it is easy to derive that the sequence

$$0 \to \Gamma(X, \Gamma_0(\mathcal{F})) \to \Gamma(X, \mathcal{F}) \to \Gamma(X, \mathcal{F}_{CM}) \to 0$$

of global sections is also exact and that $\Gamma(X, \Gamma_0(\mathcal{F})) = \prod_{x \in X} \Gamma_0(\mathcal{F})_x$ is a finite-dimensional k-vector space.

In case $\mathcal{F} = \mathcal{O}_X$, $\Gamma_0(\mathcal{O}_X)$ is an ideal sheaf of $\mathcal{O}_X$ and the quotient $\mathcal{O}_X/\Gamma_0(\mathcal{O}_X)$ is the structure sheaf of the C o h e n – M a c a u l a y f i c a t i o n X_{CM} of X. It is a closed subscheme of X with the same support and in particular, a Cohen–Macaulay curve. Since $\Gamma_0(\mathcal{O}_X) \cdot \mathcal{F} \subseteq \Gamma_0(\mathcal{F})$ the Cohen–Macaulayfication $\mathcal{F}_{CM}$ of a coherent $\mathcal{O}_X$-module $\mathcal{F}$ can always be considered as an $\mathcal{O}_{X_{CM}}$-module. But over a Cohen–Macaulay curve Y a coherent module $\mathcal{G}$ is Cohen–Macaulay if and only if, for all the closed points $y \in Y$ of the support of $\mathcal{G}$, the stalks $\mathcal{G}_y$ are torsion-free over the 1-dimensional Cohen–Macaulay local ring $\mathcal{O}_y$.

7.B.2. Reduction and Normalization

A reduced algebraic curve over k is always Cohen–Macaulay. The reduction X_{red} *of an arbitrary curve* X *is a closed subscheme of the Cohen–Macaulayfication, since* $\Gamma_0(\mathcal{O}_X) \subseteq \mathcal{N}_X$*, where* $\mathcal{O}_X/\mathcal{N}_X = \mathcal{O}_{X_{red}}$ is the structure sheaf of X_{red} (see also Exercise 4.B.6). Finally, from the reduction X_{red} we can construct the normalization $\overline{X}_{red}$ of X_{red} with the canonical finite morphism $\overline{X}_{red} \to X_{red}$ (cf. Theorem 6.C.4). The normalization is even a regular algebraic curve, i.e. all its local rings are regular local rings, see Theorem 6.B.1. Therefore, the normalization $\overline{X}_{red}$ is even a desingularization of X_{red}.[1]) For the closed points the stalks are discrete valuation rings with finite residue fields over k and for the generic points the stalks are finitely generated field extensions of transcendence degree 1 over k, the so called function fields of the irreducible components of X_{red}. In general, $\overline{X}_{red}$ is not smooth, but, if k is a perfect field, then it is (see Corollary 6.E.17). Sometimes it is convenient to factor the canonical morphism $\overline{X}_{red} \to X_{red}$ through the sum $\biguplus_{i=1}^r X_i$, where the X_i, $i = 1, \ldots, r$, are the irreducible components of X_{red} (with their reduced structure sheaves).

---

[1]) This is a special feature in the theory of curves. In higher dimensions the normalization is, in general, not a desingularization.

Altogether, we have the following chain of finite morphisms of algebraic curves (which one always should keep in mind when studying an algebraic curve X):

$$\overline{X}_{\mathrm{red}} = \biguplus_{i=1}^{r} \overline{X}_i \longrightarrow \biguplus_{i=1}^{r} X_i \longrightarrow X_{\mathrm{red}} \longrightarrow X_{\mathrm{CM}} \longrightarrow X \, .$$

For an affine algebraic curve $X = \mathrm{Spec}\, A$, it corresponds to the following chain of finite k-algebra homomorphisms:

$$\overline{A}_{\mathrm{red}} = \prod_{i=1}^{r} \overline{A}_i \longleftarrow \prod_{i=1}^{r} A_i \longleftarrow A_{\mathrm{red}} = A/\mathfrak{n}_A \longleftarrow A_{\mathrm{CM}} = A/\Gamma_0(A) \longleftarrow A \, ,$$

where $A_i = A/\mathfrak{p}_i$ are the residue class algebras of A with respect to its minimal prime ideals $\mathfrak{p}_i$ and the $\overline{A}_i$ are their normalizations, $i = 1, \ldots, r$.

As mentioned above, in order to study an algebraic curve X, it is often very useful to simplify a problem by reducing it to curves of special type occuring in this chain of morphisms. For instance, to study coherent $\mathcal{O}_X$-module $\mathcal{F}$, one can consider the exact sequence $0 \to \Gamma_0(\mathcal{F}) \to \mathcal{F} \to \mathcal{F}_{\mathrm{CM}} \to 0$ and try to divide the problem into two parts by considering separately the module of finite length $\Gamma_0(\mathcal{F})$ and the Cohen–Macaulay module $\mathcal{F}_{\mathrm{CM}}$ over the Cohen–Macaulay curve X_{CM}. *On a regular algebraic curve coherent Cohen–Macaulay modules are even locally free*.

In the following, we are mainly interested in projective algebraic curves over a field k, i. e. curves of the type $\mathrm{Proj}\, R$, where $R = \bigoplus_{m \in \mathbb{N}} R_m$ is a positively graded algebra of finite type over $k \subseteq R_0$, $\mathrm{Dim}_k R_0 < \infty$, cf. Exercise 5.A.30. By Proposition 5.A.22, we may even assume that R is standardly graded and that $k = R_0$. Then R is of Krull dimension 2 (see Exercise 5.A.31). Moreover, every minimal prime ideal of R is of dimension 2, since X is of pure dimension 1. Furthermore, we may assume that the homogeneous maximal ideal R_+ is not an associated prime ideal of R, i. e. the Hurwitz ideal $\mathfrak{I}(R)$ of R is the zero ideal (see Exercise 5.A.6). It is easily checked that then the Cohen–Macaulayfication of $X = \mathrm{Proj}\, R$ is the projective curve $X_{\mathrm{CM}} = \mathrm{Proj}(R/\Gamma_1(R))$, where $\Gamma_1(R)$ is the homogeneous ideal generated by the elements in R which are annihilated by an ideal of dimension ≤ 1. One shows that $\Gamma_0(\mathcal{O}_X) = \widetilde{\Gamma_1(R)} \subseteq \widetilde{R} = \mathcal{O}_X$.

7.B.3. Exercise In general $R/\Gamma_1(R)$ need not be a Cohen–Macaulay ring, i. e. there is no strongly regular sequence of length 2 consisting of homogeneous elements of positive degrees in $R/\Gamma_1(R)$. Show that the graded k-algebra

$$R := k[X_0, X_1, X_2, X_3]/(X_0, X_1) \cap (X_2, X_3)$$

is not Cohen–Macaulay (use Lemma 7.A.3 (2), for example), but $\mathrm{Proj}\, R \subseteq \mathbb{P}_k^3$ is the union of the two skew lines $\mathrm{Proj}\,(k[X_0, X_1, X_2, X_3]/(X_0, X_1))$, $\mathrm{Proj}\,(k[X_0, X_1, X_2, X_3]/(X_2, X_3))$ and hence even a smooth curve. – However, prove that every associated prime ideal of $R/\Gamma_1(R)$ is a minimal prime ideal of $R/\Gamma_1(R)$.

Obviously, the reduction X_{red} of $X = \mathrm{Proj}\, R$ is the projective curve $\mathrm{Proj}\, R_{\mathrm{red}}$. Finally, the normalization $\overline{X}_{\mathrm{red}}$ is the projective regular curve $\mathrm{Proj}\, \overline{R}_{\mathrm{red}}$ (see Exercise 6.C.9). Note that, in general, the normalization $\overline{R}_{\mathrm{red}}$ of R_{red} is not standardly graded even if R_{red} is. Also the homogeneous component $\left(\overline{R}_{red}\right)_0$ of degree 0 can be bigger than k even in case $(R_{red})_0 = k$. By Lemma 7.A.3 (2) one has $\left(\overline{R}_{red}\right)_0 = \Gamma\left(\overline{X}_{red}, \mathcal{O}_{\overline{X}_{red}}\right)$.

Let $X = \operatorname{Proj} R$ be a projective algebraic curve over a field k. There exist homogeneous elements x_0, x_1 of positive degree such that $R/(x_0, x_1)$ is a zero-dimensional k-algebra. By Nakayama's lemma for graded modules (see Exercise 5.A.17) the inclusion $k[x_0, x_1] \subseteq R$ is finite and induces a finite morphism $X = \operatorname{Proj} R \to \operatorname{Proj} k[x_0, x_1]$. Further, x_0, x_1 are algebraically independent over k and $\operatorname{Proj} k[x_0, x_1] \cong \mathbb{P}_k^1$ (even if the degrees $\deg x_0$ and $\deg x_1$ do not coincide, see the comment before 5.A.7; however, it is easy to attain this situation by replacing x_0, x_1 by $x_0^{\deg x_1}, x_1^{\deg x_0}$). If k is an infinite field and if R is standardly graded, then it is even possible to choose $x_0, x_1 \in R_1$. For this one has to avoid finitely many prime ideals $\neq R_+$, i.e. finitely many proper k-subspaces of R_1.

Altogether we get *a finite morphism*

$$\varphi : X \longrightarrow \mathbb{P}_k^1$$

of projective curves.

This morphism may be described in the following way: We cover the projective line $\mathbb{P}_k^1 = \operatorname{Proj} k[T_0, T_1]$, $\deg T_0 = \deg T_1 = 1$, by the two canonical affine open subsets $V_0 := D_+(T_0) = \operatorname{Spec} k[t] \cong \mathbb{A}_k^1$ and $V_1 := D_+(T_1) = \operatorname{Spec} k[t^{-1}] \cong \mathbb{A}_k^1$, $t := T_1/T_0 \in k[T_0^{\pm 1}, T_1^{\pm 1}]$. Then $\varphi^{-1}(V_0) = \operatorname{Spec} A_0$ and $\varphi^{-1}(V_1) = \operatorname{Spec} A_1$, where $A_0 = \Gamma(\varphi^{-1}(V_0), \mathcal{O}_X)$ and $A_1 = \Gamma(\varphi^{-1}(V_1), \mathcal{O}_X)$ are finite algebras of pure dimension 1 over the polynomial algebras $k[t]$ and $k[t^{-1}]$. Over the intersection $V_{01} := V_0 \cap V_1 = \operatorname{Spec} k[t^{\pm 1}]$, we get identifications of $(A_0)_t = A_0[t^{-1}]$ and $(A_1)_{t^{-1}} = A_1[t]$ with $A_{01} = \Gamma(\varphi^{-1}(V_{01}), \mathcal{O}_X)$. *The curve X is Cohen–Macaulay if and only if both algebras A_0 and A_1 are torsion-free, i.e. free over the polynomial algebras $k[t]$ and $k[t^{-1}]$ over k, respectively.* In this case we can identify A_0 and A_1 as subalgebras of the same free $k[t^{\pm 1}]$-algebra $A_0[t^{-1}] = A_1[t]$. This gives a popular description of projective Cohen–Macaulay curves. For an example see Exercise 7.B.5(3) below.

7.B.4. Exercise Let k be an algebraically closed field. Then $A := \Gamma(U, \mathcal{O}_{\mathbb{P}_k^1})$ is a principal ideal domain for an arbitrary affine open subset $U \subseteq \mathbb{P}_k^1$, $U \neq \emptyset$. – Show that this is not true for $k := \mathbb{R}$, for instance. (**Hint:** Use $\mathbb{P}_\mathbb{R}^1 \cong \operatorname{Proj} (\mathbb{R}[X_0, X_1, X_2]/(X_0^2 - X_1^2 - X_2^2))$, cf. Example 1.B.4 (3).) More generally, show that the divisor class group (= ideal class group) of A is isomorphic to $\mathbb{Z}/\mathbb{Z} \operatorname{GCD}\left([\kappa(x_1) : k], \ldots, [\kappa(x_r) : k]\right)$ where $\{x_1, \ldots, x_r\} = \mathbb{P}_k^1 \backslash U$. In particular, the class number of A is $\operatorname{GCD}\left([\kappa(x_1) : k], \ldots, [\kappa(x_r) : k]\right)$, and A is a principal ideal domain if and only if this GCD is 1 (which is always the case if the complement of U in $\mathbb{P}_k^1$ contains a k-rational point, i.e. a point x with residue field $\kappa(x) = k$).

7.B.5. Exercise (Riemann surface of a function field) (1) The construction in Section 6.C of the normalization of a (reduced and locally Noetherian) scheme can be generalized a little bit. We shall describe this only for integral algebraic schemes X over a field k. Let $K := \mathcal{R}(X) = \mathcal{O}_\xi$ (ξ = generic point of X) be the field of rational functions of X (which is a finitely generated field extension of k with transcendence degree $\dim X$) and let $K \to L$ be a *finite* field extension of K. Show that there exists a finite morphism $f : Y \to X$ of integral algebraic k-schemes, where in addition Y is normal, such that for every non-empty open affine subset $U \subseteq X$ the k-algebra homomorphism $\Gamma(U, \mathcal{O}_X) \to \Gamma(f^{-1}(U), Y)$ is the normalization of $\Gamma(U, \mathcal{O}_X)$ in L, i.e. $\Gamma(f^{-1}(U), Y)$ is the integral closure of $\Gamma(U, \mathcal{O}_X)$ in L (which is a k-algebra of finite type by Lemma 6.C.5). If $L = K$ one gets the normalization $\overline{X}$ of X. Show that $f : Y \to X$ is uniquely determined (up to isomorphism) by the given field extension $K = \mathcal{O}_\xi \to \mathcal{O}_\eta = L$ (η = generic point of Y). If X is a projective algebraic k-scheme, then Y is also such a scheme. This morphism $f : Y \to X$ is called the **normalization of X in L**.

(2) Let K be any finitely generated field extension of the field k. Show that there exists an integral and normal projective algebraic k-scheme X with function field $\mathcal{R}(X) = K$. For example: Let $t_1, \dots, t_d$ be a transcendence basis of K over k and interpret $k(t_1, \dots, t_d)$ as the rational function field of $\mathbb{P}^d_k$. Then the normalization of $\mathbb{P}^d_k$ in K is a scheme X as wanted. The scheme X is called a **normal and projective model of the function field K**. (If $\operatorname{char} k = 0$, then even a *smooth* and projective model of K exists. This is a special case of the celebrated Theorem on the Resolution of Singularities by H. Hironaka.)

(3) Let K be a function field over k of transcendence degree 1, i.e. K is finitely generated over k (as a field) and there exists an element $t \in K$ which is transcendental over k and such that the extension $k(t) \subseteq K$ is finite. Identifying $k(t)$ with $\mathcal{R}(\mathbb{P}^1_k)$, the normalization $\varphi : X \to \mathbb{P}^1_k$ of $\mathbb{P}^1_k$ in K is an integral and normal projective algebraic curve X over k with function field $\mathcal{R}(X) = K$. $A_0 := \Gamma(\varphi^{-1}(\operatorname{Spec} k[t]), X)$ is the integral closure of $k[t]$ in K, and $A_1 := \Gamma(\varphi^{-1}(\operatorname{Spec} k[t^{-1}]), X)$ is the integral closure of $k[t^{-1}]$ in K. A_0 and A_1 are Dedekind domains. Let X_0 be the set of closed points of X. For $x \in X_0$ the stalk $\mathcal{O}_x$ is a discrete valuation ring V with $k \subseteq V$ and quotient field $Q(V) = K = \mathcal{O}_\xi$, ξ = generic point of X. Let $\mathcal{V}(K)$ be the set of these valuation rings. Show that the mapping $X_0 \to \mathcal{V}(K)$, $x \mapsto \mathcal{O}_x$, is bijective and that for every open non-empty subset $U \subseteq X$ one has $\Gamma(U, \mathcal{O}_X) = \bigcap_{x \in U \cap X_0} \mathcal{O}_x$. (If $V \in \mathcal{V}(K)$, then $t \in V$ (which implies $V = \mathcal{O}_x$ for some $x \in \operatorname{Spec} A_0$) or $t^{-1} \in V$ (which implies $V = \mathcal{O}_y$ for some $y \in \operatorname{Spec} A_1$).) Show that the integral and normal projective model X of $K = \mathcal{R}(X)$ is (up to isomorphism) uniquely determined by K and is called the **Riemann surface of the function field K.**[2] Show that the field of global sections $\mathcal{O}(X) = \Gamma(X, \mathcal{O}_X) = \bigcap_{x \in X_0} \mathcal{O}_x$ of this curve is the algebraic closure of k in K.

(4) Let $f : X \to Y$ be a finite k-morphism of the Riemann surfaces of the function fields $L := \mathcal{R}(X) = \mathcal{O}_\xi$ and $K := \mathcal{R}(Y) = \mathcal{O}_\eta$ of transcendence degree 1 over k, $\xi, \eta = f(\xi)$ the generic points of X and Y respectively. Show that f is the normalization of Y with respect to the field extension $K = \mathcal{O}_\eta \to \mathcal{O}_\xi = L$. The set of finite (= non-constant) k-morphisms $X \to Y$ corresponds bijectively to the set of k-algebra homomorphisms $\mathcal{R}(Y) \to \mathcal{R}(X)$ of the function fields. In particular, the set of finite k-morphisms $X \to \mathbb{P}^1_k$ corresponds bijectively to the set of elements $t \in \mathcal{R}(X)$, which are transcendental over k.

---

[2] This notation originates from the classical case $k = \mathbb{C}$, where a Riemann surface is a *two*-dimensional real manifold.

7.C. The Projective Line

In this section we classify the coherent modules on the projective line $\mathbb{P}^1_k = \operatorname{Proj} k[T_0, T_1]$ over a field k and prove the theorem of Riemann–Roch in this case. The general case will be proved in Section 7.D by reducing it to this special one.

As we have seen in Section 7.B every coherent module $\mathcal{F}$ over $\mathbb{P}^1_k$ gives rise to an exact sequence $0 \to \Gamma_0(\mathcal{F}) \to \mathcal{F} \to \overline{\mathcal{F}} \to 0$, where the support $\operatorname{Supp}(\Gamma_0(\mathcal{F}))$ is a finite set of closed points in $\mathbb{P}^1_k$ and $\overline{\mathcal{F}}$ is a Cohen–Macaulay module. i. e. for every closed point $x \in \operatorname{Supp} \mathcal{F}$, the $\mathcal{O}_x$-module $\overline{\mathcal{F}}_x$ is torsion-free. Since $\mathcal{O}_x$ is a discrete valuation ring for every closed point $x \in \mathbb{P}^1_k$ and the rational function field $k(T_1/T_0)$ for the generic point $x \in \mathbb{P}^1_k$, it follows: The stalks $\overline{\mathcal{F}}_x$ are free modules of the same rank for all $x \in \mathbb{P}^1_k$. We call this rank the r a n k of $\overline{\mathcal{F}}$ and denote it by $r := \operatorname{rank} \overline{\mathcal{F}} = \operatorname{rank}_{\mathcal{O}_{\mathbb{P}^1_k}} \overline{\mathcal{F}}$. We also say that $\overline{\mathcal{F}}$ is a l o c a l l y f r e e $\mathcal{O}_x$-module of rank r. For an arbitrary affine open subset $U \subseteq \mathbb{P}^1_k$, it follows that $\Gamma(U, \overline{\mathcal{F}})$ is a finite projective $\Gamma(U, \mathcal{O}_{\mathbb{P}^1})$-module. Therefore the exact sequence of $\Gamma(U, \mathcal{O}_{\mathbb{P}^1_k})$-modules $0 \to \Gamma(U, \Gamma_0(\mathcal{F})) \to \Gamma(U, \mathcal{F}) \to \Gamma(U, \overline{\mathcal{F}}) \to 0$ splits. From this one easily derives that the sequence $0 \to \Gamma_0(\mathcal{F}) \to \mathcal{F} \to \overline{\mathcal{F}} \to 0$ globally splits, i. e.

$$\mathcal{F} \cong \Gamma_0(\mathcal{F}) \oplus \overline{\mathcal{F}}.$$

(The same argument holds for a coherent sheaf on an arbitrary *regular* algebraic curve.) As aready mentioned in Section 7.B the module $\Gamma_0(\mathcal{F})$ is very simple, it is determined by the finitely many stalks of its support which are finite torsion modules over the corresponding stalks of the structure sheaf. They are described by the classical structure theorem of finite modules over principal ideal domains. The main problem is now to understand the locally free module $\overline{\mathcal{F}}$ and for this we may assume that $\mathcal{F} = \overline{\mathcal{F}}$ is itself locally free.

By Corollary 7.A.2 there exists a finite graded $k[T_0, T_1]$-module M with $\mathcal{F} \cong \tilde{M}$. Since $\mathcal{F}$ is reflexive, $\mathcal{F} = \mathcal{F}^{**}$, and since $(\tilde{M})^{**} = (M^{**})^{\sim}$, we can even assume that M is a reflexive graded $k[T_0, T_1]$-module and thus free as we will prove by using the following two lemmas.

7.C.1. Lemma *Let A be an arbitrary (commutative) ring and let M, N be A-modules. If $f, g \in A$ is a strongly regular sequence for N, then f, g is also a strongly regular sequence for the A-module $\operatorname{Hom}_A(M, N)$. In particular, if f, g is a strongly regular sequence for A, then f, g is also a strongly regular sequence for the dual M^* (and the double dual M^{**}).*

PROOF. From the exact sequence $0 \to N \xrightarrow{f} N \to N/fN \to 0$, we get an exact sequence $0 \to \operatorname{Hom}_A(M, N) \xrightarrow{\tilde{f}} \operatorname{Hom}_A(M, N) \to \operatorname{Hom}_A(M, N/fN)$, where $\widetilde{f} = \mathrm{Hom}_A(M, f)$ is the multiplication by f on $\mathrm{Hom}_A(M, N)$. Therefore $\widetilde{f}$ is injective and $\mathrm{Coker}\,\widetilde{f} \subseteq \mathrm{Hom}_A(M, N/fN)$. Further, since the multiplication by g on N/fN is injective, by the above argument the multiplication by g on $\mathrm{Hom}_A(M, N/fN)$ is injective and hence injective on the submodule $\mathrm{Coker}\,\widetilde{f}$. •

7.C.2. Lemma *Let M be a finite graded module over the γ-graded polynomial ring $P := k[T_0, \ldots, T_n]$ over a field k, $\gamma = (\deg T_0, \ldots, \deg T_n) \in (\mathbb{N}^*)^{n+1}$. Then M is free with a P-basis of homogeneous elements, i.e. $M \cong P(e_1) \oplus \cdots \oplus P(e_r)$ for some integers $e_1, \ldots, e_r \in \mathbb{Z}$, if and only if $T_0, \ldots, T_n$ is a strongly regular sequence for M. In particular, in this case $\widetilde{M} \cong \mathcal{O}_{\mathbb{P}^n_{\gamma,k}}(e_1) \oplus \cdots \oplus \mathcal{O}_{\mathbb{P}^n_{\gamma,k}}(e_r)$.*

PROOF. Since $T_0, \ldots, T_n$ is a strongly regular sequence for P, it is clear that it is a strongly regular sequence for any free P-module. Conversely, suppose that $T_0, \ldots, T_n$ is a strongly regular sequence for M. By induction on n, we shall prove that M is a free P-module with a basis of homogeneous elements. The induction starts trivially at $n = -1$. By induction hypothesis there are homogeneous elements $x_1, \ldots, x_r \in M$ of degrees $-e_1, \ldots, -e_r$ whose residue classes in $M/T_0 M$ form a basis for the P'-module $M/T_0 M \cong P'(e_1) \oplus \cdots \oplus P'(e_r)$, where $P' := k[T_1, \ldots, T_n] = P/T_0 P$. Let $F := P(e_1) \oplus \cdots \oplus P(e_r)$. Consider the following commutative diagram

$$
\begin{array}{ccccccccc}
0 & \longrightarrow & F(-\deg T_0) & \xrightarrow{\;\;T_0\;\;} & F & \longrightarrow & F/T_0 F & \longrightarrow & 0 \\
  &   & \Big\downarrow{\scriptstyle f} & & \Big\downarrow{\scriptstyle f} & & \cong\Big\downarrow{\scriptstyle \overline{f}} & & \\
0 & \longrightarrow & M(-\deg T_0) & \xrightarrow{\;\;T_0\;\;} & M & \xrightarrow{\;f\;} & M/T_0 M & \longrightarrow & 0
\end{array}
$$

where $f(\varepsilon_i) := x_i$ and $\varepsilon_i \in P(e_i)$ is the standard basis element of degree $-e_i$, $i = 1, \ldots, r$. Since T_0 is a non-zero divisor for F and for M, we have $\mathrm{Ker}\,f = T_0(\mathrm{Ker}\,f)$ and $\mathrm{Coker}\,f = T_0(\mathrm{Coker}\,f)$ and hence $\mathrm{Ker}\,f = 0$ and $\mathrm{Coker}\,f = 0$ by Nakayama's lemma for graded modules (see Exercise 5.A.17). This proves that $f : F \to M$ is an isomorphism. •

7.C.3. Exercise As a generalization of Lemma 7.C.2 prove: Let M be a graded module over the γ-graded polynomial algebra $P := A[T_0, \ldots, T_n]$ with $M_m = 0$ for all $m < m_0$, where A is an arbitrary commutative ring. If all the homogeneous components of M/P_+M are free A-modules and if $T_0, \ldots, T_n$ is a strongly regular sequence for M, then there is a family of integers e_i, $i \in I$, such that $M \cong \bigoplus_{i \in I} P(e_i)$.

Combining Lemma 7.C.1 and Lemma 7.C.2 (with the remarks before 7.C.1), we get the first part of the following so called s p l i t t i n g t h e o r e m for $\mathbb{P}^1_k$ which is fundamental for the study of coherent modules over projective spaces (even in higher dimensions).

7.C.4. Theorem *Let $\mathcal{F}$ be a coherent locally free module of rank r over $\mathbb{P}^1_k$. Then there exist integers $e_1, \ldots, e_r \in \mathbb{Z}$, $e_1 \geq e_2 \geq \cdots \geq e_r$, such that*

$$
\mathcal{F} \cong \mathcal{O}_{\mathbb{P}^1_k}(e_1) \oplus \mathcal{O}_{\mathbb{P}^1_k}(e_2) \oplus \cdots \oplus \mathcal{O}_{\mathbb{P}^1_k}(e_r).
$$

Moreover, the sequence of integers $e_1, \ldots, e_r$ *is uniquely determined by* $\mathcal{F}$ *and* $\mathrm{Dim}_k\big(\Gamma(\mathbb{P}_k^1, \mathcal{F})\big) = \sum_{i=1}^r \mathrm{Dim}_k\big(\Gamma(\mathbb{P}_k^1, \mathcal{O}_{\mathbb{P}_k^1}(e_i))\big)$ *and for every* $e \in \mathbb{Z}$,

$$\mathrm{Dim}_k\big(\Gamma(\mathbb{P}_k^1, \mathcal{O}_{\mathbb{P}_k^1}(e))\big) = \begin{cases} e+1, & \text{if } e \geq 0, \\ 0, & \text{if } e < 0. \end{cases}$$

PROOF. It remains to prove the uniqueness of the sequence $e_1, \ldots, e_r$ and the dimension formula for $\mathcal{O}(e) := \mathcal{O}_{\mathbb{P}^1}(e)$. The dimension formula is a special case of 7.A.3 (2), see also Example 7.A.4. For the uniqueness, we first remark that for every integer $m \in \mathbb{Z}$, we have $\mathcal{F} \otimes_{\mathcal{O}} \mathcal{O}(m) \cong \bigoplus_{i=1}^r \mathcal{O}(e_i + m)$ and hence

$$\mathrm{Dim}_k \Gamma(\mathcal{F} \otimes \mathcal{O}(m)) = \sum_{i=1}^r \mathrm{Dim}_k \Gamma(\mathcal{O}(e_i + m)).$$

For $m < -e_1$, this dimension is 0; for $m = -e_1$, this dimension counts how often the integer e_1 occurs in the sequence $e_1 \geq \cdots \geq e_r$. Continuing in this way we determine the whole sequence $e_1 \geq \cdots \geq e_r$ by increasing m step by step. •

By 7.C.4 a locally free coherent module $\mathcal{F}$ of rank 1 over $\mathbb{P}_k^1$ is isomorphic to $\mathcal{O}(e)$ for a uniquely determined integer $e \in \mathbb{Z}$. This integer e is called the d e g r e e $\deg \mathcal{F}$ of $\mathcal{F}$. For a locally free module $\mathcal{F} \cong \bigoplus_{i=1}^r \mathcal{O}(e_i)$ of rank r we call the degree of the locally free module $\Lambda^r \mathcal{F} \cong \otimes_{i=1}^r \mathcal{O}(e_i) = \mathcal{O}(\sum_{i=1}^r e_i)$ of rank 1 the d e g r e e of $\mathcal{F}$, i.e.

$$\deg \mathcal{F} := \deg \Lambda^r \mathcal{F} = \sum_{i=1}^r e_i.$$

The Riemann–Roch theorem for the projective line is now an immediate consequence of the above definitions and Theorem 7.C.4:

7.C.5. Theorem of Riemann–Roch for the Projective Line *Let* $\mathcal{F}$ *be a coherent locally free module on the projective line* $\mathbb{P}_k^1$. *Then*

$$\mathrm{Dim}_k \Gamma(\mathcal{F}) - \mathrm{Dim}_k \Gamma(\mathcal{H}om_{\mathcal{O}}(\mathcal{F}, \mathcal{O}(-2))) = \deg \mathcal{F} + \mathrm{rank}\, \mathcal{F}.$$

PROOF. Because of Theorem 7.C.4, we may assume that $\mathcal{F} = \mathcal{O}(e)$ for some integer $e \in \mathbb{Z}$, i.e. $\mathrm{rank}\, \mathcal{F} = 1$ and $\deg \mathcal{F} = e$. Then $\mathcal{H}om_{\mathcal{O}}(\mathcal{F}, \mathcal{O}(-2)) = \mathcal{O}(-e-2)$. Now, the equality follows from the dimension formula in 7.C.4 by considering the two cases $e \leq -2$ and $e \geq -1$ separately. •

We remark that $\mathcal{H}om_{\mathcal{O}}(\mathcal{F}, \mathcal{O}(-2))$ is canonically isomorphic to $\mathcal{F}^* \otimes_{\mathcal{O}} \mathcal{O}(-2)$ for every coherent locally free sheaf $\mathcal{F}$. Furthermore, for $\mathbb{P}_k^1$,

$$\mathcal{O}(-2) \cong \Omega_{\mathbb{P}_k^1 | k} =: \omega_{\mathbb{P}_k^1} =: \omega$$

(see Theorem 6.E.20 and the comment after that). With these isomorphisms, the Riemann–Roch formula in 7.C.5 is:

$$\mathrm{Dim}_k \Gamma(\mathcal{F}) - \mathrm{Dim}_k \Gamma(\mathcal{F}^* \otimes_{\mathcal{O}} \omega) = \deg \mathcal{F} + \mathrm{rank}\, \mathcal{F}.$$

This formula we can formally extend to arbitrary coherent $\mathcal{O}_{\mathbb{P}_k^1}$-module $\mathcal{F} \cong \Gamma_0(\mathcal{F}) \oplus \overline{\mathcal{F}}$. To do this we set

$$\deg \mathcal{F} := \mathrm{Dim}_k \Gamma(\Gamma_0(\mathcal{F})) + \deg \overline{\mathcal{F}}, \qquad \mathrm{rank}\, \mathcal{F} := \mathrm{rank}\, \overline{\mathcal{F}}.$$

Furthermore, we set [3])

$$\chi(\mathcal{F}) := \mathrm{Dim}_k \Gamma(\mathcal{F}) - \mathrm{Dim}_k \Gamma(\mathcal{F}^* \otimes_{\mathcal{O}} \omega) = \mathrm{Dim}_k \Gamma(\mathcal{F}) - \mathrm{Dim}_k \Gamma(\mathcal{H}om_{\mathcal{O}}(\mathcal{F}, \omega)).$$

Then $\Gamma(\mathcal{F}) = \Gamma(\Gamma_0(\mathcal{F})) \oplus \Gamma(\overline{\mathcal{F}})$ and $\mathcal{F}^* = \overline{\mathcal{F}}^*$ and hence quite generally:

7.C.6. General Riemann–Roch Theorem for $\mathbb{P}^1_k$ *Let $\mathcal{F}$ be an arbitrary coherent module over the projective line $\mathbb{P}^1_k$. Then*

$$\chi(\mathcal{F}) = \deg \mathcal{F} + \mathrm{rank}\, \mathcal{F}.$$

The Riemann–Roch formula will be generalized in the next section by using this special case. In this section we already prove the following additivity formulas:

7.C.7. Lemma   *Let $0 \to \mathcal{F}' \xrightarrow{\alpha'} \mathcal{F} \xrightarrow{\alpha} \mathcal{F}'' \to 0$ be an exact sequence of coherent modules over $\mathbb{P}^1_k$. Then*

$$\mathrm{rank}\, \mathcal{F} = \mathrm{rank}\, \mathcal{F}' + \mathrm{rank}\, \mathcal{F}'', \quad \deg \mathcal{F} = \deg \mathcal{F}' + \deg \mathcal{F}'', \quad \chi(\mathcal{F}) = \chi(\mathcal{F}') + \chi(\mathcal{F}'').$$

PROOF. The additivity of the rank is trivial. Because of 7.C.6, it is enough to prove one of the last two formulas. We prove the formula for the degree (experts might prefer to prove this for the Euler–Poincaré characteristic!) which is trivial for modules of rank 0, because of the additivity of the length of modules.

First assume, $\mathcal{F}' \cong \mathcal{O}(e')$ and $\mathcal{F} \cong \mathcal{O}(e)$ are locally free of rank 1. Then, by the following Exercise 7.C.8, $e - e' \geq 0$ and $\mathcal{F}'' \cong \mathcal{F}/\mathcal{F}' \cong (P/FP)^{\sim}(e)$, where F is a non-zero homogeneous polynomial of degree $e - e'$ in $P = k[T_0, T_1]$. It follows $\deg \mathcal{F}'' = \mathrm{Dim}_k \Gamma(\mathcal{F}'') = \mathrm{Dim}_k \Gamma((P/FP)^{\sim}) = e - e'$.

Secondly assume $\mathcal{F}'$ and $\mathcal{F}$ are locally free of the same rank r. Then we have an exact sequence $0 \to \Lambda^r \mathcal{F}' \to \Lambda^r \mathcal{F} \to \Lambda^r(\mathcal{F})/\Lambda^r(\mathcal{F}') \to 0$. By definition of the degree and the above case we need to show $\mathrm{Dim}_k \Gamma(\mathcal{F}'') = \sum_{x \in \mathbb{P}^1_k} \mathrm{Dim}_k \mathcal{F}''_x = \Gamma(\mathcal{G}) = \sum_{x \in \mathbb{P}^1_k} \mathrm{Dim}_k \mathcal{G}_x$, where $\mathcal{G} := \Lambda^r(\mathcal{F})/\Lambda^r(\mathcal{F}')$. But, even $\mathrm{Dim}_k \mathcal{F}''_x = \mathrm{Dim}_k \mathcal{G}_x$ for any $x \in \mathbb{P}^1_k$ which is a particular case of the following well known length formula: *For an injective endomorphism $\psi : M \to M$ of a free module M of finite rank over a discrete valuation ring A, we have $\ell_A(\mathrm{Coker}\, \psi) = \ell_A(A/\mathrm{Det}\, \psi)$. This formula is true even if A is replaced by an arbitrary local Cohen–Macaulay local ring of dimension* 1. We will use this generalization eventually (for instance in Exercise 7.D.5).

Now assume all $\mathcal{F}'$, $\mathcal{F}$ and $\mathcal{F}''$ are locally free of ranks r', r and r'', respectively. Then there is a canonical isomorphism $\Lambda^r \mathcal{F} \cong \Lambda^{r'} \mathcal{F}' \otimes \Lambda^{r''} \mathcal{F}''$, i. e. $\mathcal{O}(\deg \mathcal{F}) \cong \mathcal{O}(\deg \mathcal{F}') \otimes \mathcal{O}(\deg \mathcal{F}'') = \mathcal{O}(\deg \mathcal{F}' + \deg \mathcal{F}'')$.

---

[3]) In the following $\Gamma(\mathcal{F})$ is the cohomology group $\mathrm{H}^0(\mathbb{P}^1_k, \mathcal{F})$ and $\Gamma(\mathcal{F}^* \otimes_{\mathcal{O}} \omega)$ is the k-dual of the cohomology group $\mathrm{H}^1(\mathbb{P}^1_k, \mathcal{F})$. The last equality is a special case of the so called Serre's duality theorem. The higher cohomology groups $\mathrm{H}^i(\mathbb{P}^1_k, \mathcal{F})$, $i > 1$, vanish. So the following definition of $\chi(\mathcal{F})$ is an imitation of the usual Euler–Poincaré characteristic of a coherent sheaf $\mathcal{F}$. But in these lectures we do not discuss any cohomology groups (besides $\mathrm{H}^0(-) = \Gamma(-)$).

Finally, we consider the general case. By simple diagram chasing we get the following three exact sequences:

$$0 \to \Gamma_0(\mathcal{F}') \xrightarrow{\alpha_0'} \Gamma_0(\mathcal{F}) \xrightarrow{\alpha_0} \Gamma_0(\mathcal{F}'') \to \operatorname{Coker}\alpha_0 \to 0,$$

$$0 \to \operatorname{Ker}\overline{\alpha} \longrightarrow \overline{\mathcal{F}} \xrightarrow{\overline{\alpha}} \overline{\mathcal{F}''} \to 0, \quad 0 \to \overline{\mathcal{F}'} \xrightarrow{\overline{\alpha'}} \operatorname{Ker}\overline{\alpha} \longrightarrow \operatorname{Ker}\overline{\alpha}/\operatorname{Im}\overline{\alpha'} \to 0$$

and an isomorphism $\operatorname{Coker}\alpha_0 \cong \operatorname{Ker}\overline{\alpha}/\operatorname{Im}\overline{\alpha'}$. From this, using the earlier cases already proved, we get successively the following equations which prove the additivity of the degree:

$$\deg\overline{\mathcal{F}} = \deg(\operatorname{Ker}\overline{\alpha}) + \deg\overline{\mathcal{F}''} = \deg\overline{\mathcal{F}'} + \deg(\operatorname{Coker}\alpha_0) + \deg\overline{\mathcal{F}''}$$

$$= \deg\overline{\mathcal{F}'} + \deg\Gamma_0(\mathcal{F}') - \deg\Gamma_0(\mathcal{F}) + \deg\Gamma_0(\mathcal{F}'') + \deg\overline{\mathcal{F}''}. \qquad \bullet$$

7.C.8. Exercise Let $X := \mathbb{P}_k^n = \operatorname{Proj} P$, $P := k[T_0, \dots, T_n]$ standardly graded, $n \geq 1$. Any $\mathcal{O}_X$-module homomorphism $\varphi : \mathcal{O}_X(e) \to \mathcal{O}_X(f)$, $e, f \in \mathbb{Z}$, i. e. any homomorphism $\varphi : P(e)^{\sim} \to P(f)^{\sim}$ is induced by the multiplication $P(e) \to P(f)$ with a homogeneous polynomial $F \in P_{f-e}$ of degree $f - e$. In particular, $\mathcal{O}_X(f)/\operatorname{Im}\varphi \cong (P/FP)^{\sim}(f)$. (**Hint:** Use the isomorphism $\mathcal{H}om(\mathcal{O}_X(e), \mathcal{O}_X(f)) \cong \mathcal{O}_X(-e) \otimes_{\mathcal{O}_X} \mathcal{O}(f) = \mathcal{O}_X(f-e)$.)

7.C.9. Exercise Let $P := k[T_0, T_1]$ with $\gamma_i := \deg T_i > 0$, $i = 0, 1$. Determine the degree of the locally free sheaf $P(m)^{\sim}$, $m \in \mathbb{Z}$, over $\operatorname{Proj} P \cong \mathbb{P}_k^1$. (See the commentary before Proposition 5.A.10. – One may reduce to the case GCD $(\gamma_0, \gamma_1) = 1$.) In particular, show that $\deg P(-\gamma_0 - \gamma_1)^{\sim} = -2$.

7.D. Riemann–Roch Theorem for General Curves

Let $\mathcal{F}$ be a coherent module on a projective algebraic curve X over a field k. As in Section 7.B we fix a finite morphism $\varphi : X \to \mathbb{P}_k^1$ and consider the direct image $\varphi_*\mathcal{F}$ which is a coherent $\mathcal{O}_{\mathbb{P}_k^1}$-module (see 6.E.11). Moreover, by the definition of the direct image, we have

$$\Gamma(X, \mathcal{F}) = \Gamma(\mathbb{P}_k^1, \varphi_*\mathcal{F}).$$

By the Riemann–Roch Theorem 7.C.6 for $\mathbb{P}_k^1$,

$$\chi(\varphi_*\mathcal{F}) = \operatorname{Dim}_k\Gamma(\mathbb{P}_k^1, \varphi_*\mathcal{F}) - \operatorname{Dim}_k\Gamma(\mathbb{P}_k^1, \mathcal{H}om_{\mathcal{O}_{\mathbb{P}_k^1}}(\varphi_*\mathcal{F}, \omega_{\mathbb{P}_k^1}))$$

$$= \deg\varphi_*\mathcal{F} + \operatorname{rank}\varphi_*\mathcal{F}.$$

Our aim is to interpret this formula in terms which depend only on X. At first we look at the $\mathcal{O}_{\mathbb{P}_k^1}$-module $\mathcal{H}om_{\mathcal{O}_{\mathbb{P}_k^1}}(\varphi_*\mathcal{F}, \omega_{\mathbb{P}_k^1})$. For this we recall the following general result: Let B be an A-algebra (with structure homomorphism $A \to B$) and let M be an A-module and N be a B-module. Then there is a canonical isomorphism

$$\operatorname{Hom}_A(N, M) \xrightarrow{\;\sim\;} \operatorname{Hom}_B(N, \operatorname{Hom}_A(B, M))$$

of B-modules given by $\alpha \xmapsto{\;\sim\;} (n \mapsto (b \mapsto \alpha(bn)))$.

For any coherent $\mathcal{O}_{\mathbb{P}_k^1}$-module $\mathcal{G}$, we define a coherent $\mathcal{O}_X$-module $\varphi^!\mathcal{G}$ by

$$\Gamma(\varphi^{-1}(V), \varphi^!\mathcal{G}) := \operatorname{Hom}_{\Gamma(V,\mathbb{P}_k^1)}(\Gamma(\varphi^{-1}(V), \mathcal{O}_X), \Gamma(V, \mathcal{G}))$$

for any affine open subset $V \subseteq \mathbb{P}_k^1$. Then $\varphi_*\varphi^!\mathcal{G} = \mathcal{H}om_{\mathcal{O}_{\mathbb{P}_k^1}}(\varphi_*\mathcal{O}_X, \mathcal{G})$ by definition and, more generally, for any coherent $\mathcal{O}_X$-module $\mathcal{F}$, we obtain the following identification of $\mathcal{O}_{\mathbb{P}_k^1}$-modules (or even of $(\varphi_*\mathcal{O}_X)$-modules) using the above canonical isomorphisms for modules:

$$\mathcal{H}om_{\mathcal{O}_{\mathbb{P}_k^1}}(\varphi_*\mathcal{F}, \mathcal{G}) \xrightarrow{\sim} \varphi_*\mathcal{H}om_{\mathcal{O}_X}(\mathcal{F}, \varphi^!\mathcal{G}) \,.$$

Of course, this lifting construction can be done for any finite morphism $\varphi : X \to Y$ of algebraic curves over k and coherent modules $\mathcal{F}$ and $\mathcal{G}$ over X and Y, respectively.[4]) The reader should not confound it with the pull back defined in Exercise 6.E.12. The following transitivity property of this construction is obvious using the module isomorphisms mentioned above: *If $\varphi : X \to Y$ and $\psi : Y \to Z$ are finite morphisms of algebraic curves and if $\mathcal{H}$ is a coherent $\mathcal{O}_Z$-module, then* $(\psi\varphi)^!\mathcal{H} = \varphi^!(\psi^!\mathcal{H})$.

If we set

$$\omega_X := \varphi^!\omega_{\mathbb{P}_k^1} \,,$$

we get a canonical isomorphism

$$\mathcal{H}om_{\mathcal{O}_{\mathbb{P}_k^1}}(\varphi_*\mathcal{F}, \omega_{\mathbb{P}_k^1}) \xrightarrow{\sim} \varphi_*\mathcal{H}om_{\mathcal{O}_X}(\mathcal{F}, \omega_X) \,,$$

and for $\chi(\varphi_*\mathcal{F})$ the following expression in terms of objects over X

$$\chi(\varphi_*\mathcal{F}) = \operatorname{Dim}_k\Gamma(X, \mathcal{F}) - \operatorname{Dim}_k\Gamma(X, \mathcal{H}om_{\mathcal{O}_X}(\mathcal{F}, \omega_X)) =: \chi(\mathcal{F}) \,,$$

which we now use as the definition of $\chi(\mathcal{F})$. We keep in mind the equalities $\Gamma\big(X, \mathcal{H}om_{\mathcal{O}_X}(\mathcal{F}, \omega_X)\big) = \operatorname{Hom}_{\mathcal{O}_X}(\mathcal{F}, \omega_X) = \operatorname{Hom}_{\mathcal{O}_{\mathbb{P}_k^1}}(\varphi_*\mathcal{F}, \omega_{\mathbb{P}_k^1})$ *and*

$$\chi(\mathcal{F}) = \chi(\varphi_*\mathcal{F})$$

for every coherent $\mathcal{O}_X$-module $\mathcal{F}$. The additivity of χ for coherent $\mathcal{O}_{\mathbb{P}_k^1}$-modules (see 7.C.7) implies *the additivity of χ for coherent $\mathcal{O}_X$-modules.* The coherent $\mathcal{O}_X$-module ω_X is called the c a n o n i c a l or d u a l i z i n g m o d u l e of the projective curve $X = (X, \mathcal{O}_X)$. The uniqueness of ω_X will be seen later in 7.E.21. For a non-empty open subset $U \subseteq X$ we call the restriction $\omega_U := \omega_X|U$ the canonical or dualizing module of the curve U (which is not projective if U is not closed).

Now we discuss the concept of degree for arbitrary coherent $\mathcal{O}_X$-modules. As before, for a coherent $\mathcal{O}_X$-module $\mathcal{F} = \Gamma_0(\mathcal{F})$ with finite support we define

$$\deg \mathcal{F} := \operatorname{Dim}_k \Gamma(\mathcal{F}) = \sum_{x \in X} \operatorname{Dim}_k \mathcal{F}_x \,.$$

---

[4]) The operation $\varphi^!$ can be defined in much more general situations. We shall use it only for finite morphisms of algebraic curves over a field with the proof of Theorem 7.E.21 as an exception.

Then $\chi(\mathcal{F}) = \deg \mathcal{F} = \deg \varphi_* \mathcal{F} = \chi(\varphi_* \mathcal{F})$ for such a module $\mathcal{F}$. – Next we consider the case of a locally free sheaf $\mathcal{G}$ of rank 1 over a reduced and irreducible (i. e. integral) projective curve Y with a finite morphism

$$\psi : Y \to \mathbb{P}_k^1 .$$

(We will apply this case to the irreducible components X_i of the given curve X with their reduced structures and the compositions $\psi_i : X_i \hookrightarrow X \to \mathbb{P}_k^1$.) Let η be the generic point of Y and let $\mathcal{R}(Y) = \mathcal{R}_\eta = \mathcal{O}_\eta$ be the rational function field of Y. We consider the embedding $\mathcal{G} \hookrightarrow \mathcal{R} \otimes_\mathcal{O} \mathcal{G}$ of $\mathcal{G}$ into the constant sheaf $\mathcal{R} \otimes_\mathcal{O} \mathcal{G}$ with $\Gamma(V, \mathcal{R} \otimes_\mathcal{O} \mathcal{G}) = \mathcal{R}_\eta \otimes_{\mathcal{O}_\eta} \mathcal{G}_\eta = \mathcal{G}_\eta$ for all non-empty open subsets $V \subseteq Y$. The space $\mathcal{G}_\eta$ of r a t i o n a l   s e c t i o n s of $\mathcal{G}$ is a vector space of dimension one over the function field $\mathcal{R}_\eta = \mathcal{O}_\eta$. Choose a non-zero element $s \in \mathcal{G}_\eta$, i. e. an $\mathcal{O}_\eta$-basis of $\mathcal{G}_\eta$. Then for every closed point $y \in Y$, $\mathcal{G}_y = \mathcal{O}_y m_y s$ with an element $m_y \in Q(\mathcal{O}_y)^\times = \mathcal{R}_y^\times = \mathcal{R}_\eta^\times$, which is uniquely determined up to a unit in $\mathcal{O}_y$. We write $m_y = f_y / g_y$ with $f_y, g_y \in \mathcal{O}_y \setminus \{0\}$ and set

$$\delta_y := \delta_y(\mathcal{G}, s) := \mathrm{Dim}_k(\mathcal{O}_y / \mathcal{O}_y g_y) - \mathrm{Dim}_k(\mathcal{O}_y / \mathcal{O}_y f_y) \in \mathbb{Z} .$$

Then δ_y is independent of the choice of the representation f_y / g_y of m_y. This is a special case of the following simple observation: If A is a local Cohen- Macaulay ring of dimension one and if $f, g, f', g' \in A$ are non-zero divisors with $f/g = f'/g' \in Q(A)^\times$, then $\ell_A(A/Ag) - \ell_A(A/Af) = \ell_A(A/Ag') - \ell_A(A/Af') .$[5]) Obviously, for all but a finite number of closed points $y \in Y$, we have $\mathcal{G}_y = \mathcal{O}_y s$ and, in particular, $\delta_y = 0$. Therefore, we can define

$$\deg \mathcal{G} := \sum_{y \in Y_0} \delta_y(\mathcal{G}) = \sum_{y \in Y_0} \delta_y(\mathcal{G}, s) \in \mathbb{Z} ,$$

where $Y_0 := Y \setminus \{\eta\}$ denotes here the set of closed points of Y.

For $Y = \mathbb{P}_k^1$ and $\mathcal{G} = \mathcal{O}(e)$ this definition coincides with the definition of Section 7.C, i.e. $e = \sum_{x \in (\mathbb{P}_k^1)_0} \delta_x(\mathcal{O}(e))$. For the proof of this equality we choose an arbitrary non-zero rational section $s = m T_0^e$ of $\mathcal{O}(e)$, where $m = f(t)/g(t)$, $f(t), g(t) \in k[t]$, $t := T_1/T_0$, is a rational function on $\mathbb{P}_k^1$. Then, by definition,

$$\sum_{x \in D_+(T_0)_0} \delta_x(\mathcal{O}(e), s) = \mathrm{Dim}_k \, k[t]/k[t]f - \mathrm{Dim}_k \, k[t]/k[t]g = \deg_{k[t]} f - \deg_{k[t]} g$$

since $T_0^e = g(t)s/f(t)$ is a base section of $\mathcal{O}(e)$ on $D_+(T_0)$, and

$$\delta_\infty(\mathcal{O}(e), s) = e + \deg_{k[t]} g - \deg_{k[t]} f ,$$

since $T_1^e = t^e g(t)s/f(t)$ is a base section of $\mathcal{O}(e)$ at the point $\infty \in D_+(T_1)$. Now, adding both equations, we get the required equality.

By definition,

$$\deg \mathcal{G} = 0$$

if $\mathcal{G}$ is a (not only locally free but) free $\mathcal{O}_Y$-module of rank 1 over (the reduced and irreducible curve) Y. (Choose for s a global base section of $\mathcal{G}$.)

---

[5]) Note that $A/Afg' = A/Af'g$ and that $\mathrm{Dim}_k(\mathcal{O}_y/\mathcal{O}_y f_y) = [\kappa(y) : k] \ell_{\mathcal{O}_y}(\mathcal{O}_y/\mathcal{O}_y f_y)$. Sometimes, the difference $\ell_A(A/Af) - \ell_A(A/Ag)$ (observe the opposite signs) is called the o r d e r of the element $f/g \in Q(A)^\times$ and is denoted by $\nu(f/g)$. It defines a group homomorphism $\nu : Q(A)^\times \to \mathbb{Z}$.

7.D.1. Exercise (1) Let Y be an integral projective curve. Show that $\deg(\mathcal{G} \otimes_{\mathcal{O}_X} \mathcal{H}) = \deg \mathcal{G} + \deg \mathcal{H}$ for locally free sheaves $\mathcal{G}, \mathcal{H}$ of rank 1 over Y.

(2) Let $\varphi : X \to Y$ be a finite morphism of integral projective curves and let $\mathcal{G}$ be a locally free sheaf of rank 1 over Y. Show that $\deg \varphi^* \mathcal{G} = [\mathcal{R}(X) : \mathcal{R}(Y)] \cdot \deg \mathcal{G}$, where the field extension $\mathcal{R}(Y) \to \mathcal{R}(X)$ is induced by φ. (**Hint:** For the definition of the pull back $\varphi^* \mathcal{G}$ see Exercise 6.E.12. – Use the equation $\ell_A(B/Bf) = [Q(B) : Q(A)] \cdot \ell_A(A/Af)$ for a finite extension $A \to B$ of Noetherian integral domains of Krull dimension 1 and a non-zero element $f \in A$, see Exercise 6.C.1 (1).)

We can now prove the following important lemma for the degree $\deg \mathcal{G}$ which, in particular, proves that it is independent of the choice of the section $s \in \mathcal{G}_\eta$.

7.D.2. Lemma *For any locally free sheaf $\mathcal{G}$ of rank 1 over a reduced and irreducible (i. e. integral) projective curve Y with a finite morphism $\psi : Y \to \mathbb{P}_k^1$, we have:*

$$\deg \psi_* \mathcal{G} = \deg \mathcal{G} + \deg \psi_* \mathcal{O}_Y \quad and \quad \chi(\mathcal{G}) = \deg \mathcal{G} + \chi(\mathcal{O}_Y).$$

PROOF. The second equality follows from the first one and from 7.C.6 (using $\chi(\mathcal{G}) = \chi(\psi_* \mathcal{G})$ and $\operatorname{rank} \psi_* \mathcal{G} = \operatorname{rank} \psi_* \mathcal{O}_Y$):

$$\chi(\mathcal{G}) = \chi(\psi_* \mathcal{G}) = \deg \psi_* \mathcal{G} + \operatorname{rank} \psi_* \mathcal{G} = \deg \mathcal{G} + \deg \psi_* \mathcal{O}_Y + \operatorname{rank} \psi_* \mathcal{O}_Y$$
$$= \deg \mathcal{G} + \chi(\psi_* \mathcal{O}_Y) = \deg \mathcal{G} + \chi(\mathcal{O}_Y).$$

In order to prove the first equality we write (with the notations as above)

$$\deg \mathcal{G} := \sum_{x \in (\mathbb{P}_k^1)_0} \sum_{y \in \psi^{-1}(x)} \delta_y(\mathcal{G}).$$

Let $r := \operatorname{rank} \psi_* \mathcal{G} = \operatorname{rank} \psi_* \mathcal{O}_Y = [\mathcal{O}_{Y,\eta} : \mathcal{O}_{\mathbb{P}_k^1, \xi}] = [\mathcal{R}(Y) : \mathcal{R}(\mathbb{P}_k^1)]$, where η and ξ are the generic points of Y and $\mathbb{P}_k^1$ respectively. Since we have already proved above that for $\mathbb{P}_k^1$ both definitions of the degree coincide, it suffices to show the equality

$$\delta_x(\Lambda^r(\psi_* \mathcal{G})) = \left(\sum_{y \in \psi^{-1}(x)} \delta_y(\mathcal{G}) \right) + \delta_x(\Lambda^r(\psi_* \mathcal{O}_Y)).$$

for every closed point $x \in \mathbb{P}_k^1$.

Let $U \subseteq \mathbb{P}_k^1$ be a non-empty open affine subset with $x \in U$, $A := \Gamma(U, \mathbb{P}_k^1)$ and $\mathfrak{p} = \mathfrak{p}_x \in \operatorname{Spm} A \subseteq \operatorname{Spec} A$. Then $V := \psi^{-1}(U)$ is an open affine subset of Y and $B := \Gamma(V, \mathcal{O}_Y)$ is a finite torsion-free (i. e. projective) algebra of rank r over the (Dedekind) domain A. In the same way, $N := \Gamma(V, \mathcal{G})$ is a projective module of rank one over B and a projective module of rank r over A.[6] We have $\mathcal{O}_\xi = Q(A), \mathcal{O}_\eta = Q(B) = Q(A) \otimes_A B$ and $\mathcal{G}_\eta = Q(B) \otimes_B N = Q(A) \otimes_A N$. Let $s = n/g \in \mathcal{G}_\eta$, $n \in N$, $g \in A$, $g \neq 0$, be the given rational section of $\mathcal{G}$

---

[6] We may choose for U one of the two canonical affine open subsets of $\mathbb{P}_k^1$. Then A is a principal ideal domain and B and N are finite free A-modules of rank r. See also Exercise 7.B.4.

for computing $\deg \mathcal{G}$. Choose an $\mathcal{O}_\xi$-basis $m_1, \ldots, m_r$ of $\mathcal{O}_\eta = Q(B)$. Then $m_\rho n/g$, $\rho = 1, \ldots, r$, is an $\mathcal{O}_\xi$-basis of $\mathcal{G}_\eta$, i.e. an $\mathcal{O}_\xi$-basis of $(\psi_* \mathcal{G})_\xi$ and $m_1 \wedge \cdots \wedge m_r$ and $(m_1 n/g) \wedge \cdots \wedge (m_r n/g) = (m_1 n \wedge \cdots \wedge m_r n)/g^r$ are $\mathcal{O}_\xi$-bases of $(\Lambda^r \psi_* \mathcal{O}_Y)_\xi$ and $(\Lambda^r \psi_* \mathcal{G})_\xi$, respectively. We use these rational sections to compute the degrees of the locally free sheaves $\Lambda^r(\psi_* \mathcal{O}_Y)$ and $\Lambda^r(\psi_* \mathcal{G})$ of rank one over $\mathbb{P}^1_k$, respectively.

Let $N_\mathfrak{p} = B_\mathfrak{p} m s$, $m = b/f$, $b \in B$, $f \in A \setminus \{0\}$, and let $b_\sigma := \sum_{\rho=1}^r a_{\rho\sigma} m_\rho / h$, $\sigma = 1, \ldots, r$, $a_{\rho\sigma}, h \in A$, be an $A_\mathfrak{p}$-basis of $B_\mathfrak{p}$. Then we have $b_1 \wedge \cdots \wedge b_r = \Delta \cdot (m_1 \wedge \cdots \wedge m_r)/h^r$ and $(b_1 m s) \wedge \cdots \wedge (b_r m s) = \Delta (m_1 b n) \wedge \cdots \wedge (m_r b n)/h^r f^r g^r = \Delta \cdot N b \cdot (m_1 n \wedge \cdots \wedge m_r n)/h^r f^r g^r$, $\Delta := \mathrm{Det}(a_{\rho\sigma})$, and $N : Q(B) \to Q(A)$ denotes the norm function. Now, by definition, using the notations $d := [\kappa(x) : k]$ and $\ell(-) := \ell_{A_\mathfrak{p}}(-)$, we get:

$$\delta_x(\Lambda^r(\psi_* \mathcal{G}), (m_1 n \wedge \cdots \wedge m_r n)/g^r)$$
$$= d\big(\ell(A_\mathfrak{p}/A_\mathfrak{p} h^r) + \ell(A_\mathfrak{p}/A_\mathfrak{p} f^r) - \ell(A_\mathfrak{p}/A_\mathfrak{p} \Delta) - \ell(A_\mathfrak{p}/A_\mathfrak{p} N b)\big),$$

$$\delta_x(\Lambda^r(\psi_* \mathcal{O}_Y), m_1 \wedge \cdots \wedge m_r) = d\big(\ell(A_\mathfrak{p}/A_\mathfrak{p} h^r) - \ell(A_\mathfrak{p}/A_\mathfrak{p} \Delta)\big),$$

$$\sum_{y \in \psi^{-1}(x)} \delta_y(\mathcal{G}, s) = d\big(\ell(B_\mathfrak{p}/B_\mathfrak{p} f) - \ell(B_\mathfrak{p}/B_\mathfrak{p} b)\big) = d\big(\ell(A_\mathfrak{p}/A_\mathfrak{p} f^r) - \ell(A_\mathfrak{p}/A_\mathfrak{p} N b)\big).$$

The first equality in the last line is a consequence of the Chinese remainder theorem and the second one uses $\ell(B_\mathfrak{p}/B_\mathfrak{p} b) = \ell(A_\mathfrak{p}/A_\mathfrak{p} N b)$, which is a special case of the well-known and very important length formula already mentioned in the proof of Lemma 7.C.7. Now, addition yields the required equality.   •

To define the degree in the general case, let $\mathcal{F}$ be an arbitrary coherent $\mathcal{O}_X$-module over the projective curve X with a finite morphism $\varphi : X \to \mathbb{P}^1_k$. Let X_i denote the irreducible components of X with their reduced structures, furthermore let $\xi_i \in X_i$ be their generic points and $\varphi_i : X_i \to \mathbb{P}^1_k$ the compositions of the closed embeddings $X_i \to X$ with φ (which are also finite), $i \in I$.

We represent $X = \mathrm{Proj}\, R$, where R is a standardly graded k-algebra with $R_0 = k$ and whose minimal prime ideals $\mathfrak{p}_i$ are all of dimension 2 (see 7.B). They correspond to the generic points ξ_i, $i \in I$, of X and their closures $X_i := \overline{\{\xi_i\}}$ are the irreducible components of X. By 7.A.2 there exists a finite graded R-module M with $\mathcal{F} \cong \widetilde{M}$. The module M has a composition series $0 = M_0 \subseteq M_1 \subseteq \cdots \subseteq M_r = M$ such that $M_j/M_{j-1} \cong (R/\mathfrak{q}_j)(-m_j)$ with homogeneous prime ideals $\mathfrak{q}_j$ and $m_j \in \mathbb{Z}$, $j = 1, \ldots, r$. From this we get the chain $0 = \mathcal{F}_0 \subseteq \mathcal{F}_1 \subseteq \cdots \subseteq \mathcal{F}_r = \mathcal{F}$, $\mathcal{F}_j := \widetilde{M}_j$, and its quotients $\mathcal{F}_j/\mathcal{F}_{j-1} \cong \widetilde{(R/\mathfrak{q}_j)}(-m_j)$, $j = 1, \ldots, r$. For a fixed $i \in I$, the number of $j \in \{1, \ldots r\}$ with $\mathfrak{q}_j = \mathfrak{p}_i$ is an invariant of $\mathcal{F}$, it coincides with the lengths $\ell_{\mathcal{O}_{\xi_i}}(\mathcal{F}_{\xi_i})$ and we put

$$\mathrm{rank}_{X_i} \mathcal{F} := \ell_{\mathcal{O}_{\xi_i}}(\mathcal{F}_{\xi_i}).$$

For such a j the quotient $\mathcal{F}_j/\mathcal{F}_{j-1} \cong \mathcal{O}_{X_i}(-m_j)$ is a locally free sheaf of rank 1 over X_i and $\varphi_*(\mathcal{F}_j)/\varphi_*(\mathcal{F}_{j-1}) = \varphi_*(\mathcal{F}_j/\mathcal{F}_{j-1}) = (\varphi_i)_*(\mathcal{F}_j/\mathcal{F}_{j-1})$.

For a j with $\dim R/\mathfrak{q}_j = 1$ the quotient $\mathcal{F}_j/\mathcal{F}_{j-1}$ is supported at the closed point of X corresponding to $\mathfrak{q}_j$ (note $\mathcal{F}_j/\mathcal{F}_{j-1} = 0$ if $\mathfrak{q}_j = R_+$). Now, we define

$$\deg \mathcal{F} := \sum_{j=1}^r \deg(\mathcal{F}_j/\mathcal{F}_{j-1}) \,.$$

Then we get immediately the general Riemann–Roch formula for curves which also shows that $\deg \mathcal{F}$ is well defined and, in particular, independent of the representing module M and the choice of its composition series [7]):

7.D.3. Theorem of Riemann–Roch for Projective Curves *Let X be a projective algebraic curve over the field k with irreducible components X_i (equipped with their reduced structures), $i \in I$, and let $\mathcal{F}$ be a coherent $\mathcal{O}_X$-module. Then*

$$\chi(\mathcal{F}) = \mathrm{Dim}_k \Gamma(\mathcal{F}) - \mathrm{Dim}_k \Gamma(\mathcal{H}om_{\mathcal{O}_X}(\mathcal{F}, \omega_X)) = \deg \mathcal{F} + \sum_{i \in I} (\mathrm{rank}_{X_i} \mathcal{F}) \cdot \chi(\mathcal{O}_{X_i}) \,.$$

PROOF. Using the above definitions and notations we have $\chi(\mathcal{F}) = \chi(\varphi_* \mathcal{F}) = \sum_{j=1}^r \chi(\varphi_*(\mathcal{F}_j/\mathcal{F}_{j-1})) = \sum_{j=1}^r \chi(\mathcal{F}_j/\mathcal{F}_{j-1})$. If $\mathcal{F}_j/\mathcal{F}_{j-1}$ is of finite support, then $\chi(\mathcal{F}_j/\mathcal{F}_{j-1}) = \deg(\mathcal{F}_j/\mathcal{F}_{j-1})$. If $\mathcal{F}_j/\mathcal{F}_{j-1}$ is locally free of rank 1 over X_i, then $\chi(\mathcal{F}_j/\mathcal{F}_{j-1}) = \deg(\mathcal{F}_j/\mathcal{F}_{j-1}) + \chi(\mathcal{O}_{X_i})$ by 7.D.2. Now, use the definition of $\deg \mathcal{F}$. •

An immediate corollary of 7.D.3 is:

7.D.4. Corollary (R i e m a n n ' s   i n e q u a l i t y) *In the situation of 7.D.3 we have*

$$\mathrm{Dim}_k \Gamma(\mathcal{F}) \geq \deg \mathcal{F} + \sum_{i \in I} (\mathrm{rank}_{X_i} \mathcal{F}) \, \chi(\mathcal{O}_{X_i}) \,.$$

Observe that the term $\sum_{i \in I} (\mathrm{rank}_{X_i} \mathcal{F}) \, \chi(\mathcal{O}_{X_i})$ is a constant for all coherent $\mathcal{O}_X$-modules $\mathcal{F}$ with fixed ranks $\mathrm{rank}_{X_i} \mathcal{F}, i \in I$.

The Riemann–Roch formula in 7.D.3 implies the additivity of the degree function. If $0 \to \mathcal{F}' \to \mathcal{F} \to \mathcal{F}'' \to 0$ *is an exact sequence of coherent $\mathcal{O}_X$-modules, then*

$$\deg \mathcal{F} = \deg \mathcal{F}' + \deg \mathcal{F}'' \,,$$

which is already a consequence of the definition of the degree function.

7.D.5. Exercise (R e d u c e d   d e g r e e) For the degree of a coherent sheaf $\mathcal{F}$ on a projective algebraic curve, in the literature one may find other definitions which do not coincide necessarily with our conventions. For instance: Let X be a projective *Cohen–Macaulay* curve. A coherent $\mathcal{O}_X$-module $\mathcal{F}$ is called a m o d u l e   o f   r a n k $n = \mathrm{rank}_{\mathcal{O}_X} \mathcal{F}, n \in \mathbb{N}$, if for every generic point ξ of X the stalk $\mathcal{F}_\xi$ is a free $\mathcal{O}_\xi$-module of rank n, i.e. $\mathcal{R}_X \otimes_{\mathcal{O}_X} \mathcal{F} \cong \mathcal{R}_X^n$. For such a module the r e d u c e d   d e g r e e $\mathrm{redeg} \,\mathcal{F}$ is defined as

$$\mathrm{redeg} \, \mathcal{F} := \deg \mathcal{F} - (\mathrm{rank}_{\mathcal{O}_X} \mathcal{F}) \deg \mathcal{O}_X \,.$$

It is as deg and rank an additive function for coherent $\mathcal{O}_X$-modules with rank.

If X is an integral projective curve, then $\deg \mathcal{O}_X = 0$ and hence $\mathrm{redeg} \, \mathcal{F} = \deg \mathcal{F}$ for all coherent $\mathcal{O}_X$-modules $\mathcal{F}$. In general, the reduced degree is normalized in such a way that

---

[7]) Observe that for deriving this independence we are not using the fact that ω_X is independent of the choice of the finite morphism $X \to \mathbb{P}^1_k$.

$\operatorname{redeg} \mathbb{O}_X = 0$. For a coherent module $\mathcal{F}$ with rank, prove the following Riemann–Roch formula:

$$\chi(\mathcal{F}) = \operatorname{redeg} \mathcal{F} + (\operatorname{rank}_{\mathbb{O}_X} \mathcal{F})\, \chi(\mathbb{O}_X)\,.$$

Every coherent Cohen–Macaulay $\mathbb{O}_X$-module of rank $n > 0$ [8]) is isomorphic to an $\mathbb{O}_X$-submodule of the (quasi-coherent) $\mathbb{O}_X$-module $\mathcal{R}_X^n$. Let $\mathcal{F} \subseteq \mathcal{R}_X^n$ be such a submodule. Then $\mathcal{F}_x = \mathbb{O}_x^n$ for all points $x \in X$ up to a finite set S of closed points of X. For a point $s \in S$ one can write $\mathcal{F}_s = \mathcal{F}_s'/g_s$ with $\mathcal{F}_s' \subseteq \mathbb{O}_s^n$ and a non-zero divisor $g_s \in \mathbb{O}_s$. Show that

$$\operatorname{redeg} \mathcal{F} = \sum_{s \in S} \left(n \cdot \operatorname{Dim}_k(\mathbb{O}_s/g_s\mathbb{O}_s) - \operatorname{Dim}_k(\mathbb{O}_s^n/\mathcal{F}_s') \right)\,.$$

(**Hint:** Let $s \in S$ be fixed. There exist coherent $\mathbb{O}_X$-modules $\mathcal{F}', \widetilde{\mathcal{F}} \subseteq \mathcal{R}_X^n$ with $\mathcal{F}_x' = \widetilde{\mathcal{F}}_x = \mathcal{F}_x$ for all $x \in X, x \ne s$ and $\mathcal{F}_s' = g_s\mathcal{F}_s, \widetilde{\mathcal{F}}_s = \mathbb{O}_s^n$ and short exact sequences $0 \to \mathcal{F}' \to \widetilde{\mathcal{F}} \to \widetilde{\mathcal{F}}/\mathcal{F}' \to 0$ and $0 \to \mathcal{F}' \to \mathcal{F} \to \mathcal{F}/\mathcal{F}' \to 0$. Now use induction on the cardinality of S and Exercise 6.C.1 (1).) If $\mathcal{F}$ is even locally free of rank n then $\operatorname{redeg} \mathcal{F} = \operatorname{redeg} \Lambda^n \mathcal{F}$. (**Hint:** To prove this use the general length formula mentioned in the proof of 7.C.7 for local Cohen–Macaulay rings of dimension 1.) – Note that for a finite set $S \subseteq X_0$ (= set of closed points of X) every collection $(\mathcal{F}_s)_{s \in S}$ of finite $\mathbb{O}_s$-submodules $\mathcal{F}_s \subseteq \mathcal{R}_s^n$ of rank n defines a coherent Cohen–Macaulay subsheaf $\mathcal{F} \subseteq \mathcal{R}_X^n$ of rank n (with $\mathcal{F}_x = \mathbb{O}_x^n$ for $x \notin S$). Describe the sections of $\mathcal{R}_X^n$ which are also sections of such a sheaf $\mathcal{F}$.

With the notations as in the (last) Exercise 7.D.5, if X is a regular (= normal) and connected curve and if $n = 1$, then $\mathcal{F}_s = \pi_s^{-\nu_s}\mathbb{O}_s$ with $\nu_s \in \mathbb{Z}$, where π_s is a uniformizing parameter of the discrete valuation ring $\mathbb{O}_s, s \in S$. In this case, $\mathcal{F} \subseteq \mathcal{R}_X$ is called the locally free sheaf of rank 1 (or the invertible sheaf) $\mathcal{L}(D)$ belonging to the d i v i s o r $D := \sum_{s \in S} \nu_s s$. [9]) One has

$$\operatorname{redeg} \mathcal{L}(D) = \deg \mathcal{L}(D) = \sum_{s \in S} \nu_s [\kappa(s) : k]\,.$$

This degree is also called the d e g r e e o f t h e d i v i s o r D and is denoted by $\deg D$. The non-zero sections of $\mathcal{L}(D) \subseteq \mathcal{R}(X)$ are the non-zero rational functions $f \in \mathcal{R}(X)$ with $\operatorname{div} f + D \ge 0$, where $\operatorname{div} f := \sum_{Q \in X_0} \nu_Q(f)Q$, here $\nu_Q(f)$ is the order of f in $\mathbb{O}_Q$. Note that $\deg \operatorname{div} f = 0$ because of $\deg \operatorname{div} f = \deg f\mathbb{O}_X = \deg \mathbb{O}_X = 0$.

The sum $D+E = \sum_{x \in X_0}(\nu_x + \mu_x)x$ of two divisors $D = \sum_x \nu_x x$, $E = \sum_x \mu_x x$ has degree $\deg(D + E) = \deg D + \deg E$ and corresponds to the tensor product of the invertible sheaves: $\mathcal{L}(D + E) = \mathcal{L}(D) \otimes_{\mathbb{O}_X} \mathcal{L}(E)$. $\mathcal{L}(-D)$ is the dual $\mathcal{L}(D)^* = \mathcal{H}om_{\mathbb{O}_X}(\mathcal{L}(D), \mathbb{O}_X)$ of $\mathcal{L}(D)$. All the divisors on X form the g r o u p $\operatorname{Div} X$ of (W e i l –) d i v i s o r s on X. The p r i n c i p a l d i v i s o r s $\operatorname{div} f$, $f \in \mathcal{R}(X)^\times$, form a subgroup $\operatorname{PDiv} X$ of $\operatorname{Div} X$. The factor group

$$\operatorname{Cl} X := \operatorname{Div} X/\operatorname{PDiv} X$$

is called the d i v i s o r c l a s s g r o u p of X. Show that the mapping $D \mapsto \mathcal{L}(D)$ induces an isomorphism

$$\operatorname{Cl} X \xrightarrow{\sim} \operatorname{Pic} X$$

---

[8]) With our convention in 7.B.1 a Cohen–Macaulay $\mathbb{O}_X$-module of rank 0 is the zero module.

[9]) Observe the differing signs of the ν_s in the definitions of $\mathcal{L}(D)$ and D.

from the group Cl X onto the Picard group Pic X of (the isomorphism classes of) invertible sheaves on X (with the tensor product as group multiplication). The subgroup $\text{Pic}^0 X$ of invertible sheaves of degree 0 corresponds to the special divisor class group

$$\text{Cl}^0 X = \text{Div}^0 X / \text{PDiv}\, X$$

where $\text{Div}^0 X \subseteq \text{Div}\, X$ is the subgroup of divisors of degree 0, i. e. the kernel of the group homomorphism deg $: \text{Div}\, X \to \mathbb{Z}$. Divisors $D, D' \in \text{Div}\, X$ with $\mathcal{L}(D) \cong \mathcal{L}(D')$ are called equivalent. Two divisors D and D' are equivalent if and only if their difference $D - D'$ is a principal divisor div $f \in \text{PDiv}\, X$.

7.D.6. Exercise Let $\mathcal{F}$ be a coherent Cohen–Macaulay $\mathcal{O}_Y$-module of rank 1 on an integral projective curve Y. Show that $\Gamma(\mathcal{F}) = 0$ if $\deg \mathcal{F} < 0$. (**Hint:** A non-trivial section s of $\mathcal{F}$ defines an exact sequence $0 \to \mathcal{O}_Y \to \mathcal{F} \to \mathcal{F}/\mathcal{O}_Y s \to 0$, hence $\deg \mathcal{F} = \deg (\mathcal{F}/\mathcal{O}_Y s) = \sum_{y \in Y_0} \text{Dim}_k(\mathcal{F}_y/\mathcal{O}_y s_y) \geq 0$, $Y_0 :=$ set of closed points of Y.) The proof even shows $\deg \mathcal{F} > 0$, if $\mathcal{F}$ has a non-trivial section and if $\mathcal{F} \ncong \mathcal{O}_Y$. Give examples of projective Cohen–Macaulay curves X with $\deg \mathcal{O}_X < 0$ and $\deg \mathcal{O}_X > 0$ respectively.

We want to interpret the term $\text{Dim}_k \Gamma(\mathcal{H}om_{\mathcal{O}_X}(\mathcal{F}, \omega_X))$ ($= \text{Dim}_k H^1(X, \mathcal{F})^*$, see Footnote 3) in the Riemann–Roch formula 7.D.3, in particular, for $\mathcal{F} = \mathcal{O}_X$, i. e. the genus $\text{Dim}_k \Gamma(\omega_X)$ of X. We do this in the next section. Here we add already some general remarks on the canonical (or dualizing) module ω_X of a projective curve X over k. By definition

$$\omega_{\mathbb{P}^1_k} \cong \mathcal{O}_{\mathbb{P}^1_k}(-2) \cong \Omega_{\mathbb{P}^1_k | k} \quad \text{and} \quad \omega_X \cong \varphi^! \omega_{\mathbb{P}^1_k}$$

for a finite morphism $\varphi : X \to \mathbb{P}^1_k$. From the transitivity formula $(\varphi\psi)^! \omega_{\mathbb{P}^1_k} \cong \psi^!(\varphi^! \omega_{\mathbb{P}^1_k})$ for a composition $X \to Y \to \mathbb{P}^1_k$ of finite morphisms we get the following more general result:

7.D.7. Proposition *For an arbitrary finite morphism $\psi : X \to Y$ of projective algebraic curves X, Y over a field k and a coherent $\mathcal{O}_X$-module $\mathcal{F}$ we have $\omega_X \cong \psi^! \omega_Y$, furthermore $\Gamma(X, \mathcal{F}) = \Gamma(Y, \psi_* \mathcal{F})$, $\Gamma(X, \mathcal{H}om_{\mathcal{O}_X}(\mathcal{F}, \omega_X)) \cong \Gamma(Y, \mathcal{H}om_{\mathcal{O}_Y}(\psi_* \mathcal{F}, \omega_Y))$ and, in particular, $\chi(\mathcal{F}) = \chi(\psi_* \mathcal{F})$.*

Proof. The equality $\Gamma(X, \mathcal{F}) = \Gamma(Y, \psi_* \mathcal{F})$ holds by definition of the direct image for every sheaf $\mathcal{F}$ on X. Because of $\omega_X \cong \psi^! \omega_Y$, we have $\psi_* \mathcal{H}om_{\mathcal{O}_X}(\mathcal{F}, \omega_X) \cong \psi_* \mathcal{H}om_{\mathcal{O}_X}(\mathcal{F}, \psi^! \omega_Y) \cong \mathcal{H}om_{\mathcal{O}_Y}(\psi_* \mathcal{F}, \omega_Y)$.   •

From $\varphi_* \omega_X = \mathcal{H}om_{\mathcal{O}_{\mathbb{P}^1_k}}(\varphi_* \mathcal{O}_X, \omega_{\mathbb{P}^1_k})$ for a finite morphism $\varphi : X \to \mathbb{P}^1_k$ we will deduce the following result:

7.D.8. Proposition *The canonical module ω_X of a projective algebraic curve X over a field k is a Cohen–Macaulay $\mathcal{O}_X$-module which coincides with $\iota_*(\omega_{X_{\text{CM}}})$, where $\iota : X_{\text{CM}} \hookrightarrow X$ is the canonical closed imbedding (see 7.B.1). Furthermore, $\mathcal{H}om_{\mathcal{O}_X}(\omega_X, \omega_X) = \mathcal{H}om_{\mathcal{O}_{X_{\text{CM}}}}(\omega_{X_{\text{CM}}}, \omega_{X_{\text{CM}}}) = \mathcal{O}_{X_{\text{CM}}}$.*

PROOF. Let $\varphi : X \to \mathbb{P}_k^1$ be a finite morphism and let $V \subseteq \mathbb{P}_k^1$ be an affine open subset $\neq \emptyset$ with $A := \Gamma(V, \mathbb{P}_k^1)$ and $B := \Gamma(\varphi^{-1}(V), \mathcal{O}_X)$, $B' := B_{CM} = \Gamma((\varphi\iota)^{-1}(V), \mathcal{O}_{X_{CM}})$. Since $M := \Gamma(V, \mathcal{O}_{\mathbb{P}_k^1(-2)})$ is a projective A-module of rank 1 the equalities $\Gamma(\varphi^{-1}(V), \omega_X) = \mathrm{Hom}_A(B, M) = \mathrm{Hom}_A(B', M) = \Gamma((\varphi\iota)^{-1}(V), \omega_{X_{CM}})$ hold and $\Gamma(\varphi^{-1}(V), \omega_X) = \Gamma((\varphi\iota)^{-1}(V), \omega_{X_{CM}})$ are Cohen–Macaulay modules over B and B'. Furthermore,

$$\mathrm{Hom}_{B'}(\mathrm{Hom}_A(B', M), \mathrm{Hom}_A(B', M)) = \mathrm{Hom}_A(\mathrm{Hom}_A(B', M), M) = B'$$

since B' is also a finite projective (i. e. a finite torsion-free) A-module. This proves the last equality in 7.D.8. •

Because of 7.D.8 we will identify ω_X and $\omega_{X_{CM}}$. For the computation of ω_X one can usually assume that $X = X_{CM}$ is a Cohen–Macaulay curve.

7.D.9. Exercise Let $\mathcal{F}$ be a coherent $\mathcal{O}_X$-module on the projective curve X over the field k.
(1) Show that $\mathcal{H}om_{\mathcal{O}_X}(\mathcal{H}om_{\mathcal{O}_X}(\mathcal{F}, \omega_X), \omega_X) = \mathcal{F}_{CM}$. **(Hint:** In the proof of 7.D.8 the case $\mathcal{F} = \mathcal{O}_X$ is considered. – For the Cohen–Macaulayfication $\mathcal{F}_{CM}$ see 7.B.1.)
(2) Show that $\mathrm{rank}_{X_i} \mathcal{F} = \mathrm{rank}_{X_i} \mathcal{H}om_{\mathcal{O}_X}(\mathcal{F}, \omega_X)$ for all irreducible components X_i of X. **(Hint:** Note that for a finite morphism $X \to Y := \mathbb{P}_k^1$, we have $(\mathcal{H}om_{\mathcal{O}_X}(\mathcal{F}, \omega_X))_{\xi_i} = \mathrm{Hom}_{\mathcal{O}_{Y,\eta}}(\mathcal{F}_{\xi_i}, (\omega_Y)_\eta)$, and hence $\mathrm{Dim}_{\mathcal{O}_{Y,\eta}}(\mathcal{H}om_{\mathcal{O}_X}(\mathcal{F}, \omega_X))_{\xi_i} = \mathrm{Dim}_{\mathcal{O}_{Y,\eta}} \mathcal{F}_{\xi_i}$ for every generic point ξ_i of X and the generic point η of Y.)
(3) Assume that $\mathcal{F}$ is a Cohen–Macaulay module. Then $\chi(\mathcal{F}) = -\chi(\mathcal{H}om_{\mathcal{O}_X}(\mathcal{F}, \omega_X))$ and $2\chi(\mathcal{F}) = \deg \mathcal{F} - \deg \mathcal{H}om_{\mathcal{O}_X}(\mathcal{F}, \omega_X)$, in particular, $2\chi(\mathcal{O}_X) = \deg \mathcal{O}_X - \deg \omega_X$ if X is a Cohen- Macaulay curve. **(Hint:** Use parts (1), (2) and Riemann–Roch Theorem 7.D.3.)

7.E. Genus of a Projective Curve

We still have to prove that the canonical module ω_X is uniquely determined, i. e. up to an isomorphism independent of the chosen finite morphism $X \to \mathbb{P}_k^1$. This is not so easy and we postpone the proof, cf. Theorem 7.E.21. Rather easy to prove is *the invariance of* $\mathrm{Dim}_k \Gamma(\mathcal{H}om_{\mathcal{O}_X}(\mathcal{F}, \omega_X))$ *for a coherent* $\mathcal{O}_X$-module $\mathcal{F}$ *so that* $\chi(\mathcal{F})$ *is indeed well defined*. By 7.D.3 it suffices to show that $\chi(\mathcal{O}_Y)$ is well defined for an *integral* projective curve Y over k. Let ω, ω' be canonical modules for Y with respect to finite morphisms $Y \to \mathbb{P}_k^1$. These are coherent Cohen–Macaulay $\mathcal{O}_Y$-modules of rank 1. Let $\mathcal{G}$ be any locally free sheaf of rank 1 on Y with $\deg \mathcal{G} > 0$. (Such sheaves exist!) Then $\mathcal{H}om_{\mathcal{O}_Y}(\mathcal{G}, \omega) \cong \mathcal{G}^* \otimes_{\mathcal{O}_Y} \omega$ and $\deg(\mathcal{H}om_{\mathcal{O}_Y}(\mathcal{G}, \omega)) = \deg \mathcal{G}^* + \deg \omega = -\deg \mathcal{G} + \deg \omega$ and similarly for ω'. Therefore $\Gamma(\mathcal{H}om_{\mathcal{O}_Y}(\mathcal{G}^{\otimes n}, \omega)) = 0 = \Gamma(\mathcal{H}om_{\mathcal{O}_Y}(\mathcal{G}^{\otimes n}, \omega'))$ for large n by Exercise 7.D.1 (1) and Exercise 7.D.6. It follows by 7.D.3 that, for large n,

$$\chi(\mathcal{O}_Y) = \mathrm{Dim}_k \Gamma(\mathcal{G}^{\otimes n}) - n \deg \mathcal{G}$$

for both choices ω and ω' of ω_Y.
In the formula $\chi(\mathcal{O}_X) = \mathrm{Dim}_k \Gamma(\mathcal{O}_X) - \mathrm{Dim}_k \Gamma(\omega_X)$ the space $\Gamma(\mathcal{O}_X)$ and its k-dimension are usually well understood, see for instance Example 7.A.6. More mysterious is the space $\Gamma(\omega_X)$.

7.E.1. Definition Let X be a projective algebraic curve over the field k. Then

$$g(X) := \text{Dim}_k \, \Gamma(X, \omega_X)$$

is called the (a r i t h m e t i c a l) g e n u s of X. – If L is a finitely generated field extension of k of transcendence degree 1 and if X is the Riemann surface of L, i. e. the (up to isomorphism) uniquely determined normal projective algebraic curve X over k with $\mathcal{R}(X) \cong L$ (see Exercise 7.B.5 (3)), then the genus $g(X)$ of X is also called the g e n u s o f t h e f u n c t i o n f i e l d L.

By definition

$$\chi(\mathcal{O}_X) = \text{Dim}_k \, \Gamma(\mathcal{O}_X) - g(X)$$

and the genus $g(X) = \text{Dim}_k \, \Gamma(\mathcal{O}_X) - \chi(\mathcal{O}_X)$ is always non-negative. If X is integral, $\Gamma(\mathcal{O}_X)$ is a finite field extension k' of k and X may also be considered as a projective algebraic curve over k' (or any subfield between k and k'). To distinguish both cases we write $X \mid k$ and $X \mid k'$ for X considered as a curve over k or over k' respectively. For every coherent $\mathcal{O}_X$-module $\mathcal{F}$ the equation $\text{Dim}_k \, \Gamma(\mathcal{F}) = [k' : k] \cdot \text{Dim}_{k'} \, \Gamma(\mathcal{F})$ holds and ω_X is also a dualizing sheaf of $X \mid k'$. This follows from the fact that any finite k-morphism $\varphi : X \to \mathbb{P}^1_k$ defines a commutative diagram

$$
\begin{array}{ccc}
X & \xrightarrow{\ \varphi\ } & \mathbb{P}^1_k \\
 & {\scriptstyle \tilde{\varphi}}\searrow & \downarrow{\scriptstyle \alpha} \\
 & & \mathbb{P}^1_{k'} = (\mathbb{P}^1_k)_{(k')}
\end{array}
$$

with a finite k'-morphism $\tilde{\varphi} : X \to \mathbb{P}^1_{k'}$ and $\omega_{X\mid k} \cong \varphi^!\omega_{\mathbb{P}^1_k} \cong \tilde{\varphi}^!(\alpha^!\omega_{\mathbb{P}^1_k}) \cong \tilde{\varphi}^!(\omega_{\mathbb{P}^1_{k'}}) \cong \omega_{X\mid k'}$. (The reader should check the isomorphism $\alpha^!\omega_{\mathbb{P}^1_k} \cong \omega_{\mathbb{P}^1_{k'}}$.) We get

$$g(X|k) = [k' : k] \cdot g(X|k') \quad \text{and} \quad \chi(\mathcal{O}_{X|k}) = [k' : k] \cdot \chi(\mathcal{O}_{X|k'}).$$

For an integral projective curve X, the integer

$$g_{\text{red}}(X) := g(X|\Gamma(\mathcal{O}_X)) = g(X)/\text{Dim}_k \, \Gamma(\mathcal{O}_X)$$

is sometimes called the r e d u c e d g e n u s of X. One may (and should) always assume that for such curves $\Gamma(\mathcal{O}_X)$ coincides with the base field k so that $g(X)$ and $g_{\text{red}}(X)$ coincide. Then 7.D.3 has the following classical formulation:

7.E.2. Theorem of Riemann–Roch for Integral Projective Curves *Let X be an integral projective algebraic curve over the field k with $\mathcal{O}(X) = \Gamma(\mathcal{O}_X) = k$ and let $\mathcal{F}$ be a coherent $\mathcal{O}_X$-module. Then*

$$\deg \omega_X = -2\chi(\mathcal{O}_X) = 2g(X) - 2$$

and

$$\chi(\mathcal{F}) = \text{Dim}_k \, \Gamma(\mathcal{F}) - \text{Dim}_k \, \Gamma(\mathcal{H}om_{\mathcal{O}_X}(\mathcal{F}, \omega_X)) = \deg \mathcal{F} + (\text{rank}\,\mathcal{F})(1 - g(X)).$$

The formula for $\deg \omega_X$ uses Exercise 7.D.9 (3). Note that the assumption $\Gamma(\mathcal{O}_X) = k$ is satisfied if X has a k-valued point, i.e. a (closed) point x with $\kappa(x) = k$. The best known case is a normal irreducible (= connected) projective curve X with $\Gamma(\mathcal{O}_X) = k$. For the classification of these curves cf. Exercise 7.B.5 (3), (4). Then

$$\chi(\mathcal{L}(D)) = \deg D + 1 - g(X)$$

for every divisor D on X, cf. Section 7.D just after Exercise 7.D.5. In this case ω_X is also locally free of rank 1. Every divisor K on X with $\mathcal{L}(K) \cong \omega_X$ is called a c a n o n i c a l   d i v i s o r. With this notation

$$\mathcal{H}om_{\mathcal{O}_X}(\mathcal{L}(D), \omega_X) \cong \mathcal{L}(D)^* \otimes \mathcal{L}(K) \cong \mathcal{L}(-D) \otimes \mathcal{L}(K) = \mathcal{L}(-D + K).$$

Very often one writes $\operatorname{Dim}_k D$ for $\operatorname{Dim}_k \mathcal{L}(D)$. Note that $\mathcal{L}(D) \cong \mathcal{L}(D')$ for equivalent divisors D and $D' = D + \operatorname{div} f$, $f \in \mathcal{R}(X)^\times$, hence $\operatorname{Dim}_k D = \operatorname{Dim}_k D'$ for such divisors. Now the Riemann–Roch formula reads:

$$\chi(D) := \chi(\mathcal{L}(D)) = \operatorname{Dim}_k D - \operatorname{Dim}_k(-D + K) = \deg D + 1 - g(X).$$

For any field extension $k \subseteq k'$ and for any projective algebraic curve X over k one has $\Gamma(X_{(k')}, \mathcal{F}_{(k')}) = k' \otimes_k \Gamma(X, \mathcal{F})$ by 7.A.7, where $\mathcal{F}$ is an arbitrary coherent $\mathcal{O}_X$-module. This implies that

$$g(X_{(k')}) = g(X) \qquad \text{and} \qquad \chi(\mathcal{F}_{(k')}) = \chi(\mathcal{F})$$

because of $\omega_{X_{(k')}} \cong (\omega_X)_{(k')}$. To prove this isomorphism consider a finite morphism $\varphi : X \to \mathbb{P}^1_k$. Its extension $\varphi_{(k')} : X_{(k')} \to (\mathbb{P}^1_k)_{(k')}$ is also finite and $\varphi^!_{(k')}(\omega_{\mathbb{P}^1_{k'}}) = \varphi^!_{(k')}((\omega_{\mathbb{P}^1_k})_{(k')}) = (\varphi^!(\omega_{\mathbb{P}^1_k}))_{(k')} \cong (\omega_X)_{(k')}$. Furthermore, we have

$$g(X) = g(X_{\mathrm{CM}})$$

for every projective curve X, since $\omega_X = \omega_{X_{\mathrm{CM}}}$ by 7.D.8.

7.E.3. Example (C o m p l e t e   i n t e r s e c t i o n s   a n d   P l ü c k e r   f o r m u l a) Because of $\omega_{\mathbb{P}^1_k} = \mathcal{O}_{\mathbb{P}^1_k}(-2)$ and Theorem 7.C.4, we have

$$g(\mathbb{P}^1_k) = \operatorname{Dim}_k \Gamma(\mathcal{O}_{\mathbb{P}^1_k}(-2)) = 0.$$

More generally, let $R := k[X_0, \ldots, X_n]/(F_2, \ldots, F_n)$ with the standardly graded polynomial algebra $Q := k[X_0, \ldots, X_n]$ and a strongly regular sequence $F_2, \ldots, F_n \in Q$ of homogeneous elements of positive degrees $\delta_2, \ldots, \delta_n$, cf. Example 7.A.6. Then $X :=$ Proj $R \subseteq \mathbb{P}^n_k$ is a curve which is a complete intersection in $\mathbb{P}^n_k$. To compute the genus of X, we may assume that the field k is infinite (by extending k if necessary). Then R contains a regular sequence $f_0, f_1 \in R_1$ represented by linear polynomials $F_0, F_1 \in Q_1$. The k-algebra homomorphism $P := k[T_0, T_1] \to R$, $T_1 \mapsto f_i$, $i = 0, 1$, is finite and by Lemma 7.C.2,

$$R = \bigoplus_{j=1}^r P y_j \cong \bigoplus_{j=1}^r P(-e_j)$$

is a free P-module with a P-basis y_j of homogeneous elements of degrees e_j, $j = 1, \ldots, r$. Let $c_m = \operatorname{Dim}_k \overline{R}_m$, $\overline{R} := R/(f_0, f_1)$, be the number of basis elements y_j of degree $m \in \mathbb{N}$. The series

$$\mathcal{P} = \mathcal{P}_{\overline{R}} = \sum_{m \in \mathbb{N}} c_m Z^m$$

is called the P o i n c a r é   s e r i e s of $\overline{R}$ (it is indeed a polynomial since $\overline{R}$ is finite over k) and can easily be computed. Namely, if, in general,

$$\mathcal{P}_M := \sum_{m \in \mathbb{Z}} (\text{Dim}_k M_m) \, Z^m$$

is the Poincaré series of a finite graded Q-module M and if $x \in Q_d$ is a non-zero divisor for M of positive degree d , then the exact sequence

$$0 \to M(-d) \xrightarrow{\ x\ } M \to M/xM \to 0$$

gives the equation

$$\mathcal{P}_M = \mathcal{P}_{M(-d)} + \mathcal{P}_{M/xM} = Z^d \mathcal{P}_M + \mathcal{P}_{M/xM} \quad \text{i.e.} \quad \mathcal{P}_{M/xM} = (1 - Z^d) \mathcal{P}_M \, .$$

(See also the next exercise.) Applying this result several times we get

$$\mathcal{P}_Q = \frac{1}{(1 - Z)^{n+1}} \, , \quad \mathcal{P}_R = \frac{(1 - Z^{\delta_2}) \cdots (1 - Z^{\delta_n})}{(1 - Z)^{n+1}}$$

and

$$\mathcal{P}_{\overline{R}} = (1 - Z)^2 \mathcal{P}_R = \prod_{i=2}^{n} (1 + Z + \cdots + Z^{\delta_i - 1}) \, .$$

The inclusion $P = k[T_0, T_1] \hookrightarrow R$ from above yields a finite morphism $\varphi : X = \text{Proj}\, R \to \text{Proj}\, P = \mathbb{P}_k^1$ and (cf. Example 6.E.14)

$$\varphi_* \omega_X \cong \mathcal{H}om_{\mathcal{O}_{\mathbb{P}_k^1}} (\varphi_* \mathcal{O}_X, \mathcal{O}_{\mathbb{P}_k^1}(-2)) = \text{Hom}_P(R, P(-2))^{\sim}$$

$$\cong \bigoplus_{j=1}^{r} \text{Hom}_P(P(-e_j), P(-2))^{\sim} = \bigoplus_{j=1}^{r} \mathcal{O}_{\mathbb{P}_k^1}(e_j - 2) = \bigoplus_{m \in \mathbb{N}} \mathcal{O}_{\mathbb{P}_k^1}(m - 2)^{c_m}$$

and hence, by Theorem 7.C.4,

$$g(X) = \text{Dim}_k \Gamma(\omega_X) = \text{Dim}_k \Gamma(\varphi_* \omega_X) = \sum_{m \in \mathbb{N}} c_{m+1} \, m$$

$$= \mathcal{P}_{\overline{R}}'(1) - \mathcal{P}_{\overline{R}}(1) + 1 = 1 + \mathcal{P}_{\overline{R}}(1) \left(\mathcal{P}_{\overline{R}}'(1)/\mathcal{P}_{\overline{R}}(1) - 1 \right) \, .$$

7.E.4. Proposition *For a projective complete intersection curve $X = \text{Proj}\, R$, where $R := k[X_0, \ldots, X_n]/(F_2, \ldots, F_n)$ is a graded k-algebra as described above, we have*

$$g(X) = 1 + \frac{1}{2} \delta_2 \cdots \delta_n \cdot \tau , \qquad \tau := (\delta_2 + \cdots + \delta_n) - (n + 1) \, .$$

PROOF. For any rational function $H = (1 - Z^{\alpha_1}) \cdots (1 - Z^{\alpha_m})/(1 - Z^{\beta_1}) \cdots (1 - Z^{\beta_m})$ with exponents $\alpha_1, \ldots, \alpha_m, \beta_1, \ldots, \beta_m \in \mathbb{Z} \setminus \{0\}$ the logarithmic derivative $H'(1)/H(1)$ is $\frac{1}{2} \deg H = \frac{1}{2} \left((\alpha_1 + \cdots + \alpha_m) - (\beta_1 + \cdots + \beta_m) \right)$. It suffices to check this for the simple case $H := (1 - Z^{\alpha})/(1 - Z)$, $\alpha \in \mathbb{Z} \setminus \{0\}$.   •

We remark that

$$g(X) = \text{Dim}_k R_{\tau} \, .$$

Since the polynomial $\mathcal{P}_{\overline{R}}$ is self-reciprocal, i. e. $c_m = c_{\tau + 2 - m}$, and since $\mathcal{P}_R = \mathcal{P}_{\overline{R}}/(1 - Z)^2 = \mathcal{P}_{\overline{R}} \cdot \sum_{m \in \mathbb{N}} (m + 1) Z^m$, we have indeed

$$\text{Dim}_k R_{\tau} = \sum_{m \in \mathbb{N}} c_{\tau - m} (m + 1) = \sum_{m \in \mathbb{N}} c_{m+2} (m + 1) = g(X) \, .$$

In fact, even more can be said. We mention without proof that the duality theory for complete intersections yields an isomorphism

$$\text{Hom}_P(R, P) \cong R(\tau + 2) \, , \quad \text{or equivalently} \quad \text{Hom}_P(R, P(-2)) \cong R(\tau)$$

even as graded R-modules. An R-basis of the P-dual of R of degree $-(\tau + 2)$ is given by the P-module homomorphism which maps the only P-basis element $y_j \in R$ of (maximal) degree $\tau + 2$ to 1 (and the others to 0). It follows from 7.A.3 (2) that

$$\omega_X \cong R(\tau)^{\sim} = \mathcal{O}_X(\tau) \quad \text{and} \quad \Gamma(\omega_X) \cong R_\tau .$$

The simplest case (besides the projective line $\mathbb{P}^1_k = \operatorname{Proj} k[X_0, X_1]$) is a curve $X = \operatorname{Proj} k[X_0, X_1, X_2]/(F) \subseteq \mathbb{P}^2_k$, where F is a non-zero homogeneous polynomial of positive degree δ. The equation

$$g(X) = 1 + \frac{1}{2}\delta(\delta - 3) = \binom{\delta - 1}{2}$$

for the (arithmetical) genus of X is known as the P l ü c k e r   f o r m u l a.

7.E.5. Exercise (P o i n c a r é   s e r i e s   and   H i l b e r t – S a m u e l   p o l y n o m i a l s) Let $R = R_0[x_0, \ldots, x_n]$ be a positively graded algebra over the field k with $\operatorname{Dim}_k R_0 < \infty$ and homogeneous generators $x_0, \ldots, x_n$ of positive degrees $\gamma_0, \ldots, \gamma_n$ and let $X := \operatorname{Proj} R$.

(1) Show that for a finitely generated graded R-module M, the P o i n c a r é   s e r i e s

$$\mathcal{P}_M := \sum_{m \in \mathbb{Z}} (\operatorname{Dim}_k M_m) Z^m \in \mathbb{Z}[\![Z]\!][Z^{-1}]$$

is a rational function of type $Q/(1 - Z^{\gamma_0}) \cdots (1 - Z^{\gamma_n})$ with a Laurent polynomial $Q = Q_M \in \mathbb{Z}[Z^{\pm 1}]$. (**Hint:** Use induction on n. If $n \geq 0$ and if $M(-\gamma_n) \to M$ is multiplication with x_n then the kernel K and the cokernel C are finite graded modules over the subalgebra $R_0[x_0, \ldots, x_{n-1}]$ of R and the exact sequence $0 \to K \to M(-\gamma_n) \to M \to C \to 0$ yields the equations $\mathcal{P}_K - Z^{\gamma_n}\mathcal{P}_M + \mathcal{P}_M - \mathcal{P}_C = 0$, $\mathcal{P}_K = Q_K/(1 - Z^{\gamma_0}) \cdots (1 - Z^{\gamma_{n-1}})$, $\mathcal{P}_C = Q_C/(1 - Z^{\gamma_0}) \cdots (1 - Z^{\gamma_{n-1}})$, $\mathcal{P}_M = (Q_C - Q_K)/(1 - Z^{\gamma_0}) \cdots (1 - Z^{\gamma_{n-1}})(1 - Z^{\gamma_n})$.)

(2) In the situation of (1) let $\gamma_0 = \cdots = \gamma_n = 1$. Then $\mathcal{P}_M = P + \sum_{v=0}^{d} e_v/(1 - Z)^{v+1}$ with a Laurent polynomial $P \in \mathbb{Z}[Z^{\pm 1}]$ and $e_1, \ldots, e_d \in \mathbb{Z}$. It follows that $\operatorname{Dim}_k M_m = \mathcal{H}_M(m) = \sum_{v=0}^{d} e_v \binom{m+v}{v}$ for large m (more exactly, if $m > \deg P$). In particular, asymptotically

$$\operatorname{Dim}_k M_m \sim e_d \, m^d / d!$$

if $e_d \neq 0$. The polynomial $\mathcal{H}_M(m)$ is called the H i l b e r t – S a m u e l   p o l y n o m i a l of M. (It is an object of intensive study in Commutative Algebra.) Show by induction on $\dim M$ that $\deg \mathcal{H}_M = \dim \widetilde{M} = \dim (\operatorname{Supp} \widetilde{M})$ ($= \dim M - 1$ if $M \neq 0$), where $\widetilde{M}$ is the coherent $\mathcal{O}_X$-module belonging to M (see Exercise 6.E.14). (**Hint:** Induction on $\dim M$, one may assume that $M = (R/\mathfrak{p})(m)$, where $\mathfrak{p}$ is a homogeneous prime ideal in R. Here we adopt the convention that the zero-polynomial has degree -1.)

7.E.6. Exercise (H i l b e r t – S a m u e l   p o l y n o m i a l · E u l e r – P o i n c a r é   c h a r a c t e -r i s t i c) (1) Let $R = R_0[x_0, \ldots, x_n]$ be a standardly graded k-algebra of (Krull-)dimension 2 with $\operatorname{Dim}_k R_0 < \infty$ and $x_0, \ldots, x_n \in R_1$ and let X be the projective curve $\operatorname{Proj} R$ over k. Moreover, let $\mathcal{H}_M(Z) = e_0 + e_1(Z + 1) = (e_0 + e_1) + e_1 Z$ be the Hilbert–Samuel polynomial of a finitely generated graded R- module M, cf. the previous Exercise. Show that

$$\chi\big(M(m)^{\sim}\big) = \mathcal{H}_M(m) = e_0 + e_1(m + 1)$$

for *all* $m \in \mathbb{Z}$ and $\chi\big(M(m)^{\sim}\big) = \operatorname{Dim}_k \Gamma\big(M(m)^{\sim}\big) = \operatorname{Dim}_k M_m$ ($= \mathcal{H}_M(m)$) for large m; in particular, $\chi(\widetilde{M}) = e_0 + e_1$. If the curve X is integral and $\mathcal{H}_R = e_1 Z + (e_0 + e_1)$ then

$$\chi(\mathcal{O}_X(m)) = m \deg \mathcal{O}_X(1) + \chi(\mathcal{O}_X) = e_1 m + (e_0 + e_1) \quad \text{for all} \quad m \in \mathbb{Z},$$

i.e. $e_1 = \deg \mathcal{O}_X(1)$ and $\chi(\mathcal{O}_X) = e_0 + e_1$. More generally, if X is a Cohen–Macaulay curve, then $e_1 = \text{redeg}\, \mathcal{O}_X(1)$, cf. Exercise 7.D.5. (**Hint**: One may assume that k is infinite. Then there exist elements $f_0, f_1 \in R_1$ such that the homogeneous algebra homomorphism $k[T_0, T_1] \to R$, $T_i \mapsto f_i$, $i = 0, 1$, is finite. Using the corresponding finite morphism $X \to \mathbb{P}^1_k$ one may assume that $R = k[T_0, T_1]$ and $M = (R/\mathfrak{p})(m)$, where $m \in \mathbb{Z}$ and $\mathfrak{p}$ is a homogeneous prime ideal in $R = k[T_0, T_1]$. In this case the result is obvious.)

(2) Use the result of (1) to prove once more Proposition 7.E.4.

(3) Let $Q = k[X_0, \ldots, X_n]$ be the weighted graded polynomial algebra with weights $\gamma_i = \deg X_i > 0$ for $i = 0, \ldots, n$, $R = Q/(F_2, \ldots, F_n)$, where $F_2, \ldots, F_n$ is a strongly regular sequence of homogeneous polynomials of positive degrees $\delta_2, \ldots, \delta_n$ and let $X = \text{Proj}\, R$. Then $g(X) = \text{Dim}_k R_\tau$, where $\tau := (\delta_2 + \cdots + \delta_n) - (\gamma_0 + \cdots + \gamma_n)$. (**Hint** : This generalizes the result of Proposition 7.E.4, see the remark after its proof. One may proceed similarly as in the proof of 7.E.4. For instance, if k is infinite, then choose a strongly regular sequence $f_0, f_1 \in R_\delta$, $\delta := \text{LCM}\,(\gamma_0, \ldots, \gamma_n)$. Then $\mathcal{P}_{\overline{R}} = (1 - Z^\delta)^2 \mathcal{P}_R = (1 - Z^\delta)^2 (1 - Z^{\delta_2}) \cdots (1 - Z^{\delta_n})/(1 - Z^{\gamma_0}) \cdots (1 - Z^{\gamma_n})$. – Or one may apply part (1) above using the Poincaré series $\mathcal{P}_R$ and the fact that for some appropriate $d > 0$ the Veronese transform $R^{[d]} = \bigoplus_{m \in \mathbb{N}} R_{md}$ is a standardly graded k-algebra (with $(R^{[d]})_1 := R_d$), see Exercise 5.A.25.)

7.E.7. Exercise (Geometric genus and degree of singularity) Let X be a *reduced* projective algebraic curve over the field k and let $\overline{X} = \biguplus_{i=1}^r \overline{X}_i$ be the normalisation of X, where the X_i, $i = 1, \ldots, r$, are the irreducible components of X, cf. 7.B.2. The canonical morphism $\nu : \overline{X} \to X$ is finite and the cokernel of the injection $\mathcal{O}_X \to \nu_* \mathcal{O}_{\overline{X}}$ has support in the finite set $S := S(X) := X \setminus \text{Nor}\, X$ of the closed points $x \in X$, for which $\mathcal{O}_x := \mathcal{O}_{X,x}$ is *not* normal, cf. 6.C.6. Show that

$$\chi(\mathcal{O}_{\overline{X}}) = \chi(\nu_* \mathcal{O}_{\overline{X}}) = \chi(\mathcal{O}_X) + \delta(X)$$

where $\delta(X) := \sum_{x \in S} \text{Dim}_k \left(\overline{\mathcal{O}}_x / \mathcal{O}_x\right)$ is the so called degree of singularity of X. Here $\overline{\mathcal{O}}_x = (\nu_* \mathcal{O}_{\overline{X}})_x$ is the normalization of the local ring $\mathcal{O}_x$, see example 6.C.8. The genus of the normalization $\overline{X}$ is (sometimes) called the geometric genus $g_{\text{geom}}(X)$ of X: $g_{\text{geom}}(X) := g(\overline{X})$. Show that

$$g_{\text{geom}}(X) = g(X) + \text{Dim}_k \left(\Gamma(\mathcal{O}_{\overline{X}})/\Gamma(\mathcal{O}_X)\right) - \delta(X).$$

If k is algebraically closed, then $\text{Dim}_k \left(\Gamma(\mathcal{O}_{\overline{X}})/\Gamma(\mathcal{O}_X)\right) = r - s$, *where r is the number of irreducible components of X and s is the number of connected components of X* (see Example 7.A.6), *hence*

$$g_{\text{geom}}(X) = g(X) + (r - s) - \delta(X),$$

in particular, $g_{\text{geom}}(X) = g(\overline{X}) = g(X) - \delta(X)$ *if k is algebraically closed and if X is integral.* For example, let $X = \text{Proj}\, k[X_0, X_1, X_2]/(F)$, with a homogeneous polynomial F of degree $d > 0$ without multiple factors over the algebraically closed field k, then

$$g_{\text{geom}}(X) = \binom{d - 1}{2} + (r - 1) - \delta(X),$$

where r is the number of irreducible factors of F. This formula is called the generalized or refined Plücker formula, cf. the end of Example 7.E.3. For instance, fix d distinct lines in $\mathbb{P}^2_k$ (k need not be algebraically closed). If m_P is the number of those lines which pass through the point $P \in \mathbb{P}_k^2$, then $0 = \binom{d-1}{2} + (d-1) - \sum_P \binom{m_P}{2}$, i.e. $\sum_P \binom{m_P}{2} = \binom{d}{2}$. Prove this directly (it is elementary!).

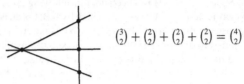

$$\binom{3}{2} + \binom{2}{2} + \binom{2}{2} + \binom{2}{2} = \binom{4}{2}$$

Another simple example is the curve $X := \operatorname{Proj} k[X_0, X_1, X_2, X_3]/(F_1, F_2) \subseteq \mathbb{P}_k^3$, where $F_1 = X_0^2 - (X_1^2 + X_2^2)$, $F_2 = X_0^2 - (aX_2^2 + X_3^2)$, $a \neq 0$ and char $k \neq 2$. (For $k = \mathbb{R}$ sketch the affine part $x_0 \neq 0$ in $\mathbb{R}^3$.) By Proposition 7.E.4 $g(X) = 1$. If $a \neq 1$, then X is smooth (i.e. an irreducible elliptic curve) and $g_{\text{geom}}(X) = g(X) = 1$. But for $a = 1$, the curve X has two singularities which are ordinary double points, hence $\delta(X) = 2$ and $g_{\text{geom}}(X) = 1 + (r-1) - 2 = r - 2$, in particular, $r \geq 2$. Indeed, $\overline{X}$ is the disjoint union of two rational curves isomorphic to $\mathbb{P}_k^1$ and $g_{\text{geom}}(X) = 0$.

7.E.8. Remark The computation of the degree of singularity $\delta(X)$ which is defined for any reduced algebraic curve X over a field k, is a purely local problem. Many special methods have been developed for this. We mention just two here.[10])

(1) Since for a closed point $x \in X$ the module $\overline{\mathcal{O}}_x/\mathcal{O}_x$ has finite length over $\mathcal{O}_x$ we can identify it with its completion $\widehat{\overline{\mathcal{O}}}_x/\widehat{\mathcal{O}}_x$. But $\widehat{\overline{\mathcal{O}}}_x = \overline{\widehat{\mathcal{O}}}_x$ is the normalization of the completion $\widehat{\mathcal{O}}_x$ of $\mathcal{O}_x$. Therefore, *the degreee of singularity* $\delta_x = \operatorname{Dim}_k(\overline{\mathcal{O}}_x/\mathcal{O}_x)$ *in the closed point* $x \in X$ *does not change if we complete the* 1-*dimensional local ring* $\mathcal{O}_x$. Very often this simplifies the computations. For instance, let $x = 0 \in \operatorname{Spec}(k[Y, Z]/(f)) \subseteq \mathbb{A}_k^2$, where $f = f_e + f_{e+1} + \cdots \in k[Y, Z]$ is a polynomial with initial homogeneous term $f_e \neq 0$ of degree $e \geq 1$. If f_e is the product of e *non-associated* linear polynomials then the completion $\widehat{\mathcal{O}}_x = k[\![Y, Z]\!]/(f)$ is isomorphic to $k[\![Y, Z]\!]/(f_m)$ with degree of singularity $\binom{e}{2}$. This degree formula follows from the example above but can also be checked directly since the normalization is the product $\prod_{i=1}^e k[\![Y, Z]\!]/(\ell_i)$, where the ℓ_i are the linear factors of f_e. Singularities with a completion of this type are called o r d i n a r y s i n g u l a r i t i e s of multiplicity e, for $e = 2$ they are called o r d i n a r y or r e g u l a r d o u b l e p o i n t s (see Example 2.C.5). Other interesting examples are the m o n o m i a l c u r v e s i n g u l a r i t i e s of Example 6.C.8. The reader should verify that the degree of singularity of the union of the m coordinate lines (with the reduced scheme structure) in affine m-space $\mathbb{A}_k^m$, $m \in \mathbb{N}^*$, is $m - 1$. For $m = 2$ it is an ordinary double point. More generally: Let $x \in X$ be a closed point of a reduced algebraic curve X over k with normalization $\nu : \overline{X} \to X$. The number of points in the fibre $\nu^{-1}(x)$, i.e. the number of closed points in the normalization $\overline{\mathcal{O}}_x$ of $\mathcal{O}_x$, is called the n u m b e r o f b r a n c h e s of X in x. Let $m > 0$ be this number. Then the completion $\widehat{\overline{\mathcal{O}}}_x = \overline{\widehat{\mathcal{O}}}_x$ is a product of m (complete) discrete valuation rings. Show that $\delta_x \geq (m-1) \cdot [\kappa(x) : k]$. The equality $\delta_x = (m-1) \cdot [\kappa(x) : k]$ holds if and only if $\mathfrak{m}_x = \mathfrak{m}_x \overline{\mathcal{O}}_x = \bigcap_{y \in \nu^{-1}(x)} \mathfrak{m}_y \ (= \mathfrak{m}_{\overline{\mathcal{O}}_x})$ and, moreover, $\kappa(y) = \kappa(x)$ for all $y \in \nu^{-1}(x)$.

(2) (B l o w i n g u p) A systematic process to construct the normalization $\nu : \overline{X} \to X$ of a reduced (not necessarily projective) algebraic curve X over a field k is the blowing

---

[10]) In some arguments we shall use in this remark the completion of a local ring which we have not introduced formally in these lectures, cf. [1] (for instance).

up described in Example 5.A.33. Let $S = S(X) = X \setminus \mathrm{Nor}\, X$ be the (finite) set of non-normal points of X. We consider S as a closed subscheme of X with its reduced scheme structure (i.e. $S \cong \mathrm{Spec}(\prod_{s \in S} \kappa(s))$) and consider the blowing up $X^{(1)} := \mathrm{Bl}_S(X)$ with the canonical surjective projection $\sigma : X^{(1)} \to X$, which by 6.B.1 is an isomorphism if and only if $X = \overline{X}$. In any case the exceptional divisor $\sigma^{-1}(S)$ is purely 1-codimensional in $X^{(1)}$ defined locally by non-zero divisors and σ defines an isomorphism $X^{(1)} \setminus \sigma^{-1}(S) \cong X \setminus S$ of open dense subsets of $X^{(1)}$ and X, respectively. In particular, $X^{(1)}$ is a reduced algebraic curve with the same algebra of global rational function as X and all the fibres of σ are finite. Therefore by 5.B.10 *the blowing up map* $\sigma : X^{(1)} \to X$ *is a finite morphism* which implies in particular, that X and $X^{(1)}$ *have the same normalization*. Hence we have the chain $\overline{X} \to X^{(1)} \to X^{(0)} = X$ of finite morphisms of reduced algebraic curves. If $\overline{X} \to X^{(1)}$ is not an isomorphism then we repeat the blowing up with $X^{(1)}$ instead of $X^{(0)}$ getting an algebraic curve $X^{(2)} = (X^{(1)})^{(1)}$ between $X^{(1)}$ and $\overline{X}$ and so on. If we denote the composition $X^{(i)} \to \cdots \to X^{(1)} \to X^{(0)}$ by $\sigma^{(i)}$ then we get the chain

$$\mathcal{O}_X = \mathcal{O}_{X^{(0)}} \subseteq \sigma_*^{(1)} \mathcal{O}_{X^{(1)}} \subseteq \cdots \subseteq \sigma_*^{(i)} \mathcal{O}_{X^{(i)}} \subseteq \cdots \subseteq \nu_* \mathcal{O}_{\overline{X}}$$

with strict inclusions as long as $X^{(i)} \not\cong \overline{X}$, i.e. $\mathrm{Dim}_k \, \Gamma(\sigma_*^{(i+1)} \mathcal{O}_{X^{(i+1)}} / \sigma_*^{(i)} \mathcal{O}_{X^{(i)}}) > 0$ for $X^{(i)} \not\cong \overline{X}$. Since $\mathrm{Dim}_k \, \Gamma(\nu_* \mathcal{O}_{\overline{X}} / \mathcal{O}_X) = \delta(X)$ is finite, this process must stop and end with $\overline{X}$ after at most $\delta(X)$ steps. We then get the formula

$$\delta(X) = \sum_{i \geq 0} \mathrm{Dim}_k \, \Gamma(\sigma_*^{(i+1)} \mathcal{O}_{X^{(i+1)}} / \sigma_*^{(i)} \mathcal{O}_{X^{(i)}}).$$

Hence, if $\overline{S} = \overline{S}(X) = \biguplus_{i \geq 0} S(X^{(i)})$ denotes the (disjoint) union of sets of non-normal points of the iterated blowing ups of X, then

$$\delta(X) = \sum_{s \in \overline{S}} \mathrm{Dim}_k (\mathrm{Bl}(\mathcal{O}_s)/\mathcal{O}_s) = \sum_{s \in \overline{S}} \ell_{\mathcal{O}_s} (\mathrm{Bl}\,(\mathcal{O}_s)/\mathcal{O}_s) \cdot [\kappa(s):k],$$

where, quite generally, $\mathrm{Bl}(A)$ denotes, for an 1-dimensional reduced Noetherian local ring A with maximal ideal $\mathfrak{m}$, the finite A-algebra $\Gamma(\mathrm{Bl}_{\{\mathfrak{m}\}}(\mathrm{Spec}\, A)) = \Gamma(\mathrm{Proj}\, A[\mathfrak{m}T]) \subseteq \overline{A}$. Note that

$$\mathrm{Bl}(A) = A[\mathfrak{m}^d/t] \subseteq \overline{A},$$

where $t \in \mathfrak{m}^d \setminus \mathfrak{m}^{d+1}$ with $d > 0$ is an element for which $\bar{t} \in \mathfrak{m}^d/\mathfrak{m}^{d+1}$ is not contained in any minimal prime ideal of the (1-dimensional) associated graded ring $\mathrm{G}_\mathfrak{m}(A) = \bigoplus_{m \geq 0} \mathfrak{m}^m/\mathfrak{m}^{m+1}$. (If $A/\mathfrak{m}$ is infinite such a t exists for $d = 1$.)

For a plane singularity s there is a simple formula for $\ell_{\mathcal{O}_s}(\mathrm{Bl}(\mathcal{O}_s)/\mathcal{O}_s)$. More generally, let s be a non-regular point of a reduced algebraic curve X over k such that the local ring $\mathcal{O}_s$ has embedding dimension $\mathrm{emdim}\, \mathcal{O}_s = 2$, i.e. $\mathfrak{m}_s$ is minimally generated by two elements. Then $\mathcal{O}_s \cong R/Rf$ where R is a regular local ring of dimension 2 with residue field $k_R = R/\mathfrak{m}_R = \mathcal{O}_s/\mathfrak{m}_s = \kappa(s)$ and $f \in \mathfrak{m}_R^2$. The largest $e = \mathrm{e}(\mathcal{O}_s)$ with $f \in \mathfrak{m}_R^e$ is called the m u l t i p l i c i t y of $\mathcal{O}_s$. For instance, if $s = 0 \in k\text{-Spec}\, k[Y, Z]/(f)$ and $f = f_e + f_{e+1} + \cdots$ with $f_e \neq 0$, then $R = k[Y, Z]_{\mathfrak{m}_0}$ and e is the multiplicity of $\mathcal{O}_s = k[Y, Z]_{\mathfrak{m}_0}/(f)$. In general, the associated graded ring $\mathrm{G}_{\mathfrak{m}_s}(\mathcal{O}_s)$ is isomorphic to $k_R[\overline{Y}, \overline{Z}]/(\overline{f})$, where $\overline{f} \in \mathfrak{m}_R^e/\mathfrak{m}_R^{e+1} \subseteq \mathrm{G}_{\mathfrak{m}_R}(R) = k[\overline{Y}, \overline{Z}]$. Here $Y, Z \in \mathfrak{m}_R \setminus \mathfrak{m}_R^2$ form a regular system of parameters of R. In particular, the multiplicity $e = \mathrm{e}(\mathcal{O}_s)$ can be read off from $\mathrm{G}_{\mathfrak{m}_s}(\mathcal{O}_s)$. The number of points in the fibre $\mathrm{Proj}\, \mathrm{G}_{\mathfrak{m}_s}(\mathcal{O}_s)$ over s of the blowing up $\mathrm{Bl}_{S(X)}(X)$ is the number of distinct prime factors (up to constants) of the homogeneous polynomial $\overline{f}(\overline{Y}, \overline{Z})$ of degree e in $k_R[\overline{Y}, \overline{Z}]$ (which gives an estimate for the number of branches of X in s). (In contrary to f the polynomial $\overline{f}$ may have multiple factors.) In the above example $s = 0 \in k\text{-Spec}\, k[Y, Z]/(f)$ one has $\overline{f} = f_e \in \mathrm{G}_{\mathfrak{m}_R}(R) = k[Y, Z]$ and $\mathrm{G}_{\mathfrak{m}_s}(\mathcal{O}_s) = k[Y, Z]/(f_e)$.

7.E.9. Proposition *Let* $\mathcal{O}_s$ *be the local ring of a point* s *of a reduced algebraic curve* X *over* k *with embedding dimension* 2 *and multiplicity* $e = \mathrm{e}(\mathcal{O}_s)$. *Then*

$$\ell_{\mathcal{O}_s}(\mathrm{Bl}(\mathcal{O}_s)/\mathcal{O}_s) = \binom{e}{2}, \quad i.\,e. \quad \mathrm{Dim}_k(\mathrm{Bl}(\mathcal{O}_s)/\mathcal{O}_s) = \binom{e}{2} \cdot [\kappa(s):k].$$

Moreover, the non-normal points of $\mathrm{Bl}_{\mathrm{S}(X)}(X)$ *over* s *also have embedding dimension* 2.

PROOF. For simplicity we replace $\mathcal{O}_s$ by its completion $A := \widehat{\mathcal{O}_s}$ with maximal ideal $\mathfrak{m} = \mathfrak{m}_s A$ and assume that k is infinite. (We can, for example, always extend k to a rational function field $k(T)$.) Then A contains an (infinite) coefficient field L, i.e. a field $L \subseteq A$ with $L \xrightarrow{\sim} A/\mathfrak{m} = \mathcal{O}_s/\mathfrak{m}_s = \kappa(s)$.[11]) Let y, z be a system of generators of $\mathfrak{m}$. The canonical homomorphism $L[\![Y, Z]\!] \to A$ with $F \mapsto F(y, z)$ is surjective with kernel (f), $f = f_e + f_{e+1} + \cdots \in L[\![Y, Z]\!]$, $f_e \neq 0$. By change of coordinates (see the proof of Lemma 1.F.1) we may assume that $f_e = Z^e + a_1(Y)Z^{e-1} + \cdots + a_e(Y)$ is monic in Z. Then $A/Ay \cong L[\![Y, Z]\!]/(Y, Z^e) = L[\![Z]\!]/(Z^e)$, hence by Nakayama's Lemma for complete local rings, A is finite (and free since it is torsion-free) over $L[\![y]\!] = L[\![Y]\!]$ with $L[\![Y]\!]$-basis $1, z, \ldots, z^{e-1}$. It follows that $A = L[\![Y]\!][z] = L[\![Y]\!][Z]/(p)$, where $p = Z^e + w_1(Y)Z^{e-1} + \cdots + w_e(Y) \in L[\![Y]\!][Z]$ is the characteristic (= minimal) polynomial of z over $L[\![Y]\!]$.[12]) Now, $\mathrm{Bl}(A)$ equals

$$\Gamma(\mathrm{Proj}\, A[\mathfrak{m}T]) = A[\mathfrak{m}/y] = L[\![y]\!][z, z/y] = L[\![Y]\!][z/Y] \cong L[\![Y]\!][U]/(q) = \bigoplus_{i=0}^{e-1} L[\![Y]\!]u^i$$

where $q = U^e + (w_1(Y)/Y)U^{e-1} + \cdots + (w_e(Y)/Y^e) \in L[\![Y]\!][U]$ is the characteristic (= minimal) polynomial of $u := z/Y$ over $L[\![Y]\!]$. It follows that

$$\ell_A(\mathrm{Bl}(A)/A) = \ell_{L[\![Y]\!]}(\mathrm{Bl}(A)/A) = \sum_{i=0}^{e-1} \ell_{L[\![Y]\!]}(L[\![Y]\!]u^i/L[\![Y]\!]z^i) = \sum_{i=0}^{e-1} i = \binom{e}{2}.$$

Since $L[\![Y]\!][Z]$ is a regular integral domain of dimension 2 (see Example 6.A.13) the localizations of $\mathrm{Bl}(A)$ at non-regular (maximal) ideals have embedding dimension 2.   •

We note that the last proof shows more generally the length formula $\ell_A(\mathrm{Bl}(A)/A) = \binom{\mathrm{e}(A)}{2}$ for an arbitrary 1-dimensional reduced Noetherian local ring A containing a field with finite normalization $\overline{A}$ and embedding dimension 2. The completion $\widehat{A} \subseteq \widehat{\overline{A}} = \overline{\widehat{A}}$ is also reduced and has a representation $\widehat{A} = L[\![Y, Z]\!]/(f)$ as above.

7.E.10. Corollary *Let* X *be a reduced algebraic curve over* k *with* $\mathrm{emdim}\,\mathcal{O}_x \leq 2$ *for all* $x \in X$ *and let* $\overline{S} = \biguplus_{i \geq 0} \mathrm{S}(X^{(i)})$ *be the (disjoint) union of the sets of non-normal points of the iterated blowing ups of* X. *Then* $\mathrm{emdim}\,\mathcal{O}_s = 2$ *for all* $s \in \overline{S}$ *and*

$$\delta(X) = \sum_{s \in \overline{S}} \binom{e_s}{2} \cdot [\kappa(s):k],$$

where $e_s = \mathrm{e}(\mathcal{O}_s)$ *denotes the multiplicity of* $\mathcal{O}_s$, $s \in \overline{S}$.

As a simple example construct the iterated blowing ups of $\mathrm{Spec}\,(k[Y, Z]/(Y^q - Z^p))$, $p, q \geq 2$, $d := \mathrm{GCD}(p, q) = 1$. What happens if $d > 1$ but still $d \neq 0$ in k? More generally, consider the blowing ups of monomial curves X as in Example 6.C.8.

---

[11]) L is not necessarily an extension of k. If k is algebraically closed then we can choose $L = k$. If the (finite) extension $k \subseteq \kappa(s)$ is separable, then we can also choose $k \subseteq L$.
[12]) Necessarily $p = \varepsilon f$ with $\varepsilon \in L[\![Y, Z]\!]^\times$. The polynomial p is called a W e i e r s t r a s s p o l y n o m i a l associated to f.

The classical (analytic) theorem of Riemann–Roch was formulated for compact (connected) Riemann surfaces (which are the smooth projective algebraic curves over $\mathbb{C}$ – this is a consequence of the analytic Riemann–Roch Theorem). Later Dedekind and Weber gave a purely algebraic proof for smooth projective algebraic curves X over an arbitrary algebraically closed field k identifying them with their function fields $\mathcal{R}(X)$, see Exercise 7.B.5. There the sheaves of differential forms $\Omega_{X|k}$ play the role of the canonical sheaves ω_X (using modern terminology). The justification for this is the following theorem which we want to prove now.

7.E.11. Theorem *Let X be a smooth projective algebraic curve over a field k. Then the canonical module ω_X is isomorphic to the $\mathcal{O}_X$-module $\Omega_{X|k}$ of Kähler differentials on X over k, i.e.*
$$\omega_X \cong \Omega_{X|k}.$$

Note that 7.E.11 says in particular that *for a smooth projective curve X over k the genus $g(X)$ can be defined as the maximal number of linearly independent global Kähler differentials on X, i.e $g(X) = \mathrm{Dim}_k\, \Gamma(\Omega_{X|k})$*. The proof of 7.E.11 needs some preparation. By Definition 6.D.22 X is smooth over k if and only if $\Omega_{X|k}$ is an invertible sheaf on X, i.e. a locally free sheaf of rank 1. Since X is the disjoint union of its irreducible (= connected) components, we may assume that X is integral with function field $\mathcal{R}(X)$. Since $\mathrm{Dim}_{\mathcal{R}(X)}\, \Omega_{\mathcal{R}(X)|k} = 1$, there is an element $t \in \mathcal{R}(X)$ such that $\Omega_{\mathcal{R}(X)|k} = \mathcal{R}(X)dt$. This implies that $\Omega_{\mathcal{R}(X)|k(t)} = 0$, i.e. that the field extension $k(t) = \mathcal{R}(\mathbb{P}^1_k) \subseteq \mathcal{R}(X)$ is finite and separable, cf. Theorem 6.D.12 (4). By Exercise 7.B.5 (4) this inclusion defines a finite morphism $\varphi : X \to \mathbb{P}^1_k$. Then by definition

$$\varphi_*\omega_X = \varphi_*\varphi^!\big(\mathcal{O}_{\mathbb{P}^1_k}(-2)\big) = \varphi_*\varphi^!\Omega_{\mathbb{P}^1_k|k} = \mathcal{H}om_{\mathcal{O}_{\mathbb{P}^1_k}}(\varphi_*\mathcal{O}_X, \Omega_{\mathbb{P}^1_k|k}).$$

We call a finite morphism $\varphi : X \to Y$ of arbitrary integral algebraic curves X, Y over k s e p a r a b l e if the finite field extension $\mathcal{R}(Y) \to \mathcal{R}(X)$ defined by φ is separable. Now, Theorem 7.E.11 is a consequence of the following more general result:

7.E.12. Theorem *Let $\varphi : X \to Y$ be a finite separable morphism of smooth integral algebraic curves over k. Then*
$$\varphi^!\Omega_{Y|k} \cong \Omega_{X|k}.$$

The isomorphism of Theorem 7.E.12 *is quite canonical* and uses a surjective t r a c e   h o m o m o r p h i s m for differential forms

$$\mathrm{tr} : \varphi_*\Omega_{X|k} \to \Omega_{Y|k}.$$

This homomorphism defines the t r a c e   f o r m for differentials

$$\varphi_*\mathcal{O}_X \times \varphi_*\Omega_{X|k} \to \Omega_{Y|k}$$

by $(s, \omega) \mapsto \mathrm{tr}(s\omega)$ for sections $s \in \Gamma(\varphi^{-1}(V), \mathcal{O}_X)$, $\omega \in \Gamma(\varphi^{-1}(V), \Omega_{X|k})$, V open in Y, and hence a homomorphism

$$\varphi_* \Omega_{X \mid k} \to \mathcal{H}om_{\mathcal{O}_Y}(\varphi_* \mathcal{O}_X, \Omega_{Y \mid k}),$$

$\omega \mapsto (s \mapsto \mathrm{tr}(s\omega))$ for s, ω as above, which is $\varphi_* \mathcal{O}_X$-linear. This homomorphism is an isomorphism and it defines the isomorphism of 7.E.12.

To perform all these constructions we may assume that $X = \mathrm{Spec}\, B$ and $Y = \mathrm{Spec}\, A$ are affine with $\mathcal{R}(X) = L = Q(B)$ and $\mathcal{R}(Y) = K = Q(A)$ as rational function fields. Since $K \subseteq L$ is a finite and separable field extension, the trace map $\mathrm{tr} = \mathrm{tr}_K^L : L \to K$ is non-trivial and defines a non-degenerate trace form $L \times L \to K$, $(x, y) \mapsto \mathrm{tr}(xy)$. Let $\Omega_{K \mid k} = K\omega$. Then $\Omega_{L \mid k} = L\omega$ since $\Omega_{L \mid K} = 0$. This allows to define the trace map

$$\mathrm{tr} : \Omega_{L \mid k} \to \Omega_{K \mid k}, \quad x\omega \mapsto \mathrm{tr}_K^L(x)\,\omega,$$

which is obviously independent of the chosen K-basis ω of $\Omega_{K \mid k}$. The modules $\Omega_{A \mid k}$ and $\Omega_{B \mid k}$ are submodules of $\Omega_{K \mid k} = K \otimes_A \Omega_{A \mid k}$ and $\Omega_{L \mid k} = L \otimes_B \Omega_{B \mid k}$, respectively (since they are torsion-free), hence $\Omega_{A \mid k} = \mathfrak{a}\,\omega$ and $\Omega_{B \mid k} = \mathfrak{b}\,\omega$ with (finitely generated) fractional ideals $\mathfrak{a} \subseteq K$ and $\mathfrak{b} \subseteq L$. Now, to complete the proof of Theorem 7.E.12 we need only:

7.E.13. Lemma *With the notations and assumptions as above, the trace map* $\mathrm{tr} :$ $\Omega_{L \mid k} \to \Omega_{K \mid k}$ *maps* $\Omega_{B \mid k} (\subseteq \Omega_{L \mid k})$ *surjectively onto* $\Omega_{A \mid k} (\subseteq \Omega_{K \mid k})$ *and the induced* A-*bilinear mapping*

$$B \times \Omega_{B \mid k} \to \Omega_{A \mid k}, \quad (x, \xi) \mapsto \mathrm{tr}(x\,\xi),$$

is a complete duality, i. e. the derived homomorphisms $\Omega_{B \mid k} \to \mathrm{Hom}_A(B, \Omega_{A \mid k})$ *and* $B \to \mathrm{Hom}_A(\Omega_{B \mid k}, \Omega_{A \mid k})$ *are isomorphisms of* B-*modules.*

PROOF. For simplicity we assume that the field k is infinite (otherwise extend k to a rational function field $k(T)$). It is enough to prove that, for every maximal ideal $\mathfrak{m} \subseteq A$, $\mathrm{tr}(\Omega_{B_{\mathfrak{m}} \mid k}) = \Omega_{A_{\mathfrak{m}} \mid k}$ and the induced $A_{\mathfrak{m}}$-bilinear mapping is a complete duality. We have $\Omega_{A_{\mathfrak{m}} \mid k} = A_{\mathfrak{m}}\,\omega$ for an A-basis ω of $\Omega_{A_{\mathfrak{m}} \mid k}$ and $\Omega_{B_{\mathfrak{m}} \mid k} = B_{\mathfrak{m}}g\,\omega$ with $g \in L^{\times}$. By Exercise 6.D.15 the finite free $A_{\mathfrak{m}}$-algebra $B_{\mathfrak{m}}$ is generated by one element $x \in B_{\mathfrak{m}}$, since the module of differentials $\Omega_{B_{\mathfrak{m}} \mid A_{\mathfrak{m}}} = B_{\mathfrak{m}}g\,\omega / B_{\mathfrak{m}}\,\omega$ (see Exercise 6.D.8) is generated by one element, i. e. $B_{\mathfrak{m}} = A_{\mathfrak{m}}[x] \cong A_{\mathfrak{m}}[X]/A_{\mathfrak{m}}[X]\,\mu$ and $L = K[x] \cong K[X]/K[X]\,\mu$, where $\mu = X^n + a_{n-1}X^{n-1} + \cdots + a_0 \in A_{\mathfrak{m}}[X]$ is the characteristic $(=$ minimal$)$ polynomial of x over $A_{\mathfrak{m}}$. It follows $\Omega_{B_{\mathfrak{m}} \mid A_{\mathfrak{m}}} = B_{\mathfrak{m}}g\omega / B_{\mathfrak{m}}\,\omega \cong B_{\mathfrak{m}}/B_{\mathfrak{m}}\,\mu'(x)$ and $\Omega_{B_{\mathfrak{m}} \mid k} = B_{\mathfrak{m}}\,\omega/\mu'(x)$. Let $b_0, \ldots, b_{n-1} \in B_{\mathfrak{m}}$ be defined by the following equation in $B_{\mathfrak{m}}[X]$: $\mu = (X - x)(b_{n-1}X^{n-1} + \cdots + b_1 X + b_0)$. By Lemma 7.E.15 (1), (5) below $b_0, \ldots, b_{n-1}$ is an $A_{\mathfrak{m}}$-basis of $B_{\mathfrak{m}}$ and $\mathrm{tr}_K^L(x^i b_j / \mu'(x)) = \delta_{ij}$ for all $i, j = 0, \ldots, n-1$. Hence $\mathrm{tr}(\Omega_{B_{\mathfrak{m}} \mid k}) = \mathrm{tr}(B_{\mathfrak{m}}/\mu'(x))\,\omega = \mathrm{tr}(\sum_j A_{\mathfrak{m}}\, b_j / \mu'(x))\,\omega = \mathrm{tr}(A_{\mathfrak{m}}\, b_0/\mu'(x))\,\omega = A_{\mathfrak{m}}\omega = \Omega_{A_{\mathfrak{m}} \mid k}$. Furthermore, the $A_{\mathfrak{m}}$-basis $1, x, \ldots, x^{n-1}$ of $B_{\mathfrak{m}}$ and the $A_{\mathfrak{m}}$-basis $b_0/\mu'(x), \ldots, b_{n-1}/\mu'(x)$ of $\Omega_{B_{\mathfrak{m}} \mid k}$ are dual with respect to the trace form. $\bullet$

If we use the $\varphi_* \mathcal{O}_X$-isomorphism $\iota : \varphi_* \omega_X = \varphi_* \Omega_{X|k} \xrightarrow{\sim} \mathcal{H}om_{\mathcal{O}_Y}(\varphi_* \mathcal{O}_X, \Omega_{Y|k}) = \mathcal{H}om_{\mathcal{O}_Y}(\varphi_* \mathcal{O}_X, \omega_Y)$ of 7.E.12 which is induced by the trace map $\varphi_* \Omega_{X|k} \to \Omega_{Y|k}$ then the trace map itself is given by $\omega \mapsto (1 \mapsto \iota_V \omega(1))$, $\omega \in \Omega_{X|k}(\varphi^{-1}(V))$, i. e. by taking the value at 1. This $\mathcal{O}_Y$-morphism $\varphi_* \omega_X = \mathcal{H}om_{\mathcal{O}_Y}(\varphi_* \mathcal{O}_X, \omega_Y) \to \omega_Y$ is defined for every finite morphism $\varphi : X \to Y$ of (projective) algebraic curves and is likewise called the t r a c e   m a p.

7.E.14. Remark (R e g u l a r   K ä h l e r   d i f f e r e n t i a l   f o r m s) Let X be a reduced and Y a smooth connected algebraic curve over k and let $\varphi : X \to Y$ be a finite separable morphism, i. e. for any generic point $\xi_i \in X$ the finite field extension $\mathcal{R}(Y) = \mathcal{O}_{Y, \varphi(\xi_i)} \hookrightarrow \mathcal{O}_{X, \xi_i}$ is separable. Then again the trace morphism

$$\mathrm{tr} : \varphi_*\big(\mathcal{R}(X) \otimes_{\mathcal{O}_X} \Omega_{X|k}\big) \to \mathcal{R}(Y) \otimes_{\mathcal{O}_Y} \Omega_{Y|k}$$

of rational differential forms is defined. Furthermore, we may define a coherent subsheaf $\widetilde{\Omega}_{X|k} \subseteq \mathcal{R}(X) \otimes_{\mathcal{O}_X} \Omega_{X|k}$ by the equation

$$\Gamma(\varphi^{-1}(V), \widetilde{\Omega}_{X|k}) = \{ \omega \in \Gamma(\varphi^{-1}(V), \mathcal{R}(X) \otimes_{\mathcal{O}_X} \Omega_{X|k}) \mid \mathrm{tr}(\Gamma(\varphi^{-1}(V), \mathcal{O}_X) \omega) \subseteq \Gamma(V, \Omega_{Y|k})\}$$

for every non-empty open affine subset $V \subseteq Y$. Then, there is a canonical $\varphi_* \mathcal{O}_X$-homomorphism $\varphi_* \widetilde{\Omega}_{X|k} \to \mathcal{H}om_{\mathcal{O}_Y}(\varphi_* \mathcal{O}_X, \Omega_{Y|k})$ defined by $\omega \mapsto (s \mapsto \mathrm{tr}(s\omega))$ for $\omega \in \Gamma(\varphi^{-1}(V), \widetilde{\Omega}_{X|k})$, $s \in \Gamma(\varphi^{-1}(V), \mathcal{O}_X)$, i. e. an $\mathcal{O}_X$-homomorphism

$$\widetilde{\Omega}_{X|k} \to \varphi^! \Omega_{Y|k}$$

which is obviously (by definition of $\widetilde{\Omega}_{X|k}$) an isomorphism. Now, let X, Y be projective. Then $\Omega_{Y|k}$ is by Theorem 7.E.11 the canonical module ω_Y of Y and we can use $\widetilde{\Omega}_{X|k}$ as a model for the canonical module ω_X of X. One can also show that this model is independent of the choosen separable morphism $X \to Y$ and hence defined for every reduced (projective) algebraic curve X which is generically (i. e. in all generic points of X) smooth over k, which, by the way, means that $\Omega_{X|k}$ is an $\mathcal{O}_X$-module of rank 1. In this case, the coherent $\mathcal{O}_X$-module $\widetilde{\Omega}_{X|k} \subseteq \mathcal{R}(X) \otimes_{\mathcal{O}_X} \Omega_{X|k}$ which is isomorphic to ω_X is called the module of r e g u l a r   (K ä h l e r)   d i f f e r e n t i a l   f o r m s on X.

The following classical lemma (due to Dedekind) was used in the proof of 7.E.13.

7.E.15. Lemma *Let A be a (commutative) ring, $f = X^n + a_{n-1}X^{n-1} + \cdots + a_0 \in A[X]$ a monic polynomial of degree $n \in \mathbb{N}^*$ and let $B = A[X]/(f) = A[x]$ be the free algebra of rank n with A-basis $1, x, \ldots, x^{n-1}$. Furthermore, let $\eta_0, \ldots, \eta_{n-1}$ be the A-dual basis of $\mathrm{Hom}_A(B, A)$ with $\eta_j(x^i) = \delta_{ij}$ for all $i, j = 0, \ldots, n-1$, and let the elements $b_0, \ldots, b_{n-1} \in B$ be defined by the equation $f = (X - x)(b_{n-1}X^{n-1} + \cdots + b_0)$ in $B[X]$. Then:*

(1) $b_0, \ldots, b_{n-1}$ *is an A-basis of B.*

(2) $\eta_j = b_j \eta_{n-1}$ *for $j = 0, \ldots, n-1$.*

(3) η_{n-1} *is a B-basis of $\mathrm{Hom}_A(B, A)$.*

(4) $\mathrm{tr}_A^B = \sum_{i=0}^{n-1} x^i \eta_i = \sum_{i=0}^{n-1} x^i b_i \eta_{n-1} = f'(x) \eta_{n-1}$.

(5) $\mathrm{tr}_A^B\big(x^i b_j / f'(x)\big) = \eta_{n-1}(x^i b_j) = b_j \eta_{n-1}(x^i) = \eta_j(x^i) = \delta_{ij}$, *if $f'(x)$ is a non-zero divisor in B.*

(6) $\mathrm{N}_A^B(f'(x)) = \det\big(\mathrm{tr}_A^B(b_i x^j)_{0 \le i, j < n}\big) = (-1)^{\binom{n}{2}} \det\big(\mathrm{tr}_A^B(x^{i+j})_{0 \le i, j < n}\big).$

PROOF. We put $a_n := 1$ and $b_n := 0$. Then $b_i = a_{i+1} + b_{i+1}x$ and hence $b_i = a_{i+1} + a_{i+2}x + \cdots + a_n x^{n-i-1}$ for $i = 0, \ldots, n - 1$ which proves (1). The equality in (2) follows by descending induction on j, using the recursion $b_{j-1} = a_j + b_j x$. By (2) η_{n-1} is a generator and hence a basis of the B-module $\mathrm{Hom}_A(B, A)$. (4) follows from (2) and from the equality $f'(x) = \sum_{i=0}^{n-1} b_i x^i$ which is deduced by differentiating the defining equation for the $b_0, \ldots, b_{n-1}$. To prove (5) replace A by $Q(A)$, B by $Q(B) = Q(A) \otimes_A B$ and use (4). To prove (6) write $b_i = \sum_{j=0}^{n-1} a_{ij} x^j$, $i = 0, \ldots, n-1$, with $a_{ij} \in A$. Then, using the formula for b_i given in the proof of part (1), we get $\det(a_{ij}) = (-1)^{\binom{n}{2}}$. From

$$f'(x)x^j = \sum_{i=0}^{n-1} \eta_i(f'(x)x^j)x^i = \sum_{i=0}^{n-1} \eta_{n-1}(f'(x)b_i x^j)x^i = \sum_{i=0}^{n-1} \mathrm{tr}(b_i x^j)x^i$$

one derives $N_A^B(f'(x)) = \det\left(\mathrm{tr}(b_i x^j)\right) = \det(a_{ij}) \cdot \det\left(\mathrm{tr}(x^{i+j})\right)$.   •

Let $X \to Y$ be a finite morphism of projective algebraic curves over k. Any formula which compares the genera of X and Y is called a R i e m a n n – H u r w i t z f o r m u l a. The discussions in Exercise 7.E.7 about the geometric genus are examples of such formulas. The case considered by Riemann and Hurwitz themselves is a finite (necessarily separable) morphism $X \to Y$ of smooth connected projective curves over $\mathbb{C}$ which is by Exercise 7.B.5 (4) uniquely determined by the finite (separable) field extension $\mathcal{R}(Y) \to \mathcal{R}(X)$. Let $\varphi : X \to Y$ be such a finite separable morphism of smooth connected projective curves, the base field k now being arbitrary. We have the exact sequence

$$0 \to \varphi^* \Omega_{Y|k} \to \Omega_{X|k} \to \Omega_{X|Y} \to 0.$$

The support of $\Omega_{X|Y}$ is a finite set of closed points of X which is called the r a m i f i c a t i o n   l o c u s   o f φ. The divisor

$$V_\varphi = V_{X|Y} := \sum_{x \in X_0} \ell_{\mathcal{O}_x}((\Omega_{X|Y})_x) \cdot x$$

and its degree

$$v_\varphi = v_{X|Y} := \deg V_{X|Y} = \sum_{x \in X_0} \ell_{\mathcal{O}_x}(\Omega_{X|Y})_x \cdot [\kappa(x) : k] = \deg \Omega_{X|Y}$$

are called the r a m i f i c a t i o n   d i v i s o r and the (t o t a l) d e g r e e   o f   r a m i - f i c a t i o n of φ, respectively. [13]) (Remember that X_0 denotes the set of closed points of a curve X.) The exact sequence above yields the equality

$$\deg \Omega_{X|k} = \deg \varphi^* \Omega_{Y|k} + \deg \Omega_{X|Y} = \deg \varphi^* \Omega_{Y|k} + v_{X|Y}.$$

Now, by 7.E.11 and Exercise 7.D.9 (3), we have $\deg \Omega_{X|k} = \deg \omega_X = -2\chi(\mathcal{O}_X)$, $\deg \Omega_{Y|k} = \deg \omega_Y = -2\chi(\mathcal{O}_Y)$, furthermore, by Exercise 7.D.1 (2), $\deg \varphi^* \Omega_{Y|k} = [\mathcal{R}(X) : \mathcal{R}(Y)] \cdot \deg \Omega_{Y|k}$. It follows:

---

[13]) "V" in $V_{X|Y}$ and "v" in $v_{X|Y}$ stand for "Verzweigung".

7.E.16. Riemann–Hurwitz Formula *Let* $\varphi : X \to Y$ *be a finite separable morphism of smooth and connected projective algebraic curves over the field* k. *Then*

$$g(X) - \text{Dim}_k \, \Gamma(\mathcal{O}_X) = [\mathcal{R}(X) : \mathcal{R}(Y)] \left(g(Y) - \text{Dim}_k \, \Gamma(\mathcal{O}_Y) \right) + \frac{1}{2} \, v_{X \mid Y} \, .$$

Note that $k \subseteq \Gamma(\mathcal{O}_Y) \subseteq \Gamma(\mathcal{O}_X)$ are finite field extensions. The calculation of the ramification divisor is easy if the field k is perfect and if φ is tamely ramified. Then all residue fields $\kappa(x), \kappa(y), x \in X_0, y \in Y_0$, are finite separable extensions of k and by Lemma 6.D.15 $(\Omega_{X \mid k})_x = \mathcal{O}_x \, d\pi_x$, $(\Omega_{Y \mid k})_y = \mathcal{O}_y \, d\pi_y$, where π_x and π_y generate the maximal ideals $\mathfrak{m}_x$ and $\mathfrak{m}_y$. For $x \in X_0$, let $e_x \in \mathbb{N}^*$ be the r a m i f i c a t i o n   e x p o n e n t   or the r a m i f i c a t i o n   i n d e x   in x defined by $\mathcal{O}_x \, \pi_{\varphi(x)} = \mathcal{O}_x \, \pi_x^{e_x}$, i. e. by $\pi_{\varphi(x)} = \pi_x^{e_x} \varepsilon_x$ with a unit $\varepsilon_x \in \mathcal{O}_x^\times$. The morphism φ is called t a m e l y   r a m i f i e d   in x if $e_x \neq 0$ in k. For such a point $x \in X_0$ $d\varepsilon_x = f_x d\pi_x$ for a suitable element $f_x \in \mathcal{O}_x$ and $d\pi_{\varphi(x)} = \varepsilon_x e_x \pi_x^{e_x-1} d\pi_x + \pi_x^{e_x} d\varepsilon_x = \widetilde{\varepsilon}_x \pi_x^{e_x-1} d\pi_x$ with a unit $\widetilde{\varepsilon}_x := \varepsilon_x e_x + \pi_x f_x \in \mathcal{O}_x^\times$, hence

$$(\Omega_{X \mid Y})_x = \mathcal{O}_x \, d\pi_x / \mathcal{O}_x \, d\pi_{\varphi(x)} = \mathcal{O}_x \, d\pi_x / \mathcal{O}_x \, \pi_x^{e_x-1} d\pi_x \, .$$

It follows: *If* k *is a perfect field and if* $X \to Y$ *is tamely ramified in all points* $x \in X_0$ *then*

$$V_{X \mid Y} = \sum_{x \in X_0} (e_x - 1) \, x \quad and \quad v_{X \mid Y} = \sum_{x \in X_0} (e_x - 1) \cdot [\kappa(x) : k] \, .$$

For an algebraically closed field of characteristic 0 we get $v_{X \mid Y} = \sum_{x \in X_0} (e_x - 1)$ for any finite morphism $X \to Y$ of smooth projective curves over k.

7.E.17. Example (T o p o l o g i c a l   g e n u s   v e r s u s   a r i t h m e t i c a l   g e n u s) Let X be a smooth connected projective curve over the field $\mathbb{C}$. By Exercise 6.E.22 the set X_0 of closed points in X carries the structure of a compact connected Riemann surface, i. e. a compact connected (complex) manifold of (complex) dimension 1 which is denoted by X^{an}. The connectedness of X^{an} (in its strong topology) follows from Theorem 6.E.24 (or more directly from its Corollary 6.E.25). We want to show that *the genus* $g(X)$ *coincides with the toplogical genus* $g_{\text{top}}(X^{\text{an}})$ *of the compact connected oriented topological surface* X^{an}, which is defined by the equation $\chi_{\text{top}}(X^{\text{an}}) = 2 - 2g_{\text{top}}(X^{\text{an}})$, where $\chi_{\text{top}}(X^{\text{an}})$ is the topological Euler–Poincaré characteristic of X^{an}. [14]) To prove this, let $\varphi : X \to \mathbb{P}^1_{\mathbb{C}}$ be a finite morphism which defines a ramified covering $\varphi^{\text{an}} : X^{\text{an}} \to (\mathbb{P}^1_{\mathbb{C}})^{\text{an}} = \overline{\mathbb{C}} = \mathbb{C} \cup \{\infty\}$ with sheet number $n := [\mathcal{R}(X) : \mathcal{R}(\mathbb{P}^1_{\mathbb{C}})]$. Let $V \subseteq X_0 = X^{\text{an}}$ be the ramification locus of φ and φ^{an} and let $W = \varphi(V) = \varphi^{\text{an}}(V)$ be its image in $\overline{\mathbb{C}}$. Then choose a triangulation of $\overline{\mathbb{C}}$ (homeomorphic to the 2-sphere S^2) which contains the set W as a part of the set of vertices and take its barycentric refinement with $V_{\overline{\mathbb{C}}}$ vertices, $E_{\overline{\mathbb{C}}}$ edges and $T_{\overline{\mathbb{C}}}$ triangles. One sees easily that this refinement can be lifted with φ^{an} to a triangulation of X^{an} with $V_{X^{\text{an}}} = n V_{\overline{\mathbb{C}}} - \sum_{x \in X^{\text{an}}} (e_x - 1)$ vertices, $E_{X^{\text{an}}} = n E_{\overline{\mathbb{C}}}$ edges and $T_{X^{\text{an}}} = n T_{\overline{\mathbb{C}}}$ triangles. Here e_x is the ramification exponent of φ (or φ^{an}) in $x \in X_0 = X^{\text{an}}$. Now, from the

---

[14]) $2g_{\text{top}}(X^{\text{an}})$ is the first Betti number of X^{an}.

equality $g_{\text{top}}(\overline{\mathbb{C}}) = g(\mathbb{P}^1_{\mathbb{C}}) = 0$ and the Riemann–Hurwitz Formula 7.E.16, the assertion results as follows:

$$2g_{\text{top}}(X) - 2 = -\chi_{\text{top}}(X^{\text{an}}) = -V_{X^{\text{an}}} + E_{X^{\text{an}}} - T_{X^{\text{an}}}$$
$$= n(-V_{\overline{\mathbb{C}}} + E_{\overline{\mathbb{C}}} - T_{\overline{\mathbb{C}}}) + \sum_{x \in X^{\text{an}}} (e_x - 1)$$
$$= n(-2) + \sum_{x \in X_0} (e_x - 1) = 2g(X) - 2.$$

7.E.18. Example (General Riemann – Hurwitz formula and discriminant ideals) For a morphism $\varphi : X \to Y$ of arbitrary integral projective curves over k the Riemann–Roch formula $\chi(\mathcal{O}_X) = \chi(\varphi_* \mathcal{O}_X) = \deg \varphi_* \mathcal{O}_X + [\mathcal{R}(X) : \mathcal{R}(Y)] \chi(\mathcal{O}_Y)$ yields the following general Riemann – Hurwitz formula:

$$g(X) - \text{Dim}_k \Gamma(\mathcal{O}_X) = [\mathcal{R}(X) : \mathcal{R}(Y)] \big(g(Y) - \text{Dim}_k \Gamma(\mathcal{O}_Y)\big) - \deg \varphi_* \mathcal{O}_X.$$

Comparing this with the formula in 7.E.16 one gets the equality

$$2 \deg \varphi_* \mathcal{O}_X = -v_{X \mid Y}$$

for a separable morphism $\varphi : X \to Y$ of smooth and connected projective curves. It is interesting to give a *direct proof* of this equality using discriminants. The $\mathcal{O}_Y$-algebra $\varphi_* \mathcal{O}_X$ is locally free of rank $n := [\mathcal{R}(X) : \mathcal{R}(Y)]$. The trace tr : $\varphi_* \mathcal{O}_X \to \mathcal{O}_Y$ induces a homomorphism

$$\text{discr}_{X \mid Y} : \Lambda^n \varphi_* \mathcal{O}_X \otimes_{\mathcal{O}_Y} \Lambda^n \varphi_* \mathcal{O}_X \to \mathcal{O}_Y$$

by $(s_1 \wedge \cdots \wedge s_n) \otimes (t_1 \wedge \cdots \wedge t_n) \mapsto \text{discr}_{X \mid Y}(s_1, \ldots, s_n; t_1, \ldots, t_n) := \det \big(\text{tr}(s_i t_j)\big)_{1 \le i,j \le n}$ for sections $s_1, \ldots, s_n; t_1, \ldots, t_n \in \Gamma(\varphi^{-1}(V), \mathcal{O}_X)$, $V \subseteq Y$ open and affine, which is called the discriminant homomorphism. It is injective because the trace is non-degenerated. (The field extension $\mathcal{R}(Y) \hookrightarrow \mathcal{R}(X)$ is finite and separable!) Its image $\mathcal{D}_{X \mid Y} = \mathcal{D}_\varphi$ in $\mathcal{O}_Y$ is called the discriminant ideal (sheaf) of φ. We claim that

$$\deg \big(\mathcal{O}_Y / \mathcal{D}_{X \mid Y}\big) = v_{X \mid Y} \ (= \deg \Omega_{X \mid Y}).$$

From the exact sequence $0 \to \Lambda^n \varphi_* \mathcal{O}_X \otimes_{\mathcal{O}_Y} \Lambda^n \varphi_* \mathcal{O}_X \to \mathcal{O}_Y \to \mathcal{O}_Y / \mathcal{D}_{X \mid Y} \to 0$ we then get the asserted result: $v_{X \mid Y} = \deg \big(\mathcal{O}_Y / \mathcal{D}_{X \mid Y}\big) = -\deg \big(\Lambda^n \varphi_* \mathcal{O}_X \otimes_{\mathcal{O}_Y} \Lambda^n \varphi_* \mathcal{O}_X\big) = -2 \deg \Lambda^n \varphi_* \mathcal{O}_X = -2 \deg \varphi_* \mathcal{O}_X$.

To prove the equality $\deg \big(\mathcal{O}_Y / \mathcal{D}_{X \mid Y}\big) = \deg \Omega_{X \mid Y}$, it suffices to show $\ell_{\mathcal{O}_y} \big(\mathcal{O}_y / (\mathcal{D}_{X \mid Y})_y\big) = \ell_{\mathcal{O}_y} \big((\varphi_* \Omega_{X \mid Y})_y\big) = \ell_{\mathcal{O}_y} \big(\Omega_{(\varphi_* \mathcal{O}_X) \mid \mathcal{O}_y}\big)$ for every closed point $y \in Y$. Let $A := \mathcal{O}_y$ and $B := (\varphi_* \mathcal{O}_X)_y$. Then $B = A[x] = A[X] / A[X] \mu$, where $\mu = X^n + a_{n-1} X^{n-1} + \cdots + a_0 \in A[X]$ is the characteristic polynomial of a generating element $x \in B$. We use the same notations as in the proof of Lemma 7.E.13 (A_{m} replaced by A and B_{m} replaced by B) and assume without loss of generality that k is infinite. Since $b_0, \ldots, b_{n-1}$ and $1, x, \ldots, x^{n-1}$ are A-bases of B, the element

$$\text{discr}\,(b_0, \ldots, b_{n-1}; 1, \ldots, x^{n-1}) = \det \big(\text{tr}(b_i x^j)\big) = N^B_A(\mu'(x))$$

(see Lemma 7.E.15 (6) for the last equality) generates $(\mathcal{D}_{X \mid Y})_y$. Furthermore, $\Omega_{B \mid A} \cong B / B \mu'(x)$, hence $\ell_A(\Omega_{B \mid A}) = \ell_A(B / B \mu'(x)) = \ell_A(A / A N^B_A \mu'(x)) = \ell_A(A / (\mathcal{D}_{X \mid Y})_y)$. This proof of the equality $-2 \deg \varphi_* \mathcal{O}_X = \deg \big(\mathcal{O}_Y / \mathcal{D}_{X \mid Y}\big) = \deg \Omega_{X \mid Y}$ for a separable morphism $\varphi : X \to Y$ uses only the facts that X and Y are normal and that the algebras $(\varphi_* \mathcal{O}_X)_y$ are cyclic over $\mathcal{O}_y$ (which, for instance, is always true if the field extensions $\kappa(y) \subseteq \kappa(x)$ are separable) but not the assumption that X and Y are smooth. We just mention without proof that even the assumption on the cyclicity of the algebras $\mathcal{O}_y \subseteq (\varphi_* \mathcal{O}_X)_y$, $y \in Y_0$, is not necessary (see, for instance, G. Scheja; U. Storch: Über Spurfunktionen bei vollständigen Durchschnitten, Jour. für die Reine und Angew. Math. **278/279**, 174-190 (1976) and [7]). Hence we have the following generalization of 7.E.16:

7.E.19. Riemann–Hurwitz Formula *Let* $\varphi : X \to Y$ *be a finite separable morphism of normal and connected projective algebraic curves over the field* k. *Then*

$$g(X) - \mathrm{Dim}_k\, \Gamma(\mathcal{O}_X) = [\mathcal{R}(X) : \mathcal{R}(Y)]\left(g(Y) - \mathrm{Dim}_k\, \Gamma(\mathcal{O}_Y)\right) + \frac{1}{2}\, v_{X|Y}\,,$$

where $v_{X|Y} := \deg \Omega_{X|Y} = \deg\left(\mathcal{O}_Y/\mathcal{D}_{X|Y}\right) = -2\, \deg \varphi_* \mathcal{O}_X$ *is again the total degree of ramification of* φ.

In the situation of 7.E.19 we have the following diagram of field extensions

$$\begin{array}{ccc} k & \subseteq\ \mathcal{O}(Y) & \subseteq\ \mathcal{O}(X) \\ & \mid \cap & \mid \cap \\ & \mathcal{R}(Y) & \subseteq\ \mathcal{R}(X) \end{array}$$

where the extension $\mathcal{R}(Y) \subseteq \mathcal{R}(X)$ is separable and where $\mathcal{O}(Y)$ (respectively $\mathcal{O}(X)$) is algebraically closed in $\mathcal{R}(Y)$ (respectively in $\mathcal{R}(X)$). Since $\mathcal{O}(Y)$ is algebraically closed in $\mathcal{R}(Y)$, the minimal polynomial of an element $s \in \mathcal{O}(X)$ over $\mathcal{O}(Y)$ coincides with the minimal polynomial of s over $\mathcal{R}(Y)$. In particular, the extension $\mathcal{O}(Y) \subseteq \mathcal{O}(X)$ is also separable and $\mathcal{O}(X)$ and $\mathcal{R}(Y)$ are linearly disjoint over $\mathcal{O}(Y)$. For, if $s \in \mathcal{O}(X)$ is a primitive element of $\mathcal{O}(X)$ over $\mathcal{O}(Y)$, then the tensor product $L := \mathcal{O}(X) \otimes_{\mathcal{O}(Y)} \mathcal{R}(Y)$ is isomorphic to $\mathcal{R}(Y)[T]/\mu\, \mathcal{R}(Y)[T]$, where $\mu \in \mathcal{O}(Y)[T]$ is the minimal polynomial of s over $\mathcal{O}(Y)$ (and over $\mathcal{R}(Y)$), and hence a field. From the chain $\mathcal{R}(Y) \subseteq L \subseteq \mathcal{R}(X)$, we get

$$\frac{[\mathcal{R}(X) : \mathcal{R}(Y)]}{[\mathcal{O}(X) : \mathcal{O}(Y)]} = [\mathcal{R}(X) : L]\,.$$

Since in the formula 7.E.19 the total degree of ramification $v_{X|Y}$ is non-negative, we get the inequality

$$g_{\mathrm{red}}(X) - 1 \geq \frac{[\mathcal{R}(X) : \mathcal{R}(Y)]}{[\mathcal{O}(X) : \mathcal{O}(Y)]}\left(g_{\mathrm{red}}(Y) - 1\right),$$

where, for an integral projective curve Z over k, $g_{\mathrm{red}}(Z) = \mathrm{Dim}_k\, \Gamma(Z, \omega_Z)/\, \mathrm{Dim}_k\, \mathcal{O}(Z)$ denotes the reduced genus of Z. In particular, the inequality $g_{\mathrm{red}}(X) \geq g_{\mathrm{red}}(Y)$ holds. If $g_{\mathrm{red}}(X) = 0 = g(X)$, *then* $g_{\mathrm{red}}(Y) = 0 = g(Y)$. This result is called L ü r o t h ' s t h e o r e m, see also Example 1.E.8 (7).

If the field k is perfect the inequality $g_{\mathrm{red}}(X) \geq g_{\mathrm{red}}(Y)$ *holds for an arbitrary finite morphism of normal and connected projective algebraic curves over k.* For the p r o o f it suffices, by the last remark, to consider the case that $\mathcal{R}(X) = \mathcal{R}(Y)[z]$ is a purely inseparable field extension of $\mathcal{R}(Y)$ of degree $p = \mathrm{char}\, k > 0$ with $z^p \in \mathcal{R}(Y) \setminus \mathcal{R}(Y)^p$. Then $z^p \notin \mathcal{O}(Y) = \mathcal{O}(Y)^p$ and from the diagram of field extensions

$$\begin{array}{ccccccc} \mathcal{R}(Y)^p & \subseteq & \mathcal{R}(X)^p & \subseteq & \mathcal{R}(Y) & \subseteq & \mathcal{R}(X) & = & \mathcal{R}(Y)[z] \\ \mid \cup & & \mid \cup & & \mid \cup \\ k(z^p) & = & k(z^p) & \subseteq & k(z) & , \end{array}$$

we have $[\mathcal{R}(X) : k(z^p)] = [\mathcal{R}(X) : k(z)] \cdot [k(z) : k(z^p)] = [\mathcal{R}(X) : \mathcal{R}(Y)] \cdot [\mathcal{R}(Y) : k(z^p)]$ which implies $[\mathcal{R}(X)^p : k(z^p)] = [\mathcal{R}(X) : k(z)] = [\mathcal{R}(Y) : k(z^p)]$ and hence $\mathcal{R}(X)^p = \mathcal{R}(Y)$ because of $\mathcal{R}(X)^p \subseteq \mathcal{R}(Y)$. It follows that X and Y are isomorphic as abstract curves over $k = k^p$ and that, in particular, $g_{\mathrm{red}}(X) = g_{\mathrm{red}}(Y)$.

Show that *for a finite morphism* $X \to Y$ *of smooth and connected projective algebraic curves over an arbitrary field* k *the inequality* $g_{\text{red}}(X) \geq g_{\text{red}}(Y)$ *holds.* (Look at the extension $X_{(\bar{k})} \to Y_{(\bar{k})}$ where $\bar{k}$ is an algebraic closure of k. The curves $X_{(\bar{k})}$ and $Y_{(\bar{k})}$ are also smooth (but not necessarily connected).)

In general, the inequality $g_{\text{red}}(X) \geq g_{\text{red}}(Y)$ *does not hold.* For example, let K be a field of characteristic $p > 2$ and let S be the standardly graded normal domain $S := k[X, Y, Z]/(T_1 X^p + T_2 Y^p + Z^p)$ over the rational function field $k := K(T_1, T_2)$.[15] By Plücker's formula (cf. Exercise 7.E.7), $Y := \text{Proj} S$ is a normal and connected projective curve over k with $\mathcal{O}(Y) = k$ and genus $g(Y) = g_{\text{red}}(Y) = (p-1)(p-2)/2 > 0$. Further, the canonical k-algebra homomorphism $S \to R$ with $R := k'[X, Y, Z]/(T_1^{1/p} X + T_2^{1/p} Y + Z)$, $k' := k(T_1^{1/p}, T_2^{1/p})$, is finite homogeneous and defines a finite k-morphism $X := \text{Proj} R = \mathbb{P}^1_{k'} \to Y$ with $g(X) = g_{\text{red}}(X) = 0$. In a similar way one constructs also examples in characteristic 2. – For further results and examples see Tate, J. : Genus Change in Inseparable Extenstions of Function Fields, Proc. Amer. Math. Soc. 3, 400-406 (1952).

7.E.20. Exercise Let k be a field of characteristic $p > 0$ and let $m \in \mathbb{N}^*$ be a positive integer which is coprime to p. Let X be the Riemann surface belonging to the function field $\mathcal{R}(X) := k(t)[Z]/(Z^p - Z - t^m) = k(t, z)$ which is an Artin–Schreier extension and, in particular, a Galois extension of the rational function field $k(t)$ (with Galois group $\mathbb{Z}_p$). The projection $X \to \mathbb{P}^1_k$ corresponding to the inclusion $k(t) \subseteq \mathcal{R}(X)$ is unramified over $\mathbb{A}^1_k = \text{Spec} k[t] \subseteq \mathbb{P}^1_k$ and *not* tamely ramified over $\infty \in \mathbb{P}^1_k$. Show that the genus $g(X)$ of X is $\frac{1}{2}(p-1)(m-1)$. (**Hint:** Use, for instance, the projection $X \to \mathbb{P}^1_k$ corresponding to the radical extension $k(z) \subseteq \mathcal{R}(X)$ and observe that $t^m - (Z^p - Z) = t^m - \prod_{a \in \mathbb{F}_p}(Z - a)$.) Show that the fields $k(t)[Z]/(Z^p - Z - t^{mp^r})$ have the same genus as X for any $r \in \mathbb{N}$.

At the end we want to show that the canonical module $\omega_X = \varphi^! \Omega_{\mathbb{P}^1_k | k}$ for a given projective algebraic curve X over a field k is well-defined, i. e. independent (up to isomorphism) of the chosen finite morphism $\varphi : X \to \mathbb{P}^1_k$. We know already that the dimension $\text{Dim}_k \, \Gamma(\mathcal{H}om_{\mathcal{O}_X}(\mathcal{F}, \omega_X))$ which occurs in the Riemann–Roch formula is indeed independent of φ (see the beginning of the present section). In the proof of the independence of ω_X itself we shall use little more homological algebra than before in these lectures, especially the first extension functors Ext^1 and $\mathcal{E}xt^1$ will occur. We hope that the reader will grasp an idea how to define canonical modules in more complicated situations.

7.E.21. Theorem *Let* X *be a projective algebraic curve over a field* k *and let* $\varphi, \psi : X \to Y := \mathbb{P}^1_k$ *be two finite* k*-morphisms. Then*

$$\varphi^! \omega_Y \cong \psi^! \omega_Y, \qquad \omega_Y := \Omega_{\mathbb{P}^1_k | k} \cong \mathcal{O}_{\mathbb{P}^1_k}(-2).$$

---

[15]) The normality of S can be verified, for instance, in the following way: The singular locus of the K-algebra $A := K[T_1, T_2, X, Y, Z]/(F)$, $F := T_1 X^p + T_2 Y^p + Z^p$, is, by Definition 6.D.22, the zero set of the residue classes of the partial derivatives $\partial F/\partial T_1$, $\partial F/\partial T_2$, $\partial F/\partial X$, $\partial F/\partial Y$, $\partial F/\partial Z$, i. e. of X^p, Y^p, which is of codimension 2 in $\text{Spec} A$. Since A is a complete intersection the normality of A follows from the normality criterion 6.B.4. Now, since A is normal, S is normal too.

PROOF. First of all, using 7.D.8, without loss of generality we may assume that X is Cohen–Macaulay. Then the $\mathcal{O}_Y$-algebras $\varphi_*\mathcal{O}_X$ and $\psi_*\mathcal{O}_X$ are locally free over $\mathcal{O}_Y$. _The isomorphism_ $\varphi^!\omega_Y \cong \psi^!\omega_Y$ _is quite canonical_, to construct it we compare both $\mathcal{O}_X$-modules $\varphi^!\omega_Y$ and $\psi^!\omega_Y$ with a third $\mathcal{O}_X$-module which is defined symmetrically by using φ and ψ in the following way: Let $\alpha := (\varphi, \psi) : X \to Y \times Y$ (with $\times := \times_k$) be the morphism defined by the equations $p_1 \circ \alpha = \varphi$ and $p_2 \circ \alpha = \psi$, where p_1, p_2 are the projections of $Y \times Y$ onto its components. The morphism α is also a finite morphism (since, for instance, $\alpha = (\varphi \times \psi) \circ \Delta_X$ is the composition of the diagonal morphism $\Delta_X : X \to X \times X$ (which is even a closed embedding) and the product $\varphi \times \psi : X \times X \to Y \times Y$ of the finite morphisms φ, ψ). Since the dimensions of X and $Y \times Y$ differ by 1, for a coherent $\mathcal{O}_{Y \times Y}$- module $\mathcal{G}$ we define the coherent $\mathcal{O}_X$-module $\alpha^!\mathcal{G}$ by the equation $\alpha_*\alpha^!\mathcal{G} = \mathcal{E}xt^1_{\mathcal{O}_{Y \times Y}}(\alpha_*\mathcal{O}_X, \mathcal{G})$, i.e for an arbitrary affine open subset $W \subseteq Y \times Y$

$$\Gamma(\alpha^{-1}(W), \alpha^!\mathcal{G}) = \mathrm{Ext}^1_{\Gamma(W, \mathcal{O}_{Y \times Y})}\big(\Gamma(\alpha^{-1}(W), \mathcal{O}_X), \Gamma(W, \mathcal{G})\big).$$

To complete the proof of 7.E.21 we shall show

$$\varphi^!\omega_Y \cong \alpha^!(\omega_Y \otimes \omega_Y) = (\varphi, \psi)^!(\omega_Y \otimes \omega_Y) \cong \psi^!\omega_Y.$$

Here we have used the following notation: If $\mathcal{E}$ and $\mathcal{F}$ are quasi-coherent modules over k-schemes X and Y respectively, then $\mathcal{E} \otimes \mathcal{F}$ denotes the quasi-coherent $\mathcal{O}_{X \times Y}$-module $p_X^*(\mathcal{E}) \otimes_{\mathcal{O}_{X \times Y}} p_Y^*(\mathcal{F})$. Note that, for affine open subsets $U \subseteq X$ and $V \subseteq Y$,

$$\Gamma(U \times V, \mathcal{E} \otimes \mathcal{F}) = \Gamma(U, \mathcal{E}) \otimes \Gamma(V, \mathcal{F})$$

(with $\otimes := \otimes_k$). Since α is symmetric in φ and ψ, it is enough to construct the first isomorphism $\varphi^!\omega_Y \cong \alpha^!(\omega_Y \otimes \omega_Y)$. It will suffice to show that the $\alpha_*\mathcal{O}_X$-modules $\alpha_*\varphi^!\omega_Y$ and $\alpha_*\alpha^!(\omega_Y \otimes \omega_Y) = \mathcal{E}xt^1_{\mathcal{O}_{Y \times Y}}(\alpha_*\mathcal{O}_X, \omega_Y \otimes \omega_Y)$ are isomorphic. To do this we interpret α as the composition

$$X \xrightarrow{\ \Gamma_\psi = (\mathrm{id}_X, \psi)\ } X \times Y \xrightarrow{\ \varphi \times \mathrm{id}_Y\ } Y \times Y.$$

The graph morphism Γ_ψ is a closed embedding and the base change morphism $\varphi_{(Y)} = \varphi \times \mathrm{id}_Y$ is again finite with a locally free $\mathcal{O}_{Y \times Y}$-algebra $(\varphi \times \mathrm{id}_Y)_*\mathcal{O}_{X \times Y} = (\varphi_*\mathcal{O}_X) \otimes \mathcal{O}_Y$. We have the following exact sequence of $\mathcal{O}_{X \times Y}$-modules

$$0 \to \mathcal{J} \longrightarrow \mathcal{O}_{X \times Y} \longrightarrow (\Gamma_\psi)_*\mathcal{O}_X \to 0.$$

The ideal sheaf $\mathcal{J}$ is invertible. Explicitly, if $V_+ = \mathrm{Spec}\, k[t] \subseteq Y$ is one of the canonical affine open subsets of $Y = \mathbb{P}^1_k$ and if $k[t] = \Gamma(V_+, Y) \to B_+ := \Gamma(\psi^{-1}(V_+), X)$, $t \mapsto b_+$, is the k-algebra homomorphism induced by ψ, then $\Gamma(\psi^{-1}(V_+) \times V_+, \mathcal{J})$ is generated over $\Gamma(\psi^{-1}(V_+) \times V_+, \mathcal{O}_{X \times Y}) = B_+ \otimes k[t] = B_+[t]$ by (the non-zero divisor) $t - b_+$. Therefore for any coherent $\mathcal{O}_{X \times Y}$-module $\mathcal{H}$, the coherent $\mathcal{O}_{X \times Y}$-module $\mathcal{E}xt^1_{\mathcal{O}_{X \times Y}}((\Gamma_\psi)_*\mathcal{O}_X, \mathcal{H})$ is the cokernel of the canonical restriction map $\mathcal{H} \to \mathcal{H}om_{\mathcal{O}_{X \times Y}}(\mathcal{J}, \mathcal{H})$.

For a coherent $\mathcal{O}_{Y \times Y}$-module $\mathcal{G}$, let the $\mathcal{O}_{X \times Y}$-module $(\varphi \times \mathrm{id}_Y)^!\mathcal{G}$ be defined in the usual way by $(\varphi \times \mathrm{id}_Y)_*(\varphi \times \mathrm{id}_Y)^!\mathcal{G} = \mathcal{H}om_{\mathcal{O}_{Y \times Y}}((\varphi \times \mathrm{id}_Y)_*\mathcal{O}_{X \times Y}, \mathcal{G})$. Then, applying the last result for $\mathcal{H} := (\varphi \times \mathrm{id}_Y)^!\mathcal{G}$, the $\mathcal{O}_{Y \times Y}$-module $(\varphi \times \mathrm{id}_Y)_*\mathcal{E}xt^1_{\mathcal{O}_{X \times Y}}((\Gamma_\psi)_*\mathcal{O}_X, (\varphi \times \mathrm{id}_Y)^!\mathcal{G})$ is the cokernel of $\mathcal{H}om_{\mathcal{O}_{Y \times Y}}((\varphi \times \mathrm{id}_Y)_*\mathcal{O}_{X \times Y}, \mathcal{G}) \to \mathcal{H}om_{\mathcal{O}_{Y \times Y}}((\varphi \times \mathrm{id}_Y)_*\mathcal{J}, \mathcal{G})$. But this cokernel is $\alpha_*\alpha^!\mathcal{G} = \mathcal{E}xt^1_{\mathcal{O}_{Y \times Y}}(\alpha_*\mathcal{O}_X, \mathcal{G})$, since

$$0 \to (\varphi \times \mathrm{id}_Y)_*\mathcal{J} \longrightarrow (\varphi \times \mathrm{id}_Y)_*\mathcal{O}_{X \times Y} \longrightarrow (\varphi \times \mathrm{id}_Y)_*(\Gamma_\psi)_*\mathcal{O}_X = \alpha_*\mathcal{O}_X \to 0.$$

is a locally free resolution of $\alpha_*\mathcal{O}_X$ over $\mathcal{O}_{Y \times Y}$. In particular, if $\mathcal{E}$ and $\mathcal{F}$ are coherent $\mathcal{O}_Y$-modules, then $(\varphi \times \mathrm{id}_Y)^!(\mathcal{E} \otimes \mathcal{F}) = \varphi^!\mathcal{E} \otimes \mathcal{F}$ (since $\varphi_*\mathcal{O}_X$ is locally free over $\mathcal{O}_Y$) and

$$(\varphi \times \mathrm{id}_Y)_* \mathcal{E}xt^1_{\mathcal{O}_{Y \times Y}}\big((\Gamma_\psi)_* \mathcal{O}_X, \varphi^! \mathcal{E} \otimes \mathcal{F}\big) = \alpha_* \alpha^! (\mathcal{E} \otimes \mathcal{F}).$$

Hence to finish the proof it suffices to show that, for every coherent $\mathcal{O}_X$-module $\mathcal{M}$,

$$\mathcal{E}xt^1_{\mathcal{O}_{X \times Y}}\big((\Gamma_\psi)_* \mathcal{O}_X, \mathcal{M} \otimes \omega_Y\big) = (\Gamma_\psi)_* \mathcal{M},$$

because with this we get the required equality

$$\alpha_* \alpha^! (\omega_Y \otimes \omega_Y) = (\varphi \times \mathrm{id}_Y)_* \, \mathcal{E}xt^1_{\mathcal{O}_{Y \times Y}}\big((\Gamma_\psi)_* \mathcal{O}_X, \varphi^! \omega_Y \otimes \omega_Y\big) = (\varphi \times \mathrm{id}_Y)_* (\Gamma_\psi)_* \varphi^! \omega_Y = \alpha_* \varphi^! \omega_Y.$$

Now, let the $\mathcal{O}_X$-module $\mathcal{M}$ be given and let $V_\pm = \mathrm{Spec}\, k[t^{\pm 1}] \subseteq Y = \mathbb{P}^1_k$ be the canonical affine open cover of $\mathbb{P}^1_k$. Furthermore, let $\psi^{-1}(V_\pm) = \mathrm{Spec}\, B_\pm$, $M_\pm = \Gamma(\psi^{-1}(V_\pm), \mathcal{M})$ and the morphism ψ be defined by $t^{\pm 1} \mapsto b_\pm$ on $\psi^{-1}(V_\pm)$, respectively. Then the two affine open sets $\psi^{-1}(V_\pm) \times V_\pm \subseteq X \times Y$ cover the graph $\Gamma_\psi(X)$. Using the exact sequence $0 \to \mathcal{I} \to \mathcal{O}_{X \times Y} \to (\Gamma_\psi)_* \mathcal{O}_X \to 0$ from above, the $B_\pm \otimes k[t^{\pm 1}] (= B_\pm[t^{\pm 1}])$-modules $\Gamma\big(\psi^{-1}(V_\pm) \times V_\pm, \mathcal{E}xt^1_{\mathcal{O}_{X \times Y}}((\Gamma_\psi)_* \mathcal{O}_X, \mathcal{M} \otimes \omega_Y)\big)$ are the cokernels of

$$M_\pm[t^{\pm 1}]\, dt^{\pm 1} \longrightarrow \mathrm{Hom}_{B_\pm[t^{\pm 1}]}\big(B_\pm[t^{\pm 1}](t^{\pm 1} - b_\pm), M_\pm[t^{\pm 1}]\, dt^{\pm 1}\big) = \frac{M_\pm[t^{\pm 1}]}{(t^{\pm 1} - b_\pm)}\, dt^{\pm 1}.$$

These cokernels can be identified with $M_\pm$ via the isomorphisms

$$\left[\frac{\sum_{i \geq 0} m_i t^{\pm i}\, dt^{\pm 1}}{t^{\pm 1} - b_\pm}\right] \mapsto \sum_{i \geq 0} m_i b_\pm^i, \quad m_i \in M_\pm.$$

We have to show that on $(\psi^{-1}(V_+) \times V_+) \cap (\psi^{-1}(V_-) \times V_-) = \psi^{-1}(V_+ \cap V_-) \times (V_+ \cap V_-)$ both these identifications coincide. But, $dt = -t^2\, dt^{-1}$ and, for $m_i \in \Gamma(\psi^{-1}(V_+ \cap V_-), \mathcal{M}) = (M_+)_{b_+} = (M_-)_{b_-}$ with $m_i = 0$ for almost all $i \in \mathbb{Z}$, we have

$$\frac{\sum_{i \in \mathbb{Z}} m_i t^i\, dt}{t - b_+} = \frac{-t^2 \sum_{i \in \mathbb{Z}} m_i t^i\, dt^{-1}}{t - b_+} = \frac{\sum_{i \in \mathbb{Z}} m_i b_- t^{i+1}\, dt^{-1}}{t^{-1} - b_-}$$

(since $b_+ b_- = 1$ and $t - b_+ = -t\, b_+(t^{-1} - b_-)$ on the intersections $\psi^{-1}(V_+ \cap V_-)$ and $\psi^{-1}(V_+ \cap V_-) \times (V_+ \cap V_-)$ respectively) and, finally, $\sum_{i \in \mathbb{Z}} m_i b_+^i = \sum_{i \in \mathbb{Z}} m_i b_- b_-^{-(i+1)}$ in $\Gamma(\psi^{-1}(V_+ \cap V_-), \mathcal{M})$. $\bullet$

7.E.22. Exercise In this exercise we give some indications on the local structure of the canonical sheaf ω_X of a projective curve X over a field k which we assume without loss of generality to be Cohen–Macaulay. By Proposition 7.D.8 ω_X is a Cohen–Macaulay $\mathcal{O}_X$-module too. Let $\varphi : X \to \mathbb{P}^1_k$ be a finite morphism.

(1) Let $x \in X$ and $y := \varphi(x) \in \mathbb{P}^1_k$ (x, y are not necessarily closed points). Show that $\omega_x / \mathfrak{m}_y \omega_x \cong \mathrm{Hom}_{\kappa(y)}(\mathcal{O}_x / \mathfrak{m}_y \mathcal{O}_x, \kappa(y))$.

(2) To study the module $\omega_x / \mathfrak{m}_y \omega_x$ in (1), let, more generally, L be a field and S be a finite (commutative) local L-algebra. Show that the S-module $E := \mathrm{Hom}_L(S, L)$ is an injective hull of $k_S = S / \mathfrak{m}_S$, i.e. E is an injective S-module with $\mathrm{Dim}_{k_S} \mathrm{Soc}_S E = 1$. (**Hint:** Use the canonical isomorphism $\mathrm{Hom}_S(M, E) \cong \mathrm{Hom}_L(M, L)$ for every S-module M and identify $\mathrm{Soc}_S E$ with $\mathrm{Hom}_L(k_S, L)$.) Furthermore, show that the minimal number of generators of E is the type $\tau(S)$ of S, i.e. $\mathrm{Dim}_{k_S} E / \mathfrak{m}_S E = \mathrm{Dim}_{k_S} \mathrm{Soc}_S S$. (**Hint:** For the concept of type see Exercise 6.B.17 (2). – Applying $\mathrm{Hom}_S(-, E)$ to the the exact sequence $0 \to \mathfrak{m}_S E \to E \to E / \mathfrak{m}_S E \to 0$, one gets $0 \to \mathrm{Hom}_S(E / \mathfrak{m}_S E, E) \to \mathrm{Hom}_S(E, E) = S \to \mathrm{Hom}_S(\mathfrak{m}_S E, E)$ which gives the equality $\mathrm{Soc}_S S = \mathrm{Hom}_S(E / \mathfrak{m}_S E, E)$. – The equations $\mathrm{Dim}_{k_S} \mathrm{Soc}_S E = 1$ and $\mathrm{Dim}_{k_S} E / \mathfrak{m}_S E = \mathrm{Dim}_{k_S} \mathrm{Soc}_S S$ hold for the injective

hull $E = E(k_S)$ of the residue field k_S of an arbitrary Artinian local ring S.)

(3) Show that for $x \in X$, the $\mathcal{O}_x$-module ω_x is a (maximal) Cohen–Macaulay module of type 1 and its minimal number of generators is the type $\tau(\mathcal{O}_x)$ of $\mathcal{O}_x$. (**Hint**: Use (1), (2) with $S := \mathcal{O}_x / \mathfrak{m}_y \mathcal{O}_x$, $L := \kappa(y)$ and observe that $\mathfrak{m}_y$ is a principal ideal in $\mathcal{O}_y$, $y := \varphi(x)$.)

(4) Show that for $x \in X$ the following conditions are equivalent: a) ω_x is generated by one element. b) $\omega_x \cong \mathcal{O}_x$ as $\mathcal{O}_x$-modules. c) The local ring $\mathcal{O}_x$ is Gorenstein (i. e. Cohen–Macaulay of type 1). In particular, ω_X is an invertible $\mathcal{O}_X$-module if and only if $\mathcal{O}_x$ is Gorenstein for every $x \in X$, and ω_X is an $\mathcal{O}_X$-module with rank (then necessarily of rank 1) if and only if the stalks $\mathcal{O}_\xi$ are Gorenstein for all generic points $\xi \in X$. – A curve X for which the $\mathcal{O}_X$-module ω_X is invertible is called a G o r e n s t e i n c u r v e and a curve for which ω_X has rank 1 is called a curve which is g e n e r i c a l l y G o r e n s t e i n .

It is only with the appearance of the Riemann–Roch Theorem that the theory of algebraic curves begins to get really fascinating. This theorem can be used as a starting point for a detailed study of (projective) algebraic curves. It is a classical subject and still an active area of research. After reading and understanding this book, a reader will be able to proceed to study more advanced topics in general Algebraic Geometry and Commutative Algebra. However, this study would require another series of lectures. We finish this book with the following two simple examples.

7.E.23. Example (C u r v e s o f g e n u s 0 and the p r o j e c t i v e l i n e) Let X be an integral projective algebraic curve of genus 0 over the field k with function field $\mathcal{R}(X)$. We assume without loss of generality that $k = \mathcal{O}(X) = \Gamma(\mathcal{O}_X)$. Note that this condition is always satisfied if X contains a k-rational point. Let $\nu : \overline{X} \to X$ be the normalization of X. By Exercise 7.E.7 $g(\overline{X}) = g_{\mathrm{geom}}(X) = \mathrm{Dim}_k \, \mathcal{O}(\overline{X}) - 1 - \delta(X)$. It follows: $X = \overline{X}$ if and only if $\mathcal{O}(\overline{X}) = k$, i. e. if and only if k is algebraically closed in $\mathcal{R}(X)$.

First we consider the case that $X = \overline{X}$ is normal and has a divisor D of degree 1 (see Section 7.D just after Exercise 7.D.5 for the concept of divisors). We claim that X has even a k- rational point. By Riemann's inequality $\mathrm{Dim}_k \, D \geq \deg D + 1 = 2$, i. e. there exists a non-constant rational function $f \in \mathcal{R}(X)$ such that $\mathrm{div} f + D \geq 0$, i. e. $\mathrm{div} f + D = \sum_{Q \in X_0} \nu_Q Q$, $\nu_Q \geq 0$. Because of $1 = \deg \mathrm{div} f + \deg D = \sum_{Q \in X_0} \nu_Q [\kappa(Q) : k]$, there exists $P \in X_0$ with $[\kappa(P) : k] = 1$, i. e. P is a k-rational point. Now using the divisor $D = P$, we get a non-constant rational function $t \in \mathcal{R}(X)$ with $\mathrm{div} t + P \geq 0$, i. e. t has only one pole in P which is moreover of order 1. Then t defines a morphism $t : X \to \mathbb{P}^1_k$ with $1 = [\mathcal{R}(X) : \mathcal{R}(\mathbb{P}^1_k)]$, i. e. $\mathcal{R}(X) = k(t)$, and hence $t : X \to \mathbb{P}^1_k$ is an isomorphism. Altogether we have the following characterization of $\mathbb{P}^1_k$:

7.E.24. Proposition *For an integral projective algebraic curve X over a field k, the following conditions are equivalent*: (1) $X \cong \mathbb{P}^1_k$. (2) X *is normal of genus 0 and contains a k-rational point.* (3) X *is normal of genus 0 and has a divisor of degree 1.* (4) X *is of genus 0 and contains a normal k-rational point.*

The curve $\mathrm{Proj}\,\mathbb{R}[X, Y, Z]/(X^2 + Y^2) \subseteq \mathbb{P}^2_\mathbb{R}$ is an integral projective non-normal curve over $\mathbb{R}$ of genus 0 with the (only) $\mathbb{R}$-rational point $\langle 0, 0, 1 \rangle \in \mathbb{P}^2_\mathbb{R}(\mathbb{R})$. What is its normalization?

Now, let us assume that the curve $X = \overline{X}$ is normal of genus 0 and has no divisor of degree 1 (but still with $\mathcal{O}(X) = k$). By 7.E.2 $\deg \omega_X = -2$ and hence $\deg \omega_X^* = 2$.

Let $N := -K$ be the negative of a canonical divisor K (with $\mathcal{L}(K) \cong \omega_X$). Such an N (with $\mathcal{L}(N) \cong \omega_X^*$) is called an **a n t i c a n o n i c a l   d i v i s o r**. By Riemann's inequality 7.D.4 $\mathrm{Dim}_k\, N \geq \deg N + 1 = 3$, indeed, $\mathrm{Dim}_k\, N = \mathrm{Dim}_k\, \omega_X^* = 3$, since $\deg(-N + K) = -4$ and so $\mathrm{Dim}_k(-N + K) = 0$ (by Exercise 7.D.6, for instance). Let $f \in \mathcal{R}(X)$ be a non-constant rational function with $\mathrm{div} f + N = \sum_{Q \in X_0} \nu_Q\, Q \geq 0$. Because of $\deg(\mathrm{div} f + N) = \deg N = 2$ and $[\kappa(Q) : k] \geq 2$ for all $Q \in X_0$ there exists a point $P \in X_0$ with $\mathrm{div} f + N = P$ and $[\kappa(P) : k] = 2$. Replacing N by the divisor P we get a non-constant rational function $t \in \mathcal{R}(X)$ with $\mathrm{div}\, t + P \geq 0$. Then t has only one pole in P (and one zero in a closed point $Q \neq P$) and the inclusion $k(t) \subseteq \mathcal{R}(X)$ defines a morphism $X \to \mathbb{P}_k^1$ with $[\mathcal{R}(X) : \mathcal{R}(\mathbb{P}_k^1)] = 2$, i.e., *X is the Riemann surface of a quadratic extension $\mathcal{R}(X)$ of the rational function field $k(t)$*.

Let us consider, quite generally, Riemann surfaces X belonging to an arbitrary quadratic extension L of $k(t)$ (and assume further $\mathcal{O}(X) = k$, i.e. k algebraically closed in L) with the finite projection $\varphi : X \to \mathbb{P}_k^1$ of degree $[L : k(t)] = 2$. Let char $k \neq 2$. Then $\mathcal{R}(X) = L = k(t)[u]$ where $u^2 = F$ and where $F = \pi_1 \cdots \pi_r \in k[t]$ is a polynomial of degree $d > 0$ with only simple prime factors $\pi_1, \ldots, \pi_r$. Since the $k[t]$-algebra $k[t, u] \cong k[t][U]/(U^2 - F)$ is normal [16] $\varphi^{-1}\big(\mathrm{Spec}\, k[t]\big) = \mathrm{Spec}\, k[t, u]$ and, similarly, $\varphi^{-1}\big(\mathrm{Spec}\, k[t^{-1}]\big) = \mathrm{Spec}\, k[t^{-1}, u/t^{\lceil d/2 \rceil}]$ with $(u/t^{\lceil d/2 \rceil})^2 = t^{-\varepsilon} \cdot (F(t)/t^d) \in k[t^{-1}]$, $\varepsilon := 2\lceil d/2 \rceil - d \in \{0, 1\}$. Since the field extension $k(t) \subseteq \mathcal{R}(X)$ is separable we may compute the genus $g(X)$ with the formula of 7.E.19. Because of $\Omega_{k[t,u] \mid k[t]} \cong k[t]/(F)$ and $\Omega_{k[t^{-1}, u] \mid k[t^{-1}]} \cong k[t^{-1}]/\big(t^{-\varepsilon}(F(t)/t^d)\big)$ and from 7.E.19 we get $\deg \Omega_{X \mid \mathbb{P}_k^1} = d + \varepsilon = 2\lceil d/2 \rceil$ and $2(g(X) - 1) = 4(-1) + d + \varepsilon$, hence

$$g(X) = \lceil d/2 \rceil - 1 = \lceil (d - 2)/2 \rceil.$$

We emphasize that X is in general not smooth. In fact, prove that X *is smooth if and only if the polynomial $F \in k[t]$ is separable, i.e.* $\mathrm{GCD}\,(F, F') = 1$ (which is of course always the case if the field k is perfect). In the smooth case X is called a **h y p e r e l l i p t i c   c u r v e**. The most famous of these curves are the curves X with $\mathcal{R}(X) = k(t)[u]$ where $u^2 = F$ and $F = a(t - a_1) \cdots (t - a_d) \in k[t]$ splits into *simple* linear factors. The projection $\varphi : X \to \mathbb{P}_k^1$ *of degree 2 is ramified over the points $a_1, \ldots, a_d \in k \subseteq \mathbb{P}_k^1$ and over $\infty \in \mathbb{P}_k^1$ in case d is odd.* The ramification exponent is 2 at every ramification point of φ in X, the total ramification order is $2\lceil d/2 \rceil$.

It follows that $g(X) = 0$ if and only if $d = 1$ or $d = 2$. If $d = 1$ then $X \cong \mathbb{P}_k^1$. If $d = 2$ and $F = a_2 t^2 + a_1 t + a_0$, $a_2 \neq 0$, then $X = \mathrm{Proj}\, k[U, T, V]/(U^2 - a_2 T^2 - a_1 T V - a_0 V^2)$. If the quadratic form $U^2 - a_2 T^2 - a_1 T V - a_0 V^2$ (which is of rank 3) has a non-trivial zero, i.e. if it is isotropic, then X has a k-rational point and $X \cong \mathbb{P}_k^1$.

7.E.25. Proposition *Let k be a field of characteristic $\neq 2$. The normal integral projective algebraic curves X of genus 0 over k with $\mathcal{O}(X) = k$ and with no k-rational point are the curves $X \cong \mathrm{Proj}\, k[U, T, V]/(Q)$ where Q is an anisotropic quadratic form in the three variables U, T, V over k. All these curves are smooth.*

By change of variables one may assume that $Q = U^2 - aT^2 - bV^2$, $a, b \in k^\times$. If Q is isotropic then $\mathrm{Proj}\, k[U, T, V]/(Q) \cong \mathbb{P}_k^1$. The reader may try to give the isomorphism

---

[16] If $h = f + gu$, $f, g \in k(t)$, is integral over $k[t]$ then $\mathrm{tr}_{k(t)}^L h = 2f$, $\mathrm{N}_{k(t)}^L h = f^2 - g^2 F \in k(t)$ are integral over $k[t]$ which implies $f, g \in k[t]$.

classes of the curves described in 7.E.25. The simplest non-trivial example is $k = \mathbb{R}$: *The only smooth projective algebraic curves X of genus 0 over $\mathbb{R}$ with $\mathcal{O}(X) = \mathbb{R}$ are up to isomorphism $\mathbb{P}^1_{\mathbb{R}} \cong$ Proj $\mathbb{R}[U, T, V]/(U^2+T^2-V^2)$ and Proj $\mathbb{R}[U, T, V]/(U^2+T^2+V^2)$.*

As mentioned at the end of Example 7.E.3 the canonical module ω_X of a curve X as in 7.E.25 is the $\mathcal{O}_X$-module $\mathcal{O}_X(-1) \cong \Omega_{X|k}$ which, by the way, follows also from Theorem 6.E.20. Its dual $\omega_X^* \cong \Omega_{X|k}^* \cong \mathcal{O}_X(1)$ is the anticanonical module $\mathcal{L}(N)$ which we started with. $\mathrm{Dim}_k \, \Gamma(\Omega_{X|k}^*) = 3$ follows now also from Lemma 7.A.3. The representation of X as in 7.E.25 is also called the a n t i c a n o n i c a l   r e p r e s e n t a t i o n of X.

The discussion of the case char $k = 2$ is left to the reader.

7.E.26. Example (C u r v e s   o f   g e n u s   1   a n d   e l l i p t i c   c u r v e s) Let E be an integral normal projective curve of genus 1 over the field k with $\mathcal{O}(E) = k$. We even assume that E has a k-rational point P_0.[17]) The canonical sheaf ω_X has degree $2g(E) - 2 = 0$ and $\mathrm{Dim}_k \, \Gamma(\omega_E) = g(E) = 1$. It follows that a non-zero section of ω_E has no zeros and hence

$$\omega_E \cong \mathcal{O}_E.$$

By Riemann–Roch Theorem, for the divisor $2P_0$ of degree 2, $\mathrm{Dim}_k(2P_0) = \deg 2P_0 = 2$. A non-constant rational function t with $\mathrm{div}\, t + 2P_0 \geq 0$ has a pole of order $\nu \leq 2$ in P_0 and defines a finite morphism $\varphi : E \to \mathbb{P}^1_k$ of degree $[\mathcal{R}(E) : k(t)] = \nu = 2$. *Let* char $k \neq 2$. Then, as discussed in the previous example,

$$\mathcal{R}(E) = k(t)[u] \quad \text{with} \quad u^2 = F(t),$$

where $F \in k[t]$ is a polynomial of degree $d = 3$ or $d = 4$ with only simple prime factors. Since φ has a pole of order 2 in the k-rational point P_0, $\varepsilon = 2\lceil d/2 \rceil - d$ has to be 1, hence $d = 3$, $u^2 = F(t) = a_3 t^3 + a_2 t^2 + a_1 t + a_0$, $a_3 \neq 0$, and

$$E = \text{Proj } k[T, U, V]/(U^2 V - (a_3 T^3 + a_2 T^2 V + a_1 T V^2 + a_0 V^3)).$$

That the genus of such a curve E is 1 follows also from Plücker's formula at the end of Example 7.E.3. If char $k \neq 3$, then we can also assume $a_2 = 0$. This description of E is called the W e i e r s t r a s s   n o r m a l   f o r m of E. The given k-rational point on E is $P_0 = \langle t_0, u_0, v_0 \rangle = \langle 0, 1, 0 \rangle$, usually written as ∞. If E is even smooth which is the case if and only if the polynomial F is separable over k, then E is called an e l l i p t i c   c u r v e with prescribed k-rational point $P_0 = \infty$. Note that the k-rational points are always smooth points of E (see, for example, Theorem 6.D.16).

7.E.27. Proposition *The set $E(k)$ of k-rational points on an integral normal projective curve E of genus 1 over the field k with a prescribed point $P_0 \in E(k)$ carries a canonical group structure $(E(k), P_0)$ with additive identity P_0 such that $\iota : E(k) \to \mathrm{Cl}^0 E$, $P \mapsto [P - P_0]$, is a group isomorphism from $(E(k), P_0)$ onto the special divisor class group $\mathrm{Cl}^0 E = \mathrm{Div}^0 E/\mathrm{PDiv}\, E \subseteq \mathrm{Cl}\, E = \mathrm{Div}\, E/\mathrm{PDiv}\, E \cong \mathrm{Pic}\, E$ (see Section 7.D just after Exercise 7.D.5).*

PROOF. At first we remark that because of $g(E) \neq 0$, there is no rational function $f \in \mathcal{R}(E)^\times$ which has no other poles than a pole of order 1 at a k-rational point of E.

---

[17]) For instance, $X = \text{Proj } \mathbb{R}[X_0, X_1, X_2, X_3]/(F_1, F_2) \subseteq \mathbb{P}^3_{\mathbb{R}}$, where $F_1 := X_0^2 - X_1^2 - X_2^2$ and $F_2 := X_0^2 - (X_1 - 3X_0)^2 - X_3^2$, is smooth with $\mathcal{O}(X) = \mathbb{R}$ and has no $\mathbb{R}$-rational point. By Proposition 7.E.4 $g(X) = 1$. Sketch the zero sets of F_1 and F_2, respectively, in the affine part $x_0 \neq 0$ in $\mathbb{R}^3$.

The map ι is injective. Because, if $[P - P_0] = [Q - P_0]$, $P, Q \in E(k)$, then $P - Q = \text{div} f$, $f \in \mathcal{R}(E)^\times$, is a principal divisor. If f is non-constant, i.e. $\text{div} f \neq 0$, i.e. $P \neq Q$, then f would have no other poles than a pole of order 1 in Q, a contradiction.

The image of ι is a subgroup of $\text{Cl}^0 E$. (1) $0 = \iota(P_0) \in \text{Im} \, \iota$. (2) If $P \in E(k)$, $P \neq P_0$, then $\deg(2P_0 - P) = 1$, hence $\text{Dim}_k(2P_0 - P) = 1$ by Riemann–Roch. If $f \in \mathcal{R}(E)^\times$ is a rational function with $\text{div} f + (2P_0 - P) \geq 0$ then f has no other poles than a pole of order at most 2 at P_0 and hence no other poles than a pole of order exactly 2 at P_0. Therefore $\text{div} f = P + Q - 2P_0$ for some $Q \in E(k)$ and $\iota(P) + \iota(Q) = 0$, hence $-\iota(P) = \iota(Q) \in \text{Im} \, \iota$. (3) If $P, Q \in E(k)$, $P \neq P_0$, $Q \neq P_0$, then $\deg(3P_0 - P - Q) = 1$, hence $\text{Dim}_k(3P_0 - P - Q) = 1$ by Riemann–Roch. If $f \in \mathcal{R}(E)^\times$ is a rational function with $\text{div} f + (3P_0 - P - Q) \geq 0$ then f has a zero at P and Q [18]) and no other poles than a pole of order at most 3 at P_0. If this order is 2, then $\text{div} f = P + Q - 2P_0$ and $\iota(P) + \iota(Q) = 0 \in \text{Im} \, \iota$. If this order is 3, then $\text{div} f = P + Q + R - 3P_0$ for some $R \in E(k)$, hence $\iota(P) + \iota(Q) + \iota(R) = 0$ and $\iota(P) + \iota(Q) = -\iota(R) \in \text{Im} \, \iota$ by (2).

The map ι is also surjective. Let $D = \sum_Q \nu_Q Q \in \text{Div}^0 X$ be a divisor of degree 0. Then $\sum_Q \nu_Q Q = \sum_Q \nu_Q \big(Q - [\kappa(Q):k] P_0\big)$. Therefore, the group $\text{Div}^0 X$ is generated by the divisors $Q - [\kappa(Q):k] P_0$, $Q \in E_0$, and it is enough to show that such a divisor belongs to $\text{Im} \, \iota$. Let $n := [\kappa(Q):k] \geq 2$. Then $\deg(Q - (n-1)P_0) = 1$ hence $\text{Dim}_k(Q - (n-1)P_0) = 1$. If $f \in \mathcal{R}(X)^\times$ is a rational function with $\text{div} f + Q - (n-1)P_0 \geq 0$ then f is not constant because of $n - 1 \geq 1$ and Q is the only pole of f and its order is 1. It follows $\text{div} f = (n-1)P_0 + R - Q$ for some $R \in E(k)$ and $[Q - nP_0] = [R - P_0] = \iota(R) \in \text{Im} \, \iota$.

The group structure $(E(k), P_0)$ on $E(k)$ is now defined in such way that the bijective map $\iota : E(k) \xrightarrow{\sim} \text{Cl}^0 E \cong \text{Pic}^0 E$ is a group isomorphism.   ●

It is possible to describe the group $(E(k), P_0) \cong \text{Cl}^0 E$ in purely affine concepts. We have seen above that $E \setminus \{P_0\} = \text{Spec} \, k[t, u]$ with $u^2 = f(t)$, $f(t) \in k[t]$ a polynomial of degree 3 without multiple prime factors. Then $\text{Cl}^0 E$ and hence $(E(k), P_0)$ *is canonically isomorphic to the divisor class group* $\text{Cl} \, B$ *of the Dedekind domain* $B := k[t, u]$. Namely, the projection $\text{Div}^0 E \to \text{Div} \, B$ with $\sum_{P \in E_0} \nu_P P \mapsto \sum_{P \neq P_0} \nu_P P$ is obviously an isomorphism which induces a canonical isomorphism $\text{Cl}^0 E \to \text{Cl} \, B$.

7.E.28. Remark Let E be an elliptic curve over a number field K (i.e. a finite extension of $\mathbb{Q}$) with a K-rational point P_0. The group $(E(K), P_0) \cong \text{Cl}^0 E \cong \text{Pic}^0 E$ of K-rational points of E is called the M o r d e l l – W e i l g r o u p of E. It is a subgroup of the group $(E(\mathbb{C}) = E_{(\mathbb{C})}(\mathbb{C}), P_0)$ of the $\mathbb{C}$-valued points of E. By the classical theory of elliptic functions the group $(E(\mathbb{C}), P_0) \cong \mathbb{C}/\Gamma$ where Γ is a lattice of rank 2 in $\mathbb{C}$, (see Section 16.C in [13], for example) so as a group $E(\mathbb{C})$ is isomorphic to the 2-dimensional torus group $T^2 = S^1 \times S^1$ where $S^1 \cong \mathbb{R}/\mathbb{Z}$ is the circle group of complex numbers of absolute value 1. The famous M o r d e l l – W e i l t h e o r e m says: *The group* $(E(K), P_0)$ *is finitely generated.* Therefore every K-rational point of E can be obtained from a fixed finite set of points by using the group operations in $(E(K), P_0)$ which we shall describe explicitly below. For $K = \mathbb{Q}$ this was proved by Mordell in 1922. It has been assumed without proof by Poincaré in 1901. Weil proved it for arbitrary number fields K. In particular, the points of the group $E(K)$ can be represented modulo its torsion part $t E(K)$ as a unique integral linear combination of some fixed finitely many points $P_1, \ldots, P_n$, $n := \text{rank} \, E(K)$. The

---

[18]) If $P = Q$, then this means that f has a zero of order at least 2 at P, but f has a pole of order at most 3 at P_0.

the study of their ranks, is one of the most active research areas in number theory. As a
finite subgroup of $S^1 \times S^1$ *the torsion subgroup* $\mathrm{t}E(K)$ *of* $E(K)$ *is generated by at most*
2 *elements.* But, depending on the field K, there are further restrictions for the torsion part
$\mathrm{t}E(K)$: For example, for $K = \mathbb{Q}$, B. Mazur proved in 1977: *The torsion subgroup* $\mathrm{t}E(\mathbb{Q})$
is one of the following 15 *finite groups:* $\mathbb{Z}_n$, $1 \le n \le 12, n \ne 11$, *and* $\mathbb{Z}_2 \times \mathbb{Z}_{2n}$, $1 \le n \le 4$.

For $k = \mathbb{R}$ and a real elliptic curve the group $E(\mathbb{R})$ is a closed subgroup of the torus group
$E(\mathbb{C}) = E_{(\mathbb{C})}(\mathbb{C}) \cong T^2 = S^1 \times S^1$ (in the strong topology). There are two possibilities
for the group $E(\mathbb{R})$: If $E(\mathbb{R})$ is a parabola campaniformis cum ovali, i. e. if E is given by
an equation of type $u^2 = F(t)$ where $F(t) \in \mathbb{R}[t]$ is of degree 3 with three distinct real
zeros, then apparently $E(\mathbb{R}) \cong S^1 \times \mathbb{Z}_2$. If $E(\mathbb{R})$ is a parabola pura, i. e. if E is given by
an equation of type $u^2 = F(t)$ where $F(t) \in \mathbb{R}[t]$ is of degree 3 with only one simple
real zero, then $E(\mathbb{R}) \cong S^1$. It follows: For an elliptic curve over a *real* number field K, the
torsion part $\mathrm{t}E(K)$ of the group $E(K)$ is cyclic or of type $\mathbb{Z}_2 \times \mathbb{Z}_{2n}$ for some $n \in \mathbb{N}^*$.

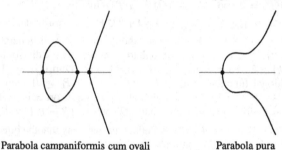

Parabola campaniformis cum ovali   Parabola pura

As already mentioned in the last remark, if char $k \ne 2$ and if E is given in the Weierstrass
normal form as above, then the group operations $\ominus$ and $\oplus$ on $E(k) = (E(k), P_0)$ can
be described explicitly. For simplicity we use the identification $(E(k), P_0) = \mathrm{Cl}\, B$ where
$B = k[t, U]/(U^2 - F)$, $F = a_3 t^3 + a_2 t^2 + a_1 t + a_0 \in k[t]$ a polynomial of degree 3
without multiple prime factors, described after the proof of 7.E.27. Then $E(k)$ is the union
of the point $P_0 = \infty$ at infinity with

$$k\text{-Spec}\, B = \{(t, u) \in k^2 \mid u^2 = F(t)\} \subseteq k^2.$$

To find *the inverse of a point* $P = (t_P, u_P) \in k\text{-Spec}\,B$, we have to find an element $f \in B$
with $\mathrm{div}_B f = P + Q \in \mathrm{Div}\, B$ for some $Q \in k\text{-Spec}\, B$. But for $f := t - t_P$ we have
$B/Bf = k[U]/(U^2 - F(t_P))$, hence $\mathrm{div}_B f = P + Q$ with $Q := (t_P, -u_P) \in k\text{-Spec}\, B$
and

$$\ominus P = \ominus(t_P, u_P) = (t_P, -u_P) \qquad \text{if} \quad u_P^2 = F(t_P).$$

In particular, the points of order 2 are precisely the points $(t_P, 0)$ where t_P is a (simple)
zero of F in k.

Now we want to compute *the sum of two points* $P = (t_P, u_P)$, $Q = (t_Q, u_Q) \in k\text{-Spec}\, B$.
Let $f \in B$ be the equation of the affine line passing through P and Q if $P \ne Q$ or the
tangent line to E at P if $P = Q$ (remember that P is a smooth point of E), i. e.

$$f = \begin{cases} (t_Q - t_P)(u - u_P) - (u_Q - u_P)(t - t_P), & \text{if } P \ne Q, \\ 2u_P(u - u_P) - F'(t_P)(t - t_P), & \text{if } P = Q. \end{cases}$$

If $Q = \ominus P$, i. e. $u_Q = -u_P$, then $\mathrm{div}_B f = P + Q$, i. e. $P \oplus Q = 0$ in $\mathrm{Cl}\, B$ (as we
already know). If $Q \ne \ominus P$ then $t_Q - t_P \ne 0$ if $P \ne Q$ and $u_P \ne 0$ if $P = Q$. Therefore

we can f replace by the equation $u - h$ with

$$h = h(t) := u_P + a \cdot (t - t_P), \qquad a := \begin{cases} (u_Q - u_P)/(t_Q - t_P), & \text{if } P \neq Q, \\ F'(t_P)/2u_P = u_P F'(t_P)/2F(t_P), & \text{if } P = Q. \end{cases}$$

Then $B/Bf = k[t]/(F - h^2)$ and so $\operatorname{div}_B f = P + Q + R$ with $R := (t_R, u_R) \in k\text{-Spec } B$, where t_R and u_R are defined by the equations

$$F - h^2 = a_3(t - t_P)(t - t_Q)(t - t_R) \quad \text{and} \quad u_R = h(t_R),$$

explicitly

$$t_R = \frac{a^2 - a_2}{a_3} - t_P - t_Q, \quad u_R = u_P + a \cdot (t_R - t_P).$$

It follows $P \oplus Q = \ominus R = (t_R, -u_R)$.

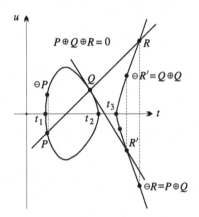

$$E(\mathbb{R}) = \{u^2 = (t - t_1)(t - t_2)(t - t_3)\} \cup \{\infty\}$$

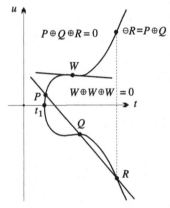

$$E(\mathbb{R}) = \{u^2 = (t - t_1)(t^2 + bt + c)\} \cup \{\infty\}$$
$$t^2 + bt + c \text{ irreducible over } \mathbb{R}$$

To describe the modifications which are necessary if char $k = 2$ is again left to the reader.

REFERENCES

[1] Atiyah, M. F. ; Macdonald, I. G. : Introduction to Commutative Algebra. Addison-Wesley, London, 1969.

[2] Eisenbud, D. : Commutative Algebra with a View Toward Algebraic Geometry. Springer, Berlin/Heidelberg, 1995.

[3] Griffith, P. ; Harris, J. : Principles of Algebraic Geometry. Wiley and Sons, New York, 1978.

[4] Grothendieck, A. ; Dieudonné, J. : Eléments de Géometrie Algébrique I-IV. Publ. Math. IHES **4**, **8**, **11**, **17**, **20**, **24**, **28**, **32** (1960-1967).

[5] Hartshorne, R. : Algebraic Geometry. Springer, Berlin/Heidelberg, 1977.

[6] Mumford, D. : Introduction to Algebraic Geometry. Harvard Lecture Notes (Published as The Red Book of Varieties and Schemes. Lect. Notes in Math. **1358**, Springer, Berlin/Heidelberg, 1988).

[7] Scheja, G. ; Storch, U. : Lokale Verzweigungstheorie. Schriftenreihe des Mathematischen Institutes der Universität Freiburg **5** (1973/74).

[8] Scheja, G. ; Storch, U. : Lehrbuch der Algebra, Teil 1, Teil 2. Teubner, Stuttgart, ^{2}1994, 1988.

[9] Scheja, G. ; Storch, U. : Regular Sequences and Resultants. Research Notes in Mathematics **8**, A. K. Peters, Natick MA, 2001.

[11] Serre, J.-P. : Local Algebra (Translated from French). Springer Monographs in Mathematics, Springer, Berlin/Heidelberg, 2000.

[12] Storch, U. : Vorlesung über Algebraische Kurven (ausgearbeitet von M. Lippa). Mathematisches Institut der Ruhr-Universität, Bochum, 1971/72.

[13] Storch, U. ; Wiebe, H. : Lehrbuch der Mathematik, Vol. IV, Analysis auf Mannigfaltigkeiten-Funktionentheorie-Funktionalanalysis. Spektrum Akademischer Verlag, Heidelberg, 2001.

[14] Zariski, O. ; Samuel, P. : Commutative Algebra, Vol. I, Vol. II, Van Nostrand, Princeton NJ, 1958, 1960.

LIST OF SYMBOLS

INDEX

BIOGRAPHY OF AUTHORS

Dilip P. Patil received B. Sc. and M. Sc. in Mathematics from the University of Pune in 1976 and 1978, respectively. From 1979 till 1992 he studied Mathematics at School of Mathematics, Tata Institute of Fundamental Research, Bombay and received Ph. D. through University of Bombay in 1989. Currently he is a Professor of Mathematics at the Departments of Mathematics and of Computer Science and Automation, Indian Institute of Science, Bangalore. He has been a Visiting Professor at Ruhr-Universität Bochum, Universität Leipzig and several universities in Europe and Canada. His research interests are mainly in Commutative Algebra and Algebraic Geometry.

Uwe Storch studied Mathematics, Physics and Mathematical Logic at the Universität Münster and Heidelberg from 1960 till 1966 and received Ph. D. from Universität Münster in 1966. In 1972 Habilitation at the Ruhr-Universität Bochum. From 1974 till 1981 and from 1981 till 2005 Full Professor at the Universität Osnabrück and at the Ruhr-Universität Bochum, respectively, holding chairs on Algebra and Geometry. Currently he is Professor Emeritus. He has been a Visiting Professor at Tata Institute of Fundamental Research, Bombay, Indian Institute of Science, Bangalore and several universities in Europe and USA. His research interests are mainly in Algebra, particularly the algebraic aspects of Complex Analytic Geometry, Commutative Algebra and Algebraic Geometry.